# Singularities in Physics and Engineering

Properties, methods, and applications

IOP Series in Advances in Optics, Photonics and Optoelectronics

## SERIES EDITOR

**Professor Rajpal S Sirohi** Consultant Scientist

### About the Editor

Rajpal S Sirohi is currently working as a faculty member in the Department of Physics, Alabama A&M University, Huntsville, Alabama (USA). Prior to this, he was a consultant scientist at the Indian Institute of Science Bangalore, and before that he was chair professor in the Department of Physics, Tezpur University, Assam. During 2000–11, he was academic administrator, being vice chancellor to a couple of universities and the director of the Indian Institute of Technology Delhi. He is the recipient of many international and national awards and the author of more than 400 papers. Dr Sirohi is involved with research concerning optical metrology, optical instrumentation, holography, and speckle phenomenon.

### About the series

Optics, photonics and optoelectronics are enabling technologies in many branches of science, engineering, medicine and agriculture. These technologies have reshaped our outlook, our way of interaction with each other and brought people closer. They help us to understand many phenomena better and provide a deeper insight in the functioning of nature. Further, these technologies themselves are evolving at a rapid rate. Their applications encompass very large spatial scales from nanometers to astronomical and a very large temporal range from picoseconds to billions of years. The series on the advances on optics, photonics and optoelectronics aims at covering topics that are of interest to both academia and industry. Some of the topics that the books in the series will cover include bio-photonics and medical imaging, devices, electromagnetics, fiber optics, information storage, instrumentation, light sources, CCD and CMOS imagers, metamaterials, optical metrology, optical networks, photovoltaics, freeform optics and its evaluation, singular optics, cryptography and sensors.

### About IOP ebooks

The authors are encouraged to take advantage of the features made possible by electronic publication to enhance the reader experience through the use of colour, animation and video, and incorporating supplementary files in their work.

### Do you have an idea of a book you'd like to explore?

For further information and details of submitting book proposals see **iopscience.org/books** or contact Ashley Gasque on **Ashley.gasque@iop.org**.

# Singularities in Physics and Engineering

## Properties, methods, and applications

**Paramasivam Senthilkumaran**
*Indian Institute of Technology Delhi*

**IOP** Publishing, Bristol, UK

ISBN 978-0-7503-1698-9 (ebook)
ISBN 978-0-7503-1696-5 (print)
ISBN 978-0-7503-1697-2 (mobi)

DOI 10.1088/978-0-7503-1698-9

Version: 20181101

IOP Expanding Physics
ISSN 2053-2563 (online)
ISSN 2054-7315 (print)

British Library Cataloguing-in-Publication Data: A catalogue record for this book is available from the British Library.

Published by IOP Publishing, wholly owned by The Institute of Physics, London

IOP Publishing, Temple Circus, Temple Way, Bristol, BS1 6HG, UK

US Office: IOP Publishing, Inc., 190 North Independence Mall West, Suite 601, Philadelphia, PA 19106, USA

*Dedicated to God*

# Contents

# Preface

Singularities are ubiquitous in nature. The theory of singularities appears in various branches of science and engineering. Hence I have started the book with a brief note on singularities in various fields. The development in the study of singularities in optics, especially phase singularities and a time-line account are provided. Important topics such as the basics of singularities, their types, properties, generation, detection, applications and the emerging research trends are covered in this book. In the main, this book concentrates on phase singularities and towards the end a chapter on polarization singularities is included. Polarization singularities can be treated as superposition of vortex fields on polarization basis. However, the singularities on non-linear media and coherence singularities are not included, since the emphasis is on waves in linear media and on electromagnetic fields and not on defining new complex functions in which similar structures can be found.

The area of optical phase singularity is relatively new and many of the books that have appeared in recent years on this topic are either collections of chapters jointly presented by multiple authors or collections of research articles published in scientific journals with a brief note from the editors. Also in many of these collections, the main emphasis is on the aspect of orbital angular momentum of light. This orbital angular momentum arises due to the phase singularities and their phase structure. The unique properties of singularities are also useful in other areas of optical engineering. Thus a book that covers the overall or broader aspects of the phase singularities is thought to be useful. Also the absence of a book with a single coherent view by a single author is another motivating factor in writing this book.

The book presents the majority of the developments in optical vortices that have happened in the recent past. It attempts to cater to research scholars who are in the beginning stage of research. This book also provides the experimentalist perspective view of singular optics. I believe that researchers working in diverse areas such as holography, interferometry, metrology, diffractive optics, polarization optics and physical optics will understand and will get a different perspective view on this subject. This book will be very useful for students who plan to pursue research in this area.

This book is aimed at introducing the subject of singular optics in a simple and easily understandable way. For those who plan to expand their research areas, this book may be very useful. The current scenario for young researchers in this area is to hunt and handpick research articles from journals to acquire the required basic expertise. While many of the interesting research articles are milestone papers in this area, they do not cater to the beginner since they are not pedagogical. To address a wider spectrum of students a gradual build-up is provided in this book. Also this book encourages the readers to think further and conceive new ideas on their own. Since the book is written by a single author, coherent, well connected and gradually developed material is presented.

Each chapter has been read by my research scholars Ruchi Rajput, Gauri Arora, Sushanta Pal and Deepa Shankar and their feedback was taken to improve/simplify the presentation at some places. Based on their feedback, new figures were added at places where they found difficulty in understanding the subject. Some figures of simulation using matlab programs were drawn by Sushanta Pal, Ruchi Rajput and Bhargava Ram.

I met my PhD supervisor Professor R S Sirohi, at a conference in the Indian Institute of Space Science and Technology, Thiruvananthapuram in November 2017 and he suggested that I should write a book on my research area for Institute of Physics Publishing. I agreed to his suggestion and wrote this book in a short span of time. But by the time I finished the manuscript in June 2018, I had learnt a lot from going through the literature. I thank Professor R S Sirohi for the good opinion he has of me and for him pushing me to do this work. I must say that this book is written from referring to the good work done by many researchers all over the world.

Many research papers have been referred to and it is realized that a large number of symbols are also required. But most readers are familiar with using certain symbols for certain variables. In an attempt to have universal symbols for the whole book, I did not want to introduce a plethora of symbols which can often be off-putting for readers. Hence the symbols in most of the book are defined when they are introduced and are used consistently within the chapter. The mostly commonly used symbols for many of the parameters are preferred.

This book I believe is self-contained and will bridge the gap between a seasoned researcher and a young, student who dreams of pursuing research on this topic in future.

# Author biography

## Paramasivam Senthilkumaran

Dr P Senthilkumaran is currently working as a full Professor in the Physics Department, Indian Institute of Technology Delhi (IITDelhi). He worked as Associate/Assistant Professor of Physics between 1996 and 2012 in IIT Guwahati (IITG) and as a lecturer in Physics between 1995 and 1996. Prior to joining IITG, he had been a senior project officer since September 1993 at IIT Madras from where he received his PhD in 1995. He is a recipient of the Young Scientist Award from Indian National Science Academy (INSA), New Delhi and Alexander von Humboldt fellowship, Germany in 1997 and 2001 respectively. He was at the University of Strathclyde, Glasgow, United Kingdom on a Royal Society London and INSA exchange fellowship in 1999 and in University of Jena, Germany during 2001–02 on a Humboldt fellowship.

He has been teaching undergraduate and postgraduate courses on basic physics, electromagnetic theory, optics and lasers, Fourier optics, holography and its applications, optical metrology and optical instrumentation. His research interests are optical beam shaping, optical phase singularities, Berry and Pancharatnam topological phases, fiber optics, holography, non-destructive testing techniques, shear interferometry, Talbot interferometry, speckle metrology and non-linear optics. He has authored/coauthored more than 100 research publications.

# Chapter 1

# Introduction

## 1.1 Singularity

Mathematically, a singularity is a point, at which a mathematical quantity is not defined or not 'well behaved' or 'blows up' [1]. Singularities can be observed in nature in many places. Its occurrence in optics is dealt with in this book. In optics, a singularity refers to a point at which some parameter(s) describing the electromagnetic field become(s) indeterminate. Even though at the singular point, things are not well defined, the neighborhood points of a singularity are characterized by very large gradients and hence there is huge interest in the study of singularities. Interestingly, people have preferred to use the singularities in diverse fields of science and engineering and there are also situations where they prefer to stay away from them. Popular ideas about interstellar/intergalactic travel through worm-holes are actually singularities.

Recent years have witnessed lots of research activities in the field of singular optics. An optical vortex, also known as an optical phase singularity is a point phase defect at which the phase is indeterminate and the amplitude is zero. This zero amplitude point draws a curve, called a thread of darkness, in space as the singular beam propagates. These dark threads can be tailored to form loops, knots, and other interesting topological structures. The spiraling Poynting vector associated with the vortices in some cases points to a backward flow of energy. Singular beams are known to carry orbital angular momentum, and in metrology they are used to lift the degeneracy in peak-valley detection. Special properties of singular beams further lead to many more novel applications in areas as diverse as micro-particle manipulation, astrophysics, information technology and processing, beam and polarization shaping, microscopy, and optical testing. Further, singular optics also deals with the presence of singularity in any of the parameters that describe a light field such as polarization vortex, coherence vortex and so on.

doi:10.1088/978-0-7503-1698-9ch1

## 1.2 Singularities in science and engineering

A vortex is a spatial point in the wavefield around which there is continuous circulation of a physical quantity. Vortices are natural occurrences of wavefields and thus have been observed in many forms and at different length and time scales. Vortices are present in nature from the macroscopic to microscopic world. Spiral galaxies, tornados and hurricanes or whirlpools are the most common examples at macroscopic scales. At microscopic scales, vortices have been observed in superfluid He-II and in Bose–Einstein condensates. It is interesting to note that the underlying concepts and the formalism of the theory of singularity are common in diverse fields.

1. *Big Bang* The origin of the Universe is from a Big Bang from a singularity. It is believed that when the laws of physics are extrapolated to a high density regime, singularities occur. After the Big Bang from the singularity there was an expansion and the Universe was formed in due course.
2. *Density* Going from the astronomical scale to the other extreme, the density of a particle is another quantity that invites singularity. The mass is defined as $m = \int \rho' d\tau$, where $d\tau$ is the volume element and since the volume of the particle is (or tends to) zero, the density $\rho'$ is supposed to blow up. Similarly the charge density of a point charge or current density in a thin wire encounters such difficulties. The delta functions are introduced to deal with them.
3. *Chronometric singularity* This singularity occurs at a point at which time cannot be measured or described. Consider the North and South Poles of the Earth at which all the longitudes meet. Since time zones are assigned based on the longitudes, it leads to the problem of defining time at the poles and this singularity is called a chronometric singularity.
4. *Cyclones* Depending on their size and the location that they occur, cyclones go by different names—hurricanes, typhoons, storms and depressions. The large swirling winds, occurring due to the coriolis force of the rotating earth form air circulations with opposite handedness in northern and southern hemispheres. In cyclones, the airflow around the low-pressure area is circulatory and this air circulation happens in a region extending over hundreds of kilometers (figure 1.1). At the center of the vortex, the core or the eye of the storm, the air is still or has very low velocity.

   On the strength scale, the weakest is a tropical depression. Stronger than this is the tropical storm. Even stronger is a hurricane/typhoon in the northern hemisphere or a cyclone in the southern hemisphere.
5. *Tornados* This is another example of a vortex in which a rapidly rotating column of air extends from the surface of the Earth to the sky. Referred as twisters or whirlwinds, they take the form of funnel shaped rotating air columns.
6. *Wing-tip vortices* Flat wing tips in an airplane produce wing-tip vortices which are rotating air columns which are produced when the airplane flies through air. The vortices formed from the tip of the left and right wings have opposite handedness. Water condensation or freezing in the cores of

**Figure 1.1.** The astronauts on the space shuttle Columbia photographed this southern hemisphere cyclone named Daniella, as it gathered force off the east coast of Madagascar in the Indian Ocean in December 1996. Being in the southern hemisphere it spins clockwise, just the opposite of the northern hemisphere hurricanes and typhoons. Figure Courtesy: NASA.

the vortices, where the air pressure is low, sometimes makes these vortices visible in a clear sky (figure 1.2). They are lift-induced vortices and are associated with induced drag. To reduce the induced drag in modern aircraft, winglets are used. Wake turbulence produced by the wing-tip vortices of huge aircraft during take-off can flip a small aircraft going through this turbulence, if it is not dissipated.

7. *Flying insect* In a flying insect, a leading-edge vortex is formed. This is a region of rapidly moving, low-pressure air created at the leading edge of an insect's wing as it flies [2]. The effect is as if the insect had a small balloon tied to each wing, helping it to stay aloft. The combination of the wing-tip vortex and downwash behind the wing may stabilize the leading-edge vortex and keep the insect airborne (figure 1.3).

**Figure 1.2.** Smoke streams from on-board generators of a 747 jumbo jet to present a visual picture of the magnitude of wake vortices created by large aircraft. This 747 is the same one acquired by NASA and later modified to carry space shuttles. (NASA Photo ECN-4242) Credit: Dryden Flight Research Center, NASA.

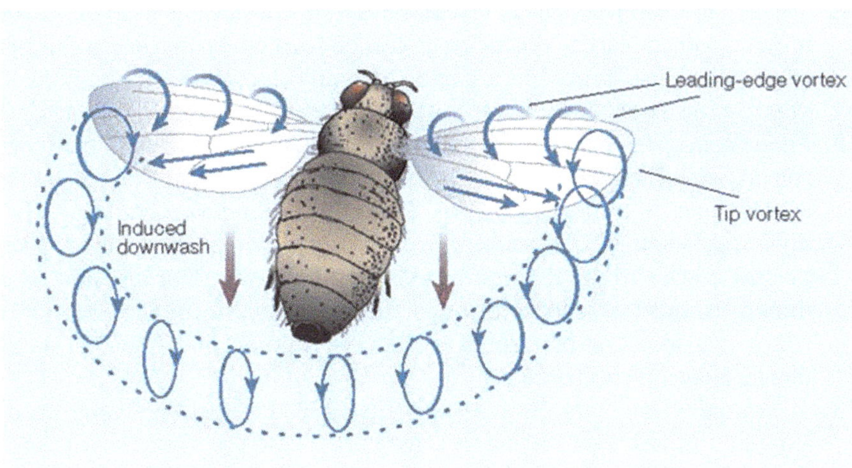


**Figure 1.3.** Patterns of airflow during the downstroke of a hovering insect. Reprinted by permission from [2] Springer Nature Copyright (2001).

8. *Toroidal vortex* This is a ring-donut like structure, made of circulating fluid. The vortex core forms a ring and has a swirling donut shape as shown in figure 1.4. These are stable structures and at some critical speed, the ring

**Figure 1.4.** Toroidal vortex by *Delphinapterus leucas*. This *Delphinapterus leucas* magic ring image has been obtained by the author(s) from the Wikimedia website where it was made available (by SR EXR [Public domain], from Wikimedia Commons) under a CC BY-SA 3.0 licence. It is included within this article on that basis. It is attributed to [SR EXR].

enlarges suddenly and breaks apart. Vortex rings are known to be formed in artillery fire, smoke signals, volcano eruptions, inside the human heart and in many other places. Vortex rings can be formed by pushing a fluid from an enclosed space through a narrow opening. Dolphins and humans form vortex rings in this way.

9. *Flying helicoptor* The formation of vortices by the rotors plays an important role in helicopters. In forward flight, the downwash from the main rotor of a helicopter is transformed into a pair of trailing vortices. But when a hovering helicopter descends, a vortex ring state can be formed. The air pushed downward by the rotor, raises up outside the rotor disc and is again drawn in and back down through the rotor thus forming a vortex ring, which is also called as toroidal vortex. Formation of a toroidal vortex is dangerous as it spoils the lift.
10. *Whirlpools* These are circular currents of water and can be formed by stirring a liquid, like stirring in a tea cup. These are vortices, rotating the fluid in clockwise or anti-clockwise sense that can be seen on the surface of the liquid. In a laminar flow, any obstruction or an obstruction moved in a static fluid can form series of vortices rotating in both directions and they are called Karman vortex streets.
11. *Circulations* Similar circulations of matter that result in the formation of circulating or spiraling patterns are abundant in nature. The spiral galaxies, the red spot on Jupiter, vortices formed by propellers, rotating fan blades,

airflow around buildings are a few examples. The list is endless, but only a few are given above to highlight the ubiquitous nature of vortices in different fields.

For all the above examples of vortex formation, there are many illustrations available on the internet and interested readers can enjoy viewing them.

## 1.3 Acoustic vortex

Although dislocations were first studied in acoustics, reports on the study of acoustic vortices [3] can be traced only to 1999. The existence and properties of screw dislocations in a wavefield were first identified in the context of acoustics. But subsequent research was on optical screw dislocations, where the similarities between the structure of screw dislocations and vortices in fluids were noticed. Indeed optical vortices interact with one another similar to vortices in fluid. Acoustic vortices are wavefields with a screw dislocation in the phase wavefront and an amplitude null on the axis.

$$P_{in} = P(r, \theta) \exp \{i(l\theta - \omega t)\} \tag{1.1}$$

where $(r, \theta)$ are polar coordinates, $\omega$ is angular frequency, and $P$ is an axisymmetric field [4]. The helicity is defined by $l$, its sign the handedness or chirality of the vortex wavefront. An acoustic vortex also has angular momentum [5] like an optical vortex.

## 1.4 Singularities in optics

There are different types of wavefront defects. A phase singularity is a point phase defect and its occurrence in linear optics is briefly discussed here. These singularities can also be formed in a non-linear medium but their generation, detection, properties, propagation and applications are not discussed in this book. The theory of singularity in physics appears in diverse areas such as superfluidity of liquid helium, Abrikosov lattice, quantum Hall effect, big bang theory, quantum chaos, fluid dynamics, global weather patterns, optics, Bose–Einstein condensates and so on. In optics, the theory of singularities appears in diverse areas such as adaptive optics, astronomy, optical testing, wave optical engineering, optical tweezers and traps, atom optics, orbital angular momentum, vortex crystals, optical communication, laser mode, atmospheric optics, phase retrieval, vortex solitons, vortex metrology, diffuser synthesis, image processing and in many other areas. Apart from phase singularities, there are also polarization singularities, coherence singularities, Stokes singularities and so on. Maximum emphasis is given to phase singularities and the underlying theories can also be extended to other types.

## 1.5 Amplitude, phase and polarization

An electromagnetic wave is characterized by amplitude, phase and polarization. In optics, techniques are developed to have precise control over any of these parameters to shape or develop a field distribution. Phase elements and polarization elements for example can be used to control the amplitude and intensity distribution at a desired plane. The electric field of a coherent electromagnetic plane wave can be given by

$$U(x, y, z, t) = U_0(x, y)\hat{n} \exp \{i(\vec{k} \cdot \vec{r} - \omega t)\} \tag{1.2}$$

The field $U(x, y, z, t)$ is complex, $U_0(x, y)$ is the amplitude at a point $(x, y)$ and $\hat{n}$ is the direction of the electric field representing the polarization vector. The phase of the wave is given by $\{(\vec{k} \cdot \vec{r} - \omega t)\}$. Here $\vec{k}$ is the propagation vector and $\omega$ is the angular frequency. Most of the time we are interested in the amplitude, phase and polarization variation in a transverse plane, namely the observation plane and it is normally taken as the $xy$ plane. The nominal direction of propagation is the $z$ direction. The time dependent phase term $\omega t$ can be safely ignored when discussing monochromatic waves. In scalar optics the polarization vector $\hat{n}$ is dropped and the amplitude and phase distributions at a particular plane kept at $z = z_0$ (say) are of interest. The propagation constant $k$ is also called the angular spatial frequency, analogous to angular temporal frequency $\omega$. Unlike $\omega$ which is a scalar, the propagation vector $\vec{k}$ has components along the coordinate axes and hence spatial frequencies along $x$, $y$ and $z$ directions can be defined. Constant phase surfaces $\phi(x, y, z)$ = constant are wavefronts or phase contour surfaces. The phase gradient vector is normal to the phase contour surface and is equal to $\vec{k}$. The direction of $\vec{k}$ also represents the direction of energy transport as the Poynting vector $\vec{S} = I\nabla\phi$, where $I$ is the intensity. Wavefronts in non-singular waves are separated by $\lambda$ distance apart.

In a transverse plane, the amplitude $U_0(x, y)$, the phase $\phi(x, y)$ and the polarization $\hat{n}(x, y)$ can be spatially varying, and non-uniform distribution in any of them leads to diffraction. In scalar optics, $\hat{n}(x, y)$ is uniform and polarization transformations during propagation are insignificant. Spatially varying $\hat{n}(x, y)$ is referred as inhomogeneous polarization distribution.

## 1.6 Brief historical account of optical phase singularities

In the early 1970s, wavefront dislocations and their different types were reported. The importance of these phase dislocations and their unique properties were realized in later years. The field witnessed an increasing level of interest thereafter. A brief but not exhaustive historical account on the developments in optical phase singularities is given in this section. There has been a steady growth in the research interest and number of groups working on this area. In the late 1990s and after, the number of research articles in this area increased significantly.

In complex functions represented by complex variables $f(z) = u + iv$, where $u$ and $v$ are real and imaginary parts of the function $f$, the theory of singularities has been there for a long time [6]. The circulatory motion of matter has been the subject of study in engineering and science and this is evident from many examples given in the previous section [7]. In electromagnetic waves, the idea of phase singularity was first introduced by Nye and Berry [8] in 1974. They observed that the radio echoes from the bottom of Antarctic ice sheets consist of imperfections or defects in the waves. These defects are similar to those that occur in crystal structures. Although there are different types of wavefront dislocations, the point phase dislocation, also called the screw dislocation or optical vortex, has attracted lot of attention and has become the subject of study in recent years.

In the 1980s the problems caused by these phase defects came to light [9–12] as they are found in large numbers in random fields [13]. In computational optics wherever random phases or diffused illuminations are used, the presence of optical vortices become inevitable in the computed fields and they tend to affect the iterative processes employed. Experimentally the presence of edge dislocation in a $TEM_{01}$ (transverse electromagnetic) laser mode was reported [14] in 1983. Diffraction of the screw dislocation in wavefronts was investigated [15] in 1985. A three beam method to study the dislocation in wave pulse [16] was reported in 1987.

In 1990, a method of generation of a radially polarized beam was reported [17]. This beam, even though having a plane wavefront, is actually made up of two helical wavefronts in superposition. In scalar optical fields, some reports on the properties of phase singular beams appeared during this time [18, 19]. The use of spiral gratings [20, 21], and phase rotor filters [22] were also reported in the early 1990s where the elements that were used were capable of producing vortices. In 1992, the use of spiral zone plates for the generation of vortices [23] in a controlled way and laser beams that possess phase singularities [24, 25] were identified and the field of singular optics began to gain momentum. During this period, in 1992 Allen and his group reported the orbital angular momentum [25] carried by the helical waves. In 1993, the use of diffractive lenses [26] and mode converters [27] for vortex generation were also reported. For isolated vortex generation, binary computer generated spiral zone plates were used. An important element, namely the spiral phase plate [28], for the generation of vortices was reported in 1994, and in later years lithographically generated phase plates were used by many research groups.

One of the first reports on the detection of vortices by interference was by White *et al* [29] in 1991. Although there have been many generation methods reported from time to time, the number of detection methods were very few with one of them appearing [30] in 1996. A method based on diffraction appeared a decade later [31]. The effect of aberration, especially astigmatism [32–34] on vortices of higher charge were useful in the development of detection methods later [35–38]. Another detection method that was reported in 2007 used a Shack–Hartmann sensor [39]. In 2008 the use of a lateral shear interferometer [40, 41] for vortex detection was demonstrated. Diffraction methods were reported in 2009 and afterwards [42–47]. Other methods of vortex generation were also reported in the literature. They include the use of non-spiral phase plates [48], plexi glass [49], wedge plates [50, 51], spatial light modulators [52], spatial filtering [53], laser etched mirrors [54] micro-electro-mechanical systems [55], and adaptive mirrors [56–58]. It was shown theoretically by Coullet *et al*, that by using a laser with a large Fresnel number an optical phase singularity can be generated [59].

The interest in singular optics was driven by its applications associated with its angular momentum [60–68], trapping [69–76] and rotation [77–85] of microscopic particles. Properties of vortices in non-linear media was also studied [86–90], but in this book we restrict ourselves to the linear regime only. The applications of optical vortices goes beyond harnessing the orbital angular momentum property and in this book a wide range of applications in diverse areas are described in detail.

Propagation of vortices in a non-linear medium can result in the creation of solitons [86], and the higher order vortex soliton can break up into fundamental units [87] and in this process the angular momentum is conserved. Frequency doubling of a vortex leads to doubling of orbital angular momentum (OAM) per photon. On the parametric down conversion of singular beams the angular momentum conservation is debated [91, 92]. The spatial modes, also called orbital angular momentum states, can be used to define an infinite dimension discrete Hilbert space. These modes, therefore, involve many, multi-dimensional quantum states, unlike two-dimensional quantum states involving two orthogonal polarization states of photons. The spatially separated photons namely signal and idler, resulting in the parametric down conversion of phase singular beam are entangled in their arrival times, say at the detector and in their transverse position [93].

The presence of optical vortices in speckle fields also has drawn the attention of researchers [94, 95]. On average, there is one vortex per speckle and there is little probability of the presence of optical vortices of topological charge greater than one in a speckle field [96]. Freund *et al*, studied the network of randomly distributed optical vortices in a speckle field and predicted that the random distribution of optical vortices in speckle may demonstrate unexpected topological correlations [13, 97]. They proved that adjacent vortices on a certain phase contour alternate sign, in a speckle field distribution.

## References

[1] Arfken G B and Weber H J 2005 *Mathematical Methods for Physicists* (Amsterdam: Elsevier)

[2] Lauder G V 2001 Flight of the robofly *Nature* **412** 688–9

[3] Hefner B T and Marston P L 1999 An acoustical helicoidal wave transducer with applications for the alignment of ultrasonic and underwater systems *J. Acoust. Soc. Am.* **106** 3313–6

[4] Thomas J-L and Marchiano R 2003 Pseudo angular momentum and topological charge conservation for nonlinear acoustical vortices *Phys. Rev. Lett.* **91** 244302

[5] Lekner J 2006 Acoustic beams with angular momentum *J. Acoust. Soc. Am.* **120** 3475–8

[6] Brown J W and Churchill R V 1996 *Complex Variables and Applications* (New York: McGraw-Hill)

[7] Lugt H J 1983 *Vortex Flow in Nature and Technology* (New York: Wiley)

[8] Nye J F and Berry M V 1974 Dislocation in wave trains *Proc. R. Soc. Lond. Ser.* A **336** 165–90

[9] Fienup J R 1982 Phase retrieval algorithms: a comparison *Appl. Opt.* **21** 2758–69

[10] Wyrowski F and Bryngdahl O 1988 Iterative fourier-transform algorithm applied to computer holography *J. Opt. Soc. Am.* A **5** 1058–65

[11] Fienup J R and Wackerman G C 1986 Phase retrieval stagnation problems and solutions *J. Opt. Soc. Am.* A **3** 1897–907

[12] Ghiglia D C and Pritt M D 1998 *Two-dimensional Phase Unwrapping: Theory, algorithms and Software* (New York: Wiley)

[13] Freund I, Shvartsman N and Freilkher V 1993 Optical dislocation network in highly random media *Opt. Commun.* **101** 247–64
[14] Vaughan J M V and Willetts D V 1983 Temporal and interference fringe analysis of excimer $TEM_{01}$ laser *J. Opt. Soc. Am.* **73** 1018–21
[15] Condell W J 1985 Fraunhofer diffraction from a circular annular aperture with helical phase factor *J. Opt. Soc. Am.* **2** 206–8
[16] Nicholls K W and Nye J F 1987 Three beam model for studying dislocations in wave pulse *J. Phys.* A **20** 4673–96
[17] Tidwell W C, Ford D H and Kimura W D 1990 Generating radially polarized beams interferometrically *Appl. Opt.* **29** 2234–9
[18] Bazhenov V Y, Vasnetsov M V and Soskin M S 1990 Laser beams with screw dislocation in their wavefronts *JETP Lett.* **52** 429–31
[19] Bazhenove V Y, Soskin M S and Vasnetsov M V 1992 Screw dislocations in light wavefronts *J. Mod. Opt.* **39** 985–90
[20] Szwaykowski P and Patorski K 1989 Moire fringes by evolute gratings *Appl. Opt.* **28** 4679–81
[21] Chang C W and Su D C 1991 Collimation method that uses spiral gratings and Talbot interferometry *Opt. Lett.* **16** 1783–4
[22] Khonina S N, Kotlyar V V, Shinkaryev M V, Soifer V S and Uspieniev G V 1992 The phase rotor filter *J. Mod. Opt.* **39** 1147–54
[23] Heckenberg N R, McDuff R, Smith C P and White A G 1992 Generation of optical phase singularities by computer generated hologram *Opt.Lett.* **17** 221–3
[24] Heckenberg N R, McDuff R, Smith C P, Rubinsztein Dunlop H and Wegener M J 1992 Laser beams with phase singularities *Opt. Quantum Electron.* **24** S951–62
[25] Allen L, Beijersbergen M W, Spreeuw R J C and Woerdman J P 1992 Orbital angular momentum of light and the transformation of Laguerre–Gaussian laser modes *Phys. Rev.* A **45** 8185–9
[26] Roux F S 1993 Diffractive lens with a null in the centre of its focal point *Appl. Opt.* **32** 4191–2
[27] Beijersbergen M W, Allen L, van der Veen H E L O and Woerdman J P 1993 Astigmatic laser mode converters and transfer of orbital angular momentum *Opt. Commun.* **96** 123–32
[28] Beijersbergen M W, Coerwinkel R P C, Kristensen M and Woerdman J P 1994 Helical wavefront laser beams produced with a spiral phase plate *Opt. Commun.* **112** 321–2
[29] White A G, Smith C P, Heckenberg N R, Rubinsztein-Dunlop H, Mcduff R, Weiss C O and Tamm C 1991 Interferometric measurements of phase singualrities in the output of a visible laser *J. Mod. Opt.* **38** 2531–41
[30] Padgett M J, Arlt J, Simpson N B and Allen L 1996 An experiment to observe the intensity and phase structure of Laguerre–Gaussian laser modes *Am. J. Phys.* **64** 77–83
[31] Sztul H I and Alfano R R 2006 Double-slit interference with Laguerre–Gaussian beams *Opt. Lett.* **31** 999–1001
[32] Singh R K, Senthilkumaran P and Singh K 2007 Effect of astigmatism on the diffraction of a vortex carrying beam with Gaussian background *J. Opt. A: Pure Appl. Opt.* **9** 543–54
[33] Singh R K, Senthilkumaran P and Singh K 2007 Influence of astigmatism and defocusing on the focusing of a singular beam *Opt. Commun.* **270** 128–38
[34] Bekshaev A Y and Karamoch A I 2008 Astigmatic telescopic transformation of a high-order optical vortex *Opt. Commun.* **281** 5687–96

[35] Bekshaev A Y, Soskin M S and Vasnetsov M V 2004 Transformation of higher order optical vortices upon focusing by an astigmatic lens *Opt. Commun.* **241** 237–47

[36] Vaity P, Banerji J and Singh R P 2013 Measuring the topological charge of an optical vortex by using a tilted convex lens *Phys. Lett.* A **377** 1154–6

[37] Reddy S G, Prabhakar S, Aadhi A, Banerji J and Singh R P 2014 Propagation of an arbitrary vortex pair through an astigmatic optical system and determination of its topological charge *J. Opt. Soc. Am.* A **31** 1295–302

[38] Kotlyar V V, Kovalev A A and Porfirev A P 2017 Astigmatic transforms of an optical vortex for measurement of its topological charge *Appl. Opt.* **56** 4095–104

[39] Chen M, Roux F S and Olivier J C 2007 Detection of phase singularities with a Shack-Hartmann wavefront sensor *J. Opt. Soc. Am.* A **24** 1994–2002

[40] Ghai D P, Senthilkumaran P and Sirohi R S 2008 Shearograms of optical phase singularity *Opt. Commun.* **281** 1315–22

[41] Ghai D P, Vyas S, Senthilkumaran P and Sirohi R S 2008 Detection of phase singularity using a lateral shear interferometer *Opt. Laser Eng.* **46** 419–23

[42] Moreno I, Davis J A, Melvin B, Pascoguin L, Mitry M J and Cottrell D M 2009 Vortex sensing diffraction gratings *Opt. Lett.* **34** 2927–9

[43] Ghai D P, Senthilkumaran P and Sirohi R S 2009 Single-slit diffraction of an optical beam with phase singularity *Opt. Laser Eng.* **47** 123–6

[44] Zhang N, Yuan X C and Burge R E 2010 Extending the detection range of optical vortices by Dammann vortex gratings *Opt. Lett.* **35** 3495–7

[45] Hickmann J, Fonseca E and Chavez Cerda S 2010 Unveiling a truncated optical lattice associated with a triangular aperture using light's orbital angular momentum *Phys. Rev. Lett.* **105** 053904

[46] De Araujo L E E and Anderson M E 2011 Measuring vortex charge with a triangular aperture *Opt. Lett.* **36** 787–9

[47] Fu S, Wang T, Zang S and Gao C 2016 Integrating 5x5 Dammann gratings to detect orbital angular momentum states of beams with the range of −24 to +24 *Appl. Opt.* **55** 1514–7

[48] Kim G H, Jeon J H, Ko K H, Moon H J, Lee J H and Chang J S 1997 Optical vortices produced with a non-spiral phase plate *Appl. Opt.* **36** 8614–21

[49] Rotschild C and Zommer S 2004 Adjustable spiral phase plate *Appl. Opt.* **43** 721–7

[50] Yuan X C, Ahluwalia B P S, Tao S H, Cheong W C, Zhang L S, Lin J, Bu J and Burge R E 2007 Wavelength-scalable micro-fabricated wedge for generation of optical vortex beam in optical manipulation *Appl. Phys.* B **86** 209–13

[51] Lin J, Yuan X C, Bu J, Ahluwalia B P, Sun Y Y and Burge R E 2007 Selective generation of high-order optical vortices from a single phase wedge *Opt. Lett.* **32** 2927–9

[52] Ganic D, Gan X and Gu M *et al* 2002 Generation of doughnut laser beams by use of a liquid-crystal cell with a conversion efficiency near 100% *Opt. Lett.* **27** 1351–3

[53] Guo C S, Zhang Y, Han Y J, Ding J P and Wang H T 2006 Generation of optical vortices with arbitrary shape and array via helical phase spatial filtering *Opt. Commun.* **259** 449–54

[54] Strohaber J, Scarborough T and Uiterwaal C J G J 2007 Ultrashort intense-field optical vortices produced with laser etched mirrors *Appl. Opt.* **46** 8583–90

[55] Zhou G and Chau F S 2005 Helical wave front laser beam generated with a micro-electro-mechanical systems (mems) based device *IEEE Photon. Technol. Lett.* **18** 292–4

[56] Bokyo O, Mercere T A P, Valentin C and Balcou P H 2005 Adaptive shaping of a focused intense laser beam into a doughnut mode *Opt. Commun.* **246** 131–40

[57] Ghai D P, Senthilkumaran P and Sirohi R S 2008 Adaptive helical mirror for generation of optical phase singularity *Appl. Opt.* **47** 1378–83

[58] Ghai D P 2011 Generation of optical vortices with an adaptive helical mirror *Appl. Opt.* **50** 1374–81

[59] Coullet P, Gil L and Rocca F 1989 Optical vortices *Opt.Commun.* **73** 403–8

[60] Allen L, Padgett M J and Babikar M 1999 *The orbital Angular Momentum of Light, Progress in Optics* vol 39 (Amsterdam: Elsevier)

[61] Allen L, Barnett S M and Padgett M J (ed) 2003 *Optical Angular Momentum* (Bristol: IOP Publishing)

[62] Padgett M, Courtial J and Allen L 2004 Light's orbital angular momentum *Phys. Today* **57** 35–40

[63] Beth R A 1936 Mechanical detection and measurement of the angular momentum of light *Phys. Rev.* **50** 115–27

[64] Molina-Terriza G, Recolons J, Torres J P, Torner I and Wright W M 2001 Observation of the dynamical inversion of the topological charge of an optical vortex *Phys. Rev. Lett.* **87** 23902–5

[65] O'Neil A T, MacVicar I, Allen L and Padgett M J 2002 Intrinsic and extrinsic nature of the orbital angular momentum of a light beam *Phys. Rev. Lett.* **88** 053601

[66] Bekshaev A Y 1999 Mechanical properties of the light wave with phase singularity *Proc. SPIE* **3994** 131–9

[67] Bekshaev A Y 2000 Manifestation of mechanical properties of light waves in vortex beam optical systems *Opt. Spectrosc.* **88** 904–10

[68] Tao S H, Yuan X C, Lin J, Burge R E, Tao S H, Yuan X C, Lin J and Burge R E 2006 Residue orbital angular momentum in interferenced double vortex beams with unequal topological charge *Opt. Express* **14** 535–41

[69] Ghaghan K T and Swartzlander G Jr 1996 Optical vortex trapping of particles *Opt. Lett.* **21** 827–9

[70] Heckenberg N R, Nieminen T A, Friese M E J and Rubinsztein-Dunlop H 1998 Trapping microscopic particles with singular beams *Proc. SPIE* **3487** 46

[71] Grzegorczyk T M and Kong J A 2007 Analytical prediction of stable optical trapping in optical vortices created by three TE or TM plane waves *Opt. Express* **15** 8010–8

[72] Gahagan K T and Swartzlander G A Jr 1998 Trapping of low-index micro particles in an optical vortex *J. Opt. Soc. Am.* B **15** 524–34

[73] Gahagan K T and Swartzlander G A 1999 Simultaneous trapping of low-index and high-index microparticles observed with an optical vortex trap *J. Opt. Soc. Am.* B **16** 533–7

[74] Liesener J, Reicherter M, Haist T and Tiziani H J 2000 Multi-functional optical tweezers using computer-generated holograms *Opt. Commun.* **185** 77

[75] Bradshaw D S and Andrews D L 2005 Interaction between spherical nano-particles optically trapped in Laguerre–Gaussian modes *Opt. Lett.* **30** 3039–41

[76] Lee W M and Yuan X-C 2003 Observation of three-dimensional optical stacking of micro particles using a single Laguerre–Gaussian beam *Appl. Phys. Lett.* **83** 5124–6

[77] Friese M E J, Heckenberg N R and Rubinsztein-Dunlop H 1995 Direct observation of transfer of angular momentum to absorptive particles from a laser beam with a phase singularity *Phys. Rev. Lett.* **75** 826–9

[78] Simpson B N, Dholakia K, Allen L and Padgett M J 1997 Mechanical equivalence of spin and orbital angular momentum of light: An optical spanner *Opt. Lett.* **22** 52–4

[79] Simpson N B, Allen L and Padgett M J 1996 Optical tweezers and optical spanners with Laguerre–Gaussian modes *J. Mod. Opt.* **43** 2485–92
[80] O'Neil A T and Padgett M J 2000 Three-dimensional optical confinement of micron-sized metal particles and the decoupling of the spin and orbital angular momentum within an optical spanner *Opt. Commun.* **185** 139–43
[81] Lee W M, Ahluwalia B P S, Yuan X-C, Cheong W C and Dholakia K 2005 Optical steering of high and low index micro-particles by manipulating an off-axis optical vortex *J. Opt. A: Pure Appl. Opt.* **7** 1–6
[82] Galajda P and Ormos P 2001 Complex micro-machines produced and driven by light *Appl. Phys. Lett.* **78** 249–51
[83] Ladavac K and Grier D G 2004 Microoptomechanical pumps assembled and driven by holographic optical vortex arrays *Opt. Express* **12** 1144–9
[84] Friese M E J, Rubisztein-Dunlop H, Gold J, Hagberg P and Hanstorp D 2001 Optically driven micro-machine elements *Appl. Phys. Lett.* **78** 547–9
[85] Soskin M S, Bekshaev A Y and Vanetsov M V 2004 An optical vortex as a rotating body: mechanical features of a singular light beam *J. Opt. A. Pure Appl.* **6** S170–4
[86] Rozas D and Swartzlander G A Jr 2000 Observed rotational enhancement of nonlinear optical vortices *Opt. Lett.* **25** 126–8
[87] Mcdonald G S, Syed K S and Firth W J 1996 Optical vortices in beam propagation through a self-defocusing medium *Opt. Commun.* **94** 469–76
[88] Chen Y and Atai J 1994 Dynamics of optical-vortex solitons in perturbed nonlinear media *J. Opt. Soc. Am.* B **11** 2000–3
[89] Law C T, Zhang X and Swartzlander G A Jr 2000 Wave guiding properties of optical vortex solitons *Opt. Lett.* **25** 55–7
[90] Tikhonenko V, Christou J and Luther-Davies B 1996 Three-dimensional bright spatial soliton collision and fusion in a saturable nonlinear medium *Phys. Rev. Lett.* **76** 2698–701
[91] Arlt J, Dholakia K, Allen L and Padgett M J 1999 Parametric down-conversion for light beams possessing orbital angular momentum *Phys. Rev.* A **59** 3950–2
[92] Mair A, Vaziri A, Weihs G and Zeilinger A 2001 Entanglement of the orbital angular momentum states of photon *Nature* **412** 313–6
[93] Eliel E R, Dutra S M, Nienhuis G and Woerdman J P 2001 Comment on, orbital and intrinsic angular momentum of single photons and entangled pairs of photons generated by parametric down-conversion *Phys. Rev. Lett.* **86** 5208–10
[94] Baranova N B, Zel'dovich B Y, Mamaev A V, Pilipetski N F and Shkunov V V 1981 Dislocations of the wavefront of a speckle-inhomogeneous field *JETP Lett.* **33** 195–9
[95] Staliunas K, Berrzanskis A and Jarutis V 1995 Vortex statistics in optical speckle fields *Opt. Commun.* **120** 23–8
[96] Baranova N B, Mamaev A V, Pilipetsky N F, Shkunov V V and Zel'dovich Y B 1983 Wavefront dislocations: topological limitations for adaptive systems with phase conjugation *J. Opt. Soc. Am.* **73** 525 9
[97] Freund I and Shvartsman N 1994 Wave-field phase singularities: the sign principle *Phys. Rev.* A **50** 5164–72

**IOP** Publishing

Singularities in Physics and Engineering
Properties, methods, and applications
**Paramasivam Senthilkumaran**

# Chapter 2

# Topological features

## 2.1 Introduction

To understand many of the properties in the phase distribution, knowledge of the shape of the wavefront and the topological features are important. Many of the beam characteristics such as propagation, divergence, and diffraction are dependent on the wavefront. Critical points shape up the wavefront and hence this chapter starts with the study of wavefronts.

The structure of an optical vortex wavefront, phase and amplitude distributions are studied. In field distributions, phase contours, real and imaginary zero curves are employed in the discussion on the topological features of the phase singularity. Other critical points such as extrema and saddles in fields and their coexistence with vortices are described. Critical points can also undergo disintegration and there can be byproducts. The critical points can be analyzed with the help of phase contours. The difference between phase contour, zeros of real/imaginary part of the field and bifurcation lines are made clear. Similarly there is also difference between the charge, index and order of a vortex. There are different sign rules and conservation rules for vortices in distributions. With the help of Berry's paradox, it is shown that concepts like charge and its conservation can be dispensed with and instead vortex trajectories can be used. To deal with these vortex trajectories, concepts like topological manifolds are used.

## 2.2 Wavefront shape

The phase contour surfaces of an optical field are the wavefronts. The wavefronts are usually drawn with a spacing of $\lambda$ between two consecutive surfaces. Depending on the shape of the wavefronts, the waves are named as plane, spherical or cylindrical waves. From a primary wavefront, according to Huygen's principle, the secondary wavelets lead to the construction of the next wavefront. The next wavefront, that is constructed is also of the same type, i.e. a plane wavefront leads to the construction of plane wavefront and so on.

doi:10.1088/978-0-7503-1698-9ch2

For a phase singular beam, the wavefronts have a helical shape [1–4] and are shown in figure 2.1. It is a ramp-like structure winding about the phase singular point which draws a curve in three dimensions, upon propagation. Depending on the handedness of the phase ramp, the singularity is termed as positive (anti-clockwise ramp) or negative (clockwise ramp) as shown in figure 2.2. Along the general propagation direction the distance between two consecutive constant phase surfaces is $\lambda$. But there is a little twist in the story. Due to the presence of the phase singularity the adjacent wavefronts get connected and there is a single wavefront spanning the entire space and the notion of distance between two wavefronts becomes meaningless. When a wavefront has a vortex, the wavefront extends from $-\infty$ to $+\infty$ along the propagation direction. Also construction from one wavefront to another becomes difficult to comprehend as there is going to be only one wavefront.

Imagine that travel on a circular path is performed over the wavefront surface. In a plane wave or in a spherical wave, travel between two consecutive wavefront surfaces entails a jump $\lambda$ while such a jump is not needed if the travel is carried over a helical wavefront. When multiple spatially distributed singularities are present these ramps allow forward and backward travel from one wavefront to succeeding or preceding wavefronts while maintaining the sense of (say, clockwise) direction of the travel path. To be exact, I cannot use the term different wavefronts—hence the travel results in forward and backward displacements on the same surface which is cut and stitched at various places to the preceding or succeeding wavefronts. (This can be seen later in this chapter where oriented curves are introduced for vortices. The propagation vector spirals around the vortex core and this indicates that backward flow of energy is possible.) For a vortex of higher charge, in the wavefront

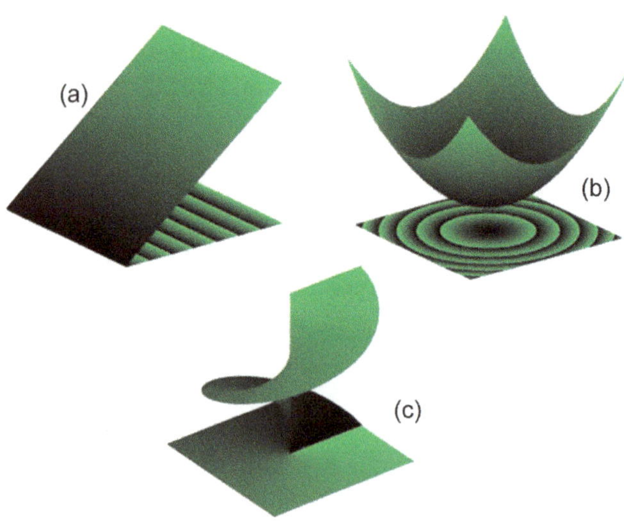


**Figure 2.1.** Phase distributions and the wavefront structures corresponding to (a) plane wave with a tilt, (b) spherical wave and (c) helical wave of charge +1. On the wavefront the phase is constant and hence if the observation plane is of the same shape as that of the wavefront, the phase distribution would have been constant. But normally phase distribution is the one observed on a plane surface (say $z = z_1$ plane if the nominal propagation direction is along $z$).

**Figure 2.2.** Phase distribution and wavefront shape of (a) a positive and (b) a negative vortex.

there are multi-start ramps starting from one surface leading to the next surface at a height of $\lambda$, as depicted in figure 2.3. This is similar to going from one floor to the next floor in a building using (multiple start ramps) any one of the many helical ramps. For example for a topological charge of 2, there are two ramps starting simultaneously from one floor and leading to the next floor and one can use either one of them to go from one floor to the other.

Instead of discussing travel from one surface to another let us consider travel over a closed path on the wavefront. For a non-singular beam, a circular closed travel path will leave a person on the same floor (same starting point) whereas a similar circular travel path on a singular beam around the singularity leads to climbing up/down by a height of $\lambda$. For a vortex of charge $l$ the height gained is $l\lambda$. When there are two vortices of opposite charges present, the evolution of the wavefront is hard to visualize as the structure continuously evolves as depicted in figure 2.4. There is global change in the shape of the wavefront as it propagates and is periodic. Individually, each of the vortices has a helical structure winding in the opposite sense to each other as the wave propagates. This can be seen by tracking the immediate neighborhood of each vortex.

## 2.3 Amplitude and phase distribution of an optical vortex beam

Now, instead of dealing with constant phase surfaces, let us examine the situation on a plane surface. For a wave traveling along the $z$ direction, the plane under consideration is an $xy$-plane. The phase at each point on this plane is given by the phase distribution. The phase distributions for a spherical wave, a plane wave traveling at an angle to the $z$ axis, and a helical wave are presented in figure 2.1.

For a non-singular beam such as a plane wave or a spherical wave, the phase distribution is such that the accumulated phase (or the phase difference between any two different spatial points) is given by

$$\Delta\phi = \int_a^b \nabla\phi \cdot dl = \phi(b) - \phi(a) \tag{2.1}$$

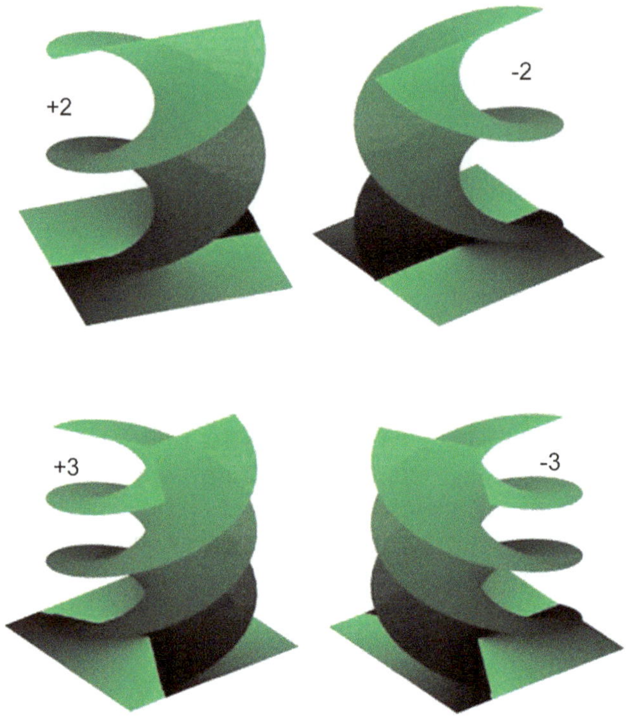


**Figure 2.3.** Vortices of higher charges and different polarities. Phase and wavefront maps.

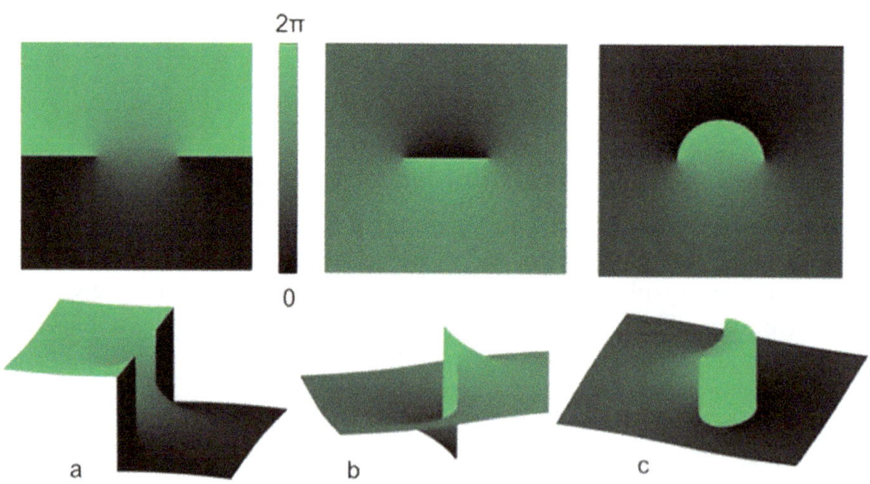


**Figure 2.4.** Vortex dipole and the wavefront structure (a) where the $2\pi$ virtual phase jump line is pointing away from the center of the dipole, (b) where the $2\pi$ virtual phase jump is between the charges and (c) at an intermediate state. This $2\pi$ jump is called 'virtual' as at this position the wavefront is connected to the next wavefront and hence there is actually no phase jump and the wavefront is continuous. The wavefront structure from (a) to (c) can evolve by constant phase shifts.

where $dl$ is the path of integration. In equation (2.1) the phase difference is independent of the path taken for computation. Further for closed path $\oint \nabla\phi \cdot dl = 0$. But for phase singular beams

$$\oint \nabla\phi \cdot dl \neq 0 \tag{2.2}$$

This means $\nabla \times \nabla\phi \neq 0$ and in the neighborhood region of the singularity the gradient field has non-zero curl or circulating phase gradient. At the point at which the singularity is located, the phase is indeterminate. For a complex field which is single valued, at every point the amplitude and phase take a single value and this undefined phase is possible only by considering the amplitude to be zero at the singular point. It is also true that at all points where the amplitude is zero, the phase is indeterminate, because the phase is given by the ratio of the imaginary to the real part of the complex field. The field is $A\exp\{i\phi\} = \xi + i\eta = A\cos\phi + iA\sin\phi$. At a null amplitude we have,

$$\tan[\phi(x, y)] = \frac{\eta}{\xi} = \frac{0}{0} \tag{2.3}$$

But an amplitude zero point does not guarantee the presence of an optical vortex at that point. For example, the Newton's ring experiment in which concentric circular interference fringes are formed due to interference of a plane and a spherical wave. The center of the fringe pattern observed in reflection is an amplitude zero point, and this happens to be an extremum point in phase but not a vortex point. Even though amplitude null makes the phase indeterminate, that point does not correspond to a wavefront with helical shape. Hence amplitude zero is a necessary condition but not sufficient condition for the existence of phase singularity. The condition given by the closed path integral equation (2.2) has to be satisfied for a vortex.

The amplitude and phase distributions of an optical vortex beam are such that at the phase singular point the phase is undefined and the amplitude is zero. Even if you start with a uniform background amplitude wave with an embedded vortex in it, (this is possible by illuminating the vortex phase plate by a uniform amplitude plane wave), during propagation at the vortex point the amplitude becomes zero due to diffraction effects. The amplitude background can be of any form but has zero value at the vortex point. For example, a Gaussian wave has non-zero amplitude at the center, but the Gaussian amplitude is modulated (multiplied) with Laguerre polynomials of any order along with an $r$ term, resulting in the zero amplitude at the vortex point at $r = 0$. Laguerre–Gaussian beams are often encountered in singular optics. Similarly the amplitude of the vortex beam can have tan hyperbolic variation and in that case the beam can be termed as **tanh vortex**. Another kind of vortex often used in discussions is the **r-vortex** in which case, the amplitude is linearly varying as a function of $r$. The linear increase in amplitude as we go radially away from the origin (vortex point) leads to larger amplitude and since the wave function is finite, this is not a physically feasible form of describing the vortex. To avoid this amplitude blow-up a Gaussian envelope can be used. Nevertheless use of such an

$r$-type vortex is common and is very helpful in understanding many of the properties of the vortex [5]. It is useful to introduce the following way of writing the complex field for a vortex. For a positive vortex at the origin, the field is given by $(x + iy)$ in which the amplitude is $r = \sqrt{x^2 + y^2}$ and the phase is $\arctan\left\{\frac{y}{x}\right\}$. This is in fact an $r$-type vortex. A negative vortex has the complex field given by $(x-iy)$, a vortex of higher charge $l$ is given by $(x \pm iy)^l$ and a vortex located at a point other than the origin is given by $(x-x_0) \pm i\ (y-y_0)$ where $(x_0, y_0)$ is the location of the phase singularity. Consider the complex function given by

$$\begin{aligned}\psi(x, y) &= (x \pm iy)^l = r^l \exp\{\pm il\phi\} \\ &= r^l[\cos(l\phi) \pm i \sin(l\phi)] = R(x, y) \pm iI(x, y)\end{aligned} \tag{2.4}$$

Here the charge of the vortex is $l$. $R$ and $I$ indicate the real and imaginary parts of the complex field respectively. By denoting the zero of the real part of the field as $Z_R$ and zero of the imaginary part of the field as $Z_I$ we will discuss various aspects of the complex field in the following sections.

## 2.4 Topological charge

The topological charge of a phase singularity is defined by

$$\oint \nabla\phi \cdot dl = l2\pi \tag{2.5}$$

where $l$ is called the topological charge of the phase singularity. It can take positive and negative integer values [6, 7]. The sign of the vortex is decided by the sign of the azimuthal phase gradient. For a positive vortex, the phase increases in an anti-clockwise sense around the singular point and for a negative vortex, the phase increase is in the clockwise sense.

There are also vortices with fractional charges and they are called fractional vortices. They do not produce a donut intensity profile, but produce a radial cut in the donut shape.

## 2.5 Phase contours and zero crossings

Phase contours in two-dimensional phase distributions are the curves on which the phase value is constant. Phase contours for a single vortex and a vortex diplole are shown in figure 2.5. The random field has a large number of vortices. The phase, phase contours and real and imaginary zero curves for a random field are shown in figure 2.6. Contours corresponding to all phase values terminate on the vortex giving it a star-like appearance. This is similar to electric field lines terminating or originating from an electric charge [5]. Because of the similar structure, the phrase 'charge' is chosen to describe the strength of the vortices. Consider four phase contours at phase values of 0, $\frac{\pi}{2}$, $\pi$ and $\frac{3\pi}{2}$ that terminate on a vortex point. Phase values of zero and $\pi$ correspond to zero of the imaginary part ($Z_I$) of the wave function and the phase values of $\frac{\pi}{2}$ and $\frac{3\pi}{2}$ correspond to zero of the real part ($Z_R$) of the wave function. Termination of phase contours can be seen as the zeros of the real

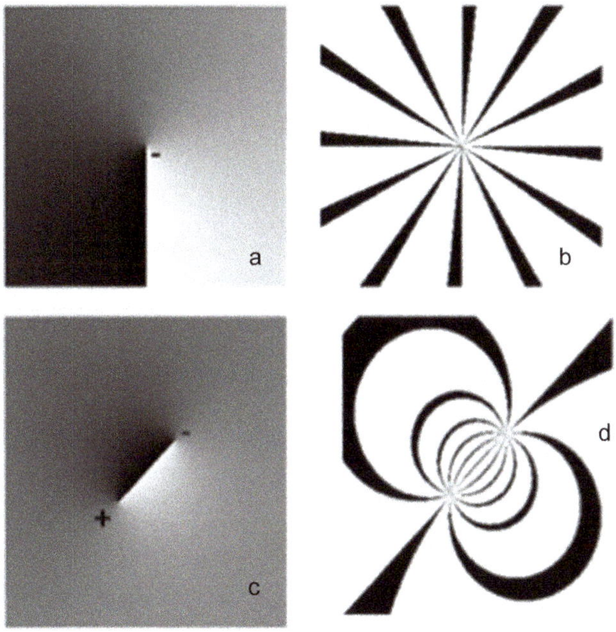


**Figure 2.5.** Phase (a) and (c) and phase contours (b) and (d) of a vortex and a vortex dipole.

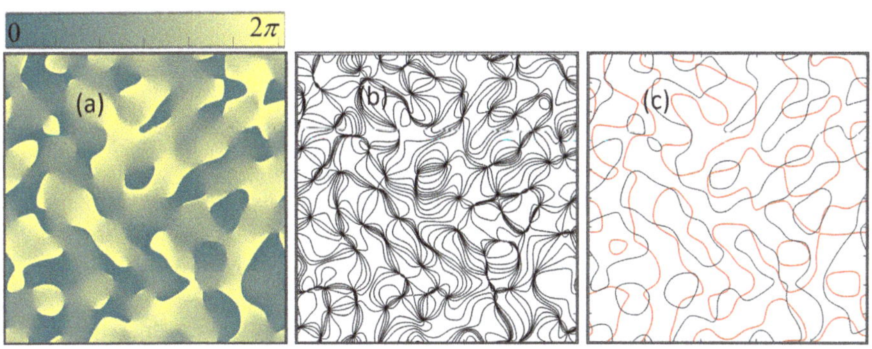


**Figure 2.6.** (a) Random phase distribution, (b) phase contours and (c) curves of real and imaginary zeros ($Z_R$ and $Z_I$) for the phase distribution shown in (a).

and imaginary part of the wave function crossing at the vortex point. A phase contour at phase value zero and a phase contour at phase value $\pi$ terminating on a vortex point is seen as an imaginary zero curve which is continuously passing through the vortex point. Hence by using the continuous curves of $Z_I$ and $Z_R$ we can analyze the critical points. The distinction between the phase contours and the real and imaginary zero curves are depicted in figure 2.7, in which phase contours are shown by the solid curves, and the $Z_R$ and $Z_I$ curves are given by broken line curves.

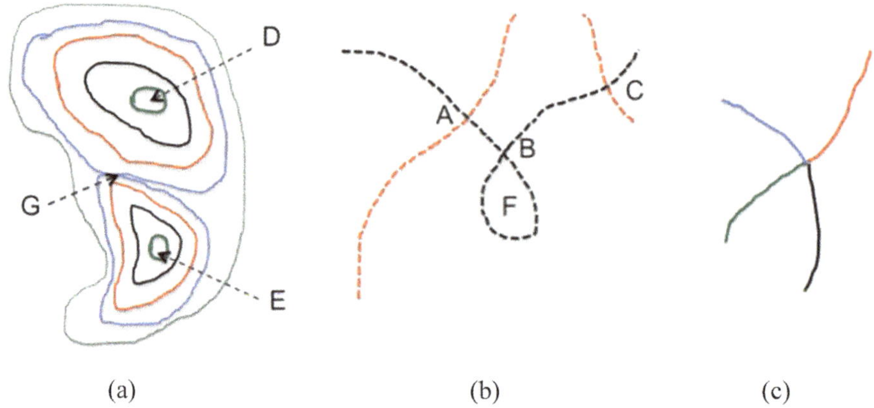


**Figure 2.7.** Real and imaginary zero contours. (a) Phase contours, each color is for different phase values, say $0, \frac{\pi}{2}, \pi$ and $\frac{3\pi}{2}$. Hence two of them are $Z_R$ and two of them are $Z_I$. (b) Zero crossing at the vortex points. Note also at the saddle point there is a zero crossing (self-intersection). (c) Zero crossing at the vortex, made of different phase contours. The locations given by D, E and F are extrema, G and B are saddle points, locations A and C are vortex points. In (a) each curve of zero ($Z_I$ and $Z_R$) is made of single phase value whereas in (b) and (c) they are not. Solid lines are phase contours and broken curves are real and imaginary zero contours.

## 2.6 Phase gradients of an optical vortex beam

The phase contours for a vortex and a vortex dipole are shown in figure 2.5. For a spherical wave, the phase contours are closed, whereas for the tilted plane wave the phase contours are equally spaced, straight lines. The phase values are taken at equal intervals to draw the contours. It can be seen in figure 2.6, that the contours are densely packed wherever the phase gradient is high. The phase gradient vector, points in the direction of maximum (rate of) change of phase and is perpendicular to the phase contour lines. For a phase contour surface, the gradient vector is normal to the surface. Examples: (1) in electrostatics, the electric field vector is normal to the equi-potential surfaces and (2) similarly in optics, the propagation vector is normal to the constant phase surfaces.

For a vortex beam, the phase distribution is shown in figure 2.5(a). The phase is constant along a radial line from the vortex core [6, 8, 9]. Since different phase contour lines emanate from the singular point and go radially outward, the phase gradient for a vortex is azimuthal and the gradient is shown in figure 2.8.

The phase gradient vectors in the transverse plane can be computed by the phase distribution of a vortex in a transverse plane. For an $r$-vortex of charge $l$, the phase distribution [10] can be written as

$$\phi(x, y) = Arg[(x + iy)^l] = \frac{l}{i} \ln\left\{\frac{x + iy}{\sqrt{x^2 + y^2}}\right\} \tag{2.6}$$

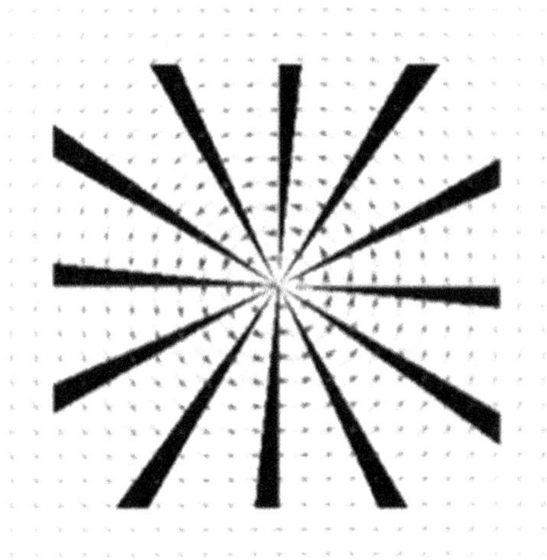

**Figure 2.8.** The transverse phase gradient field is shown superimposed on the phase contour map for a vortex.

and the phase gradient is given by

$$\nabla\phi(x, y) = (\nabla\phi)_x\hat{x} + (\nabla\phi)_y\hat{y} = l\left[\frac{x\hat{y} - y\hat{x}}{x^2 + y^2}\right] \tag{2.7}$$

Note that the gradient is circulating about the vortex point. This can be seen by taking the curl of $\nabla\phi$. This same result in polar coordinates appears elegant and easy to interpret. By noting that the phase distribution of the vortex of charge $l$ is $\phi = l\theta$, the gradient is given by

$$\nabla\phi(r, \theta) = (\nabla\phi)_r\hat{r} + (\nabla\phi)_\theta\hat{\theta} = \frac{\partial\phi}{\partial r}\hat{r} + \frac{1}{r}\frac{\partial\phi}{\partial\theta}\hat{\theta} = \frac{l}{r}\hat{\theta} \tag{2.8}$$

It can be seen that the vortex phase distribution does not have radial phase gradient as $(\nabla\phi)_r\hat{r} = 0$ and has only an azimuthal (circulating) phase gradient.

Let us now return to the phase distribution of a vortex. On the phase distribution corresponding to the phase singularity of charge ±1, any two points positioned diametrically opposite with respect to the singular point is out of phase. Because of this, at the vortex core locations the Huygen's secondary waves emanating from a point at a distance $r$ from the core will destructively interfere with the secondary waves coming from the other point which is at the same distance on the phase distribution. Any point on the vortex core along the propagation direction of the vortex is also equidistant from these two Huygen's point sources which are out of phase. Likewise for every point on the primary wavefront, there is another point on the wavefront that is out of phase and at the vortex core the secondary waves destructively interfere. Hence the vortex core is an amplitude null point.

Even if, at the beginning the wavefront has uniform amplitude, the propagation process drills an amplitude hole at the singular point. The field distribution immediately after the spiral phase plate illuminated by an uniform amplitude plane

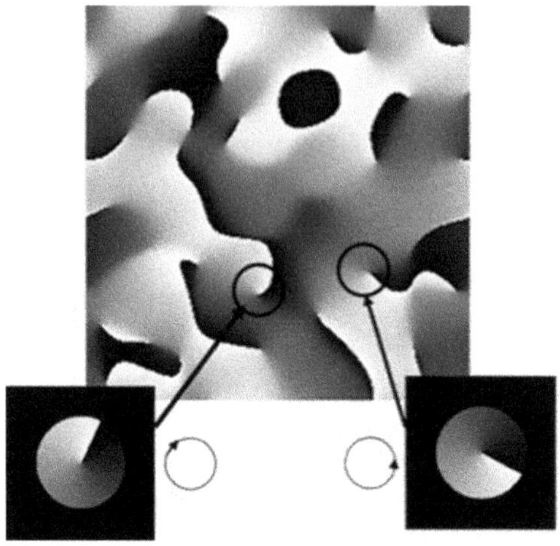

**Figure 2.9.** Random phase distribution and vortices. Most of the vortices are paired as vortex dipoles. Two of the vortices with opposite charge are cut out and shown.

wave, can have uniform amplitude with a vortex phase in it. But the uniform amplitude distribution from the phase element does not sustain as the beam propagates. Destructive interference of secondary waves at the vortex core produces an amplitude zero point at the vortex. Random fields have positively and negatively charged vortices in equal number in the form of dipoles (figure 2.9) and both the charges of the dipoles produce dark amplitude points.

### Phase gradient near zeros

Let us have a closer look at the azimuthal part of the phase gradient given in equation (2.8). The magnitude of the gradient vector is given by $\frac{l}{r}$ and this indicates that the phase gradient increases as one goes near the core and at the core $r = 0$, the gradient blows up. Also note that the phase gradient for a monochromatic wave cannot exceed the propagation constant $\frac{2\pi}{\lambda}$. Hence, near to the core the situation has to be explained using evanescent waves [11] or super oscillations [12].

The region near the core is a mysterious area, and by using an angular spectrum of plane waves, only the regions near the core [13] are considered and the propagation is studied. Studies reveal that the phase gradient near the zero of a vortex [13–16] does have a radial component of phase gradient apart from the circulating phase gradient component. This leads to a dip in the structure of a vortex wavefront near the core as depicted by figure 2.10, where the helical variation is removed for clarity.

## 2.7 Critical points

Real-valued functions of a single variable have extrema in them whereas real-valued functions of two variables can host saddles and extrema in them. But complex-valued functions of two variables can have vortices in addition to saddles and extrema in them.

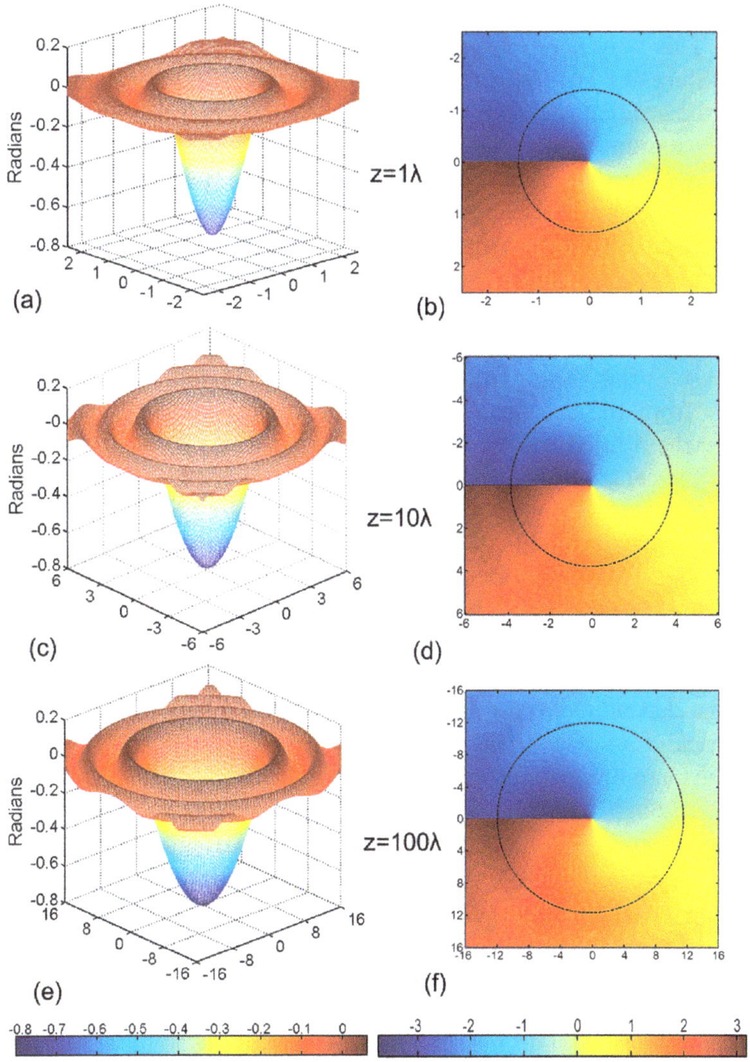


**Figure 2.10.** The structure of the wavefront near the core of a vortex. The helical phase contribution is removed to show the dip that has radial phase gradient. On the left (a, c and e) the wavefront dips after removing the helical phase are shown and on the right (b, d and f) the phase distributions are shown at different propagation distances. Reproduced with permission from [13]. Copyright (2016) by the OSA.

The critical points in an optical field distribution are the maxima, minima, saddles and singular points [17]. Phase singularity refers to the presence of singular point in the phase distribution. In a random or any other spatially varying phase distribution, it is possible to draw phase contours and phase gradient field distribution. To draw the phase contour, first consider a particular value of phase and connect all the points with this phase value on the phase distribution. Likewise for another phase value another contour can be drawn and so on. The phase of an optical field is a scalar and to construct a vector field, draw local normals to the contour surfaces.

The tips of the local normals are such that they point in the direction of increasing phase value. This way a phase gradient field can be constructed from the phase distribution. Since we have considered two-dimensional phase distribution at any given plane, the gradient field that is considered is a transverse gradient field.

A phase extremum (maximum or minimum) is always surrounded by closed phase contours. The gradient field distribution is such that the phase gradient vectors point in the direction of maximum ascent. At the extremum point the gradient (transverse gradient) vector becomes zero and in the immediate neighborhood of the extremum point the gradients are pointing inwards/outwards to the extrema. In the jargon of Fourier optics, these extremum points are zero local spatial frequency points. A wavefront that has at least one phase extremum point has at least one propagation vector that is along the optical axis of the system. Otherwise the beam is drifting away from the optical axis (e.g. as in the case of tilt). Beams that have symmetrically distributed spatial frequency components about the zero frequency can maintain the centroid of the beam along the axis of the optical system.

At the second critical point namely the saddle point, the phase contours touch each other [18] and these two touching contours correspond to the same phase value. Since contours represent the same phase value lines, at the saddle point the gradient vector disappears as in the case of an extremum point, but at the neighborhood of a saddle point the gradient vectors point towards the saddle point in certain regions and in other regions they point outwards. Basically, a saddle point can be a maximum point and/or a minimum point depending on the direction of approach. In a saddle which is normally used on the back of a horse, if one moves from the front to the back of the horse, you will go through a minimum and if you go from the left to the right side of the horse, you will have to go through a maximum point. Hence the saddle point is seen as a maximum as well as a minimum point depending on the direction of approach to the point. Saddle points radiate a pair of bifurcation lines—phase contour lines.

The extrema and saddle points have a close association with each other. The four arms of the bifurcation lines that emanate from a saddle point can be open, two arms closed at one end and two arms open at the other end, all the four arms are closed forming a double loop structure or a loop interior to the other as shown in figure 2.11. In the immediate neighborhood of a saddle point, the regions where the gradient vectors are pointing towards the saddle point are indicated by positive sign and the regions where the gradient vectors are pointing away from the saddle point are indicated by negative sign. The coexistence of extrema and saddles in possible configurations are depicted in figure 2.12.

The third critical point we consider here is a vortex point in which the phase contour lines appear to converge at, or diverge from, a point. At the vortex point since all the phase contour lines terminate, the value of phase at the vortex point is indeterminate. Because according to the definition of a contour line drawn at a particular value, at the vortex point the phase is decided by the contour line in which the point lies. But since all the contour lines of different phase values meet at the singular point, there is an ambiguity of which value of phase has to be assigned to the vortex point and hence the phase at that point becomes undefined. In a coherent monochromatic electromagnetic field at every point and at every

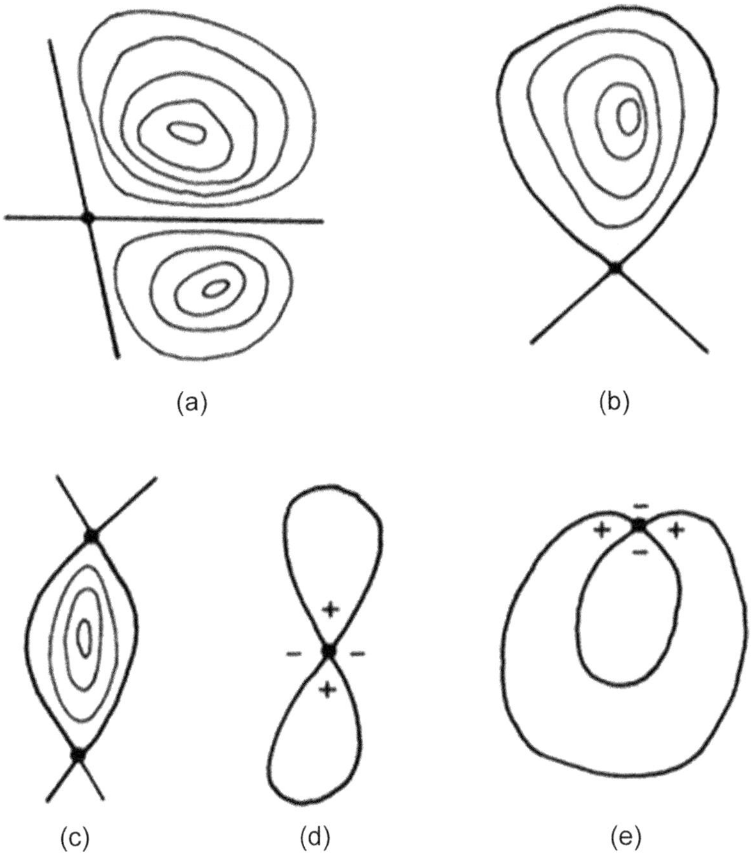


**Figure 2.11.** Possible bifurcation line configurations. Saddles and extrema. (a) Neighboring extrema must be separated by a bifurcation line. (b) Extremum embraced by the joined arms of a single saddle. This is the generic arrangement in a random phase field. (c) Two saddles join arms to embrace one extremum yielding a topologically possible but non-generic arrangement. (d) Both loops of a figure eight embrace the same type of extremum (maximum or minimum). (e) The loops of a reentrant saddle embrace extrema of opposite type. Here the exterior loop embraces a maximum and the interior loop a minimum. Reprinted with permission from [17]. Copyright (1995) by the American Physical Society.

time the phase has to have a value. Since the wave function has to be single valued, we do not have the choice of having many phase values at the singular point. Hence an easy way out of this situation is to make the electromagnetic field vanish at this point, so that the point has undefined phase value. Hence a vortex point is an amplitude null point.

The coexistence of these three critical points can be seen by considering a random phase distribution in which phase contour lines are drawn and shown in figure 2.6.

## 2.8 Zero crossings and bifurcation lines

At the lowest order saddle, four phase contours corresponding to the same phase value will touch each other and these phase contour lines are called bifurcation lines.

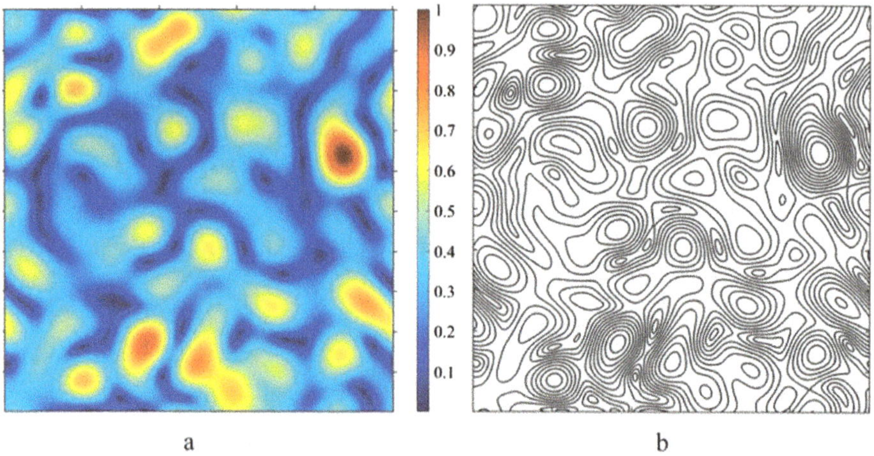


**Figure 2.12.** (a) Height distribution in a landscape. (b) Height contours. In real-valued functions, coexistence of extrema and saddle points in a distribution occur. A random positive valued function is considered here. This can be a landscape (or intensity distribution) that has peaks, valleys and saddles and the contours represent height contours. Steeper up- and down-slopes have crowded contours.

A bifurcation line can be a zero crossing (of the real/imaginary part of the wavefield) but not all zero crossings are bifurcation lines. The phase contours going through the saddle need not have the phase value corresponding to real/imaginary zero. But by adding constant phase to the phase distribution, it is possible to move the phase contours in the distribution. The phase contours corresponding to real/imaginary zeros can be moved in this way and can be made to touch at the saddle point. If the phase value is any one of $n\,\pi$, where $n$ is an integer, then the phase contour is an imaginary zero curve and if the phase value is any one of $\left\{n + \frac{1}{2}\right\}\pi$ then the phase contour is a real zero curve. The saddle can be of higher order and some examples are shown in figure 2.13 in which the curves shown are phase contours of single value and they all touch at (or pass through) the saddle point. It is possible to make all these contours be real zero curves or imaginary zero curves by adding an appropriate constant phase value to the distribution.

With these points in mind one can see that at the saddle point an imaginary zero contour can cross another imaginary zero contour line (basically they touch each other so that it appears as a crossing). Likewise at the saddle, a real zero contour can cross only another real zero contour (figure 2.14). But at a vortex point a real zero contour and an imaginary zero contour cross each other. Secondly, the zero crossing has no phase jump while going through the saddle whereas the zero crossing has a $\pi$ phase jump (charge $\pm 1$ considered for argument) while going through a vortex. Since all the phase values are present in the immediate neighborhood of a vortex, all the phase contours and hence the zero crossing of the real and imaginary part of the wave function will happen to cross each other at all times and at all phase shifts unlike in the case of saddle points.

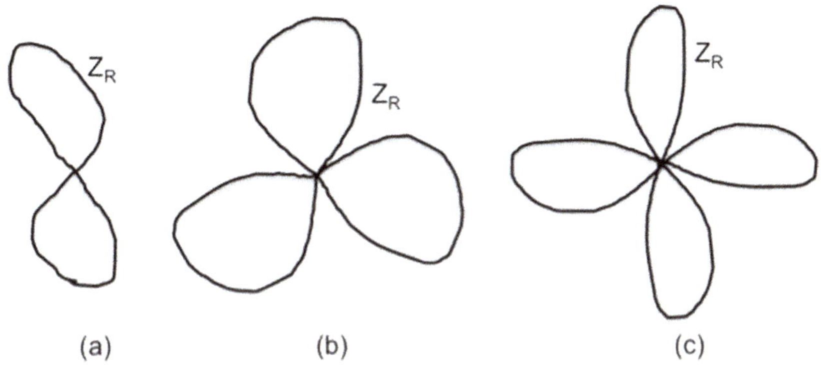


**Figure 2.13.** Saddle of order (a) one (b) two and (c) three are shown. The saddle shown in (a) has four arms, the saddle in (b) has six arms and the saddle in (c) has eight arms. Hence index the saddle in (a) is +1, (b) is +2 and (c) is +3 as all of them have closed arms.

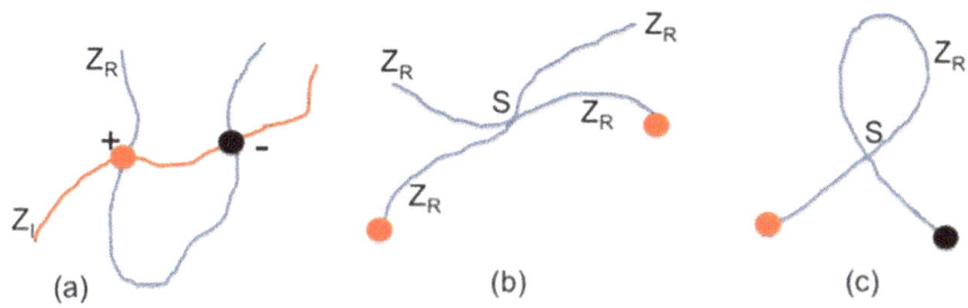


**Figure 2.14.** (a) At the vortex contours of $Z_R$ and $Z_I$ cross each other. Any two adjacent vortices on a contour of $Z_R$ have alternate signs. Any two adjacent vortices on a contour of $Z_I$ also have alternate signs. (b) Any two adjacent vortices on a $Z_R$ contour (or a $Z_I$ contour) with a open saddle in between them have same sign. (c) Any two adjacent vortices on a $Z_R$ (or a $Z_I$ contour) have alternate signs if there is a closed saddle between them. In a one-loop saddle, the bifurcation line passing between the singularities contains two saddle points as the saddle point is traversed twice by the bifurcation line.

## 2.9 Charge, order and index

The charge of a vortex is equal to the order multiplied with sign of the vortex [19]. The order is equal to the magnitude of the charge for a vortex. In terms of zeros, crossing the critical point, the order of a vortex is equal to $\frac{N_v}{2}$, where $N_v$ is the number of zeros (both real and imaginary) crossing the vortex. The order of a saddle is given by $\frac{N_s}{2} - 1$, where $N_s$ is the number of arms of the saddle. Therefore an $n$th order vortex has $n$ zero crossings, an $n$th order saddle has $n + 1$ zero crossings. Note that at a zero crossing two zeros (two $Z_R$s or two $Z_I$s for a saddle or one $Z_R$ and one $Z_I$ for a vortex) are crossing. To find the order of the saddle, consider the $n$th order vortex for example, $(x + iy)^n = r^n \cos(n\theta) + ir^n \sin(n\theta)$. If you consider the real part of the field, $R^{(n)} = r^n \cos(n\theta)$ it has a saddle point of order $n - 1$ at the origin [19]. Similarly the imaginary part of the field, $I^{(n)} = r^n \sin(n\theta)$ can be seen to contain a saddle point of order $n-1$. Hence when a vortex of order $n$ is disintegrated, there can be many saddle points appearing as byproducts.

Index is the topological index given to each of the critical points [19]. Since basic topology is beyond the scope of this book, I request readers to assume the topological index values as presented here. For both positive and negative charged vortices of any order, the topological index is +1. Extremum has a topological index +1. An open saddle of order $n$ has topological index of $-n$ and a closed saddle has an index of $+n$. A saddle that has $n+1$ zero crossings has an index of $-n$ and order $n$.

## 2.10 Sign rules

Also termed as principles, there are sign principles [20], enlarged sign principles [17] and extended sign principles [21] reported in the literature. The real and imaginary zeros of the wave functions cross each other at the singular point. On real zero contour adjacent singularities are of opposite sign (sign rule). It is also true for imaginary zero contours. But when the real (imaginary) zero contour is going through a saddle point, on either side of it singularities of the same polarity can exist. Application of the (enlarged) sign principle demonstrates that the singularities that terminate on a bifurcation line of an open arm saddle must be of the same sign. Further singularities with the same (opposite) sign terminate bifurcation lines containing an odd (even) number of saddle points.

According to the above rule, the singularities that terminate the bifurcation line of a one-loop saddle are shown to have opposite sign. As the saddle point is traversed twice in passing between the singularities, the bifurcation line contains two saddle points. These sign rules are depicted in figure 2.14.

An important topological constraint involves the signs of adjacent vortices on zero crossings. If two vortices terminate a zero crossing segment containing no saddle points then these two vortices are of opposite sign. But if there is an intervening ordinary first-order saddle, both vortices must have the same sign. This highly useful rule is called the sign principle, which in the form stated is applicable only to generic vortices of any order. It permits one to rapidly determine the relative signs of all vortices on a given set of zero crossings.

The sign principle can be extended to include the case of non-contacting, apparently isolated zero crossings. This can be done by extending either the $Z_R$ or $Z_I$ line that is crossing a vortex of known charge. If the $Z_I$ contour is taken, it is extended till it crosses the (say) $Z_R$ that runs through the vortex of unknown charge. Now the sign rule is applied in the extended segment and tracing the charges of the vortices at each of the zero crossings, the polarity of the unknown charge can be determined. This extended sign rule is explained in figure 2.15.

## 2.11 Disintegrations or explosions

Higher-order critical points are called degenerate critical points. Under the influence of a small perturbation a degenerate critical point can decay explosively into a large number of irreducible non-degenerate components [19]. The mound of debris resulting from decay of a vortex of order 10 as found in superconductors can

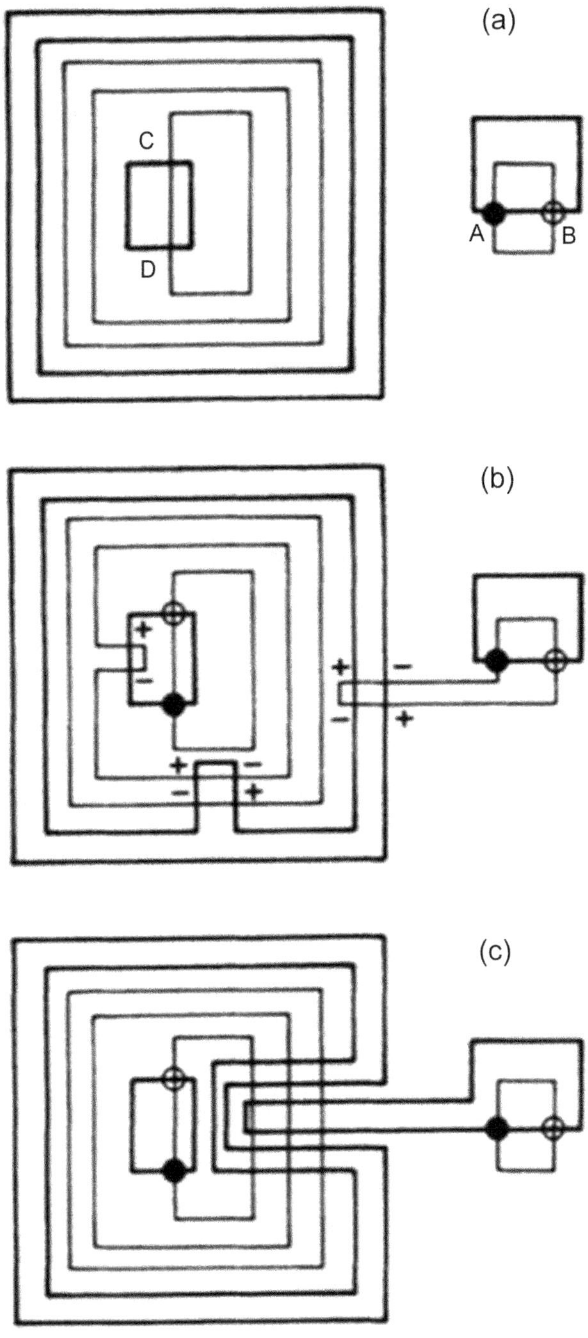


**Figure 2.15.** Application of the extended sign rule to identify the signs of enclosed vortices C and D. Thick and thin lines are lines of $Z_R$ and $Z_I$, respectively. (a) Vortex A is positive and vortex B is negative. (b) and (c) are examples of two different possibilities of contour extensions that permit the sign of the vortex C to be determined as negative and the sign of the vortex D as positive. Reprinted with permission from [21]. Copyright (1994) by the American Physical Society.

exceed $5 \times 10^6$. These explosions are within the constraints of conservation rules. The index conservation and charge conservations are discussed here.

A higher-order saddle can explode into a large number of first-order saddles and extrema. An $m$th order saddle can disintegrate into $\frac{m(m+1)}{2}$ first-order saddles and $\frac{m(m-1)}{2}$ extrema. In a phase distribution enriched with phase saddles and extrema, the phase gradient lines reach and disappear at the saddle in certain directions and leave from the saddle from other directions. At extrema the phase gradient lines normally terminate or originate. The phase gradients associated with phase vortices are circulating in nature around the singular point.

## 2.12 Charge conservation

The topological charge of the vortex is a conserved quantity. According to the law of charge conservation, the net charge within some bounded region is conserved under continuous deformation of the wave function, provided that no charges enter or leave the region by crossing the boundary. Therefore, a vortex of charge +5, after disintegration results in the final product of five vortices each with charge +1.

While the charge conservation decides the charge of each of the resulting vortices, actually the disintegration gives rise to other types of critical points appearing in the field distribution. An $n$th order vortex disintegrates into $n^2$ first-order vortices along with many saddle points and extrema. Out of these $n^2$ vortices, there are $\frac{n(n+1)}{2}$ vortices having one sign and $\frac{n(n-1)}{2}$ vortices having the opposite sign so that the net charge before and after explosion is conserved [19]. This means that a vortex of charge +6 disintegrates into 21 vortices each having charge +1 and 15 vortices of charge −1 so that the net charge before and after explosion is equal to +6.

## 2.13 Index conservation

Index conservation is more important than charge conservation. The net index of critical point(s) before and after disintegration is a conserved quantity. Consider the disintegration of an $n$th order vortex, that leads to the creation of $n^2$ first-order vortices. Since the topological index of each of the vortices irrespective of the order or sign is +1, the initial index is +1 and the byproduct $n^2$ vortices put together has index $n^2$, hence there should be $(n^2 - 1)$ saddles each with topological index (−1) according to the index theorem. Hence $+1 = n^2 - (n^2 - 1)$ and the index is conserved [19]. If the explosion creates extrema also, it is possible to have more saddles than this number to conserve the topological index.

## 2.14 Limitation on vortex density

The phase gradient of the vortex is such that at regions near to the core of the vortex very high phase gradients are possible. But having phase vortex dipoles can limit the magnitude of the rapid spatial fluctuation of phase. Accordingly, to avoid rapid spatial fluctuations in phase there exists a kind of limitation on the number of vortices in a given area. Roux [22] has shown that the net topological charge in an area cannot exceed the circumference of that area divided by the wavelength.

In speckle fields which are random vortex fields [20], the vortices are anti-correlated with their nearest neighbors and tend to be uncorrelated with vortices further apart.

The local spatial frequency which is a function of local phase gradient gives an indication of the angular spectrum of the optical field. The maximum spatial frequency of a propagating monochromatic wave (wavelength $\lambda$) is $\frac{1}{\lambda}$. If the average spatial frequency on the contour is larger than $\frac{1}{\lambda}$, at certain regions it indicates that there is a need to have evanescent waves excited in these regions that do not propagate. Such high vortex density regions are possible in engineered fields. According to Roux, if $n$ is the net enclosed topological charge of vortices in a region surrounded by a boundary of length $L$, the vortex density limitation that is applicable is given by $\frac{n}{L} < \frac{1}{\lambda}$.

## 2.15 Threads of darkness

When two beams interfere in space, the interference pattern forms bright and dark surfaces in the volume of overlap. The intersection of these surfaces on a two-dimensional observation plane appears as an interference fringe. Instead of two beams if three or more plane waves whose propagation vectors are non-coplanar interfere, interference makes the light vanish at lines rather than on surfaces. And these dark lines when observed on a two-dimensional plane, appear as dark points instead of fringes and each of these dark points are seen as vortices, where the phase is indeterminate.

These vortices extend in three dimensions as a curve, and are called the threads of darkness. In the context of sound waves, these are called curves of absolute silence. Therefore singular optics can be considered as the study of the dark side of the light.

## 2.16 Berry's paradox

Reduction of three-dimensional fields to a set of parallel two-dimensional slices is called foliation of the field [23]. These two-diminsional observation planes are called the leaves of the foliation. Normally these leaves of foliation are made in such a way that the foliation and the normal beam axis is 90°, but when this angle is not 90° the observer sees in his leaves different results. Critical foliations are in a way rules and not exceptions as the event is same as that being observed, but viewed from different angles. Isaac Freund terms this paradoxical situation [24] in which nominally equivalent, equally valid experiments produce entirely different results as a 'Berry's paradox'.

Consider the vortex trajectory as shown in figure 2.16 in which the three observation planes are indicated by A, B and C. Depending on which observation plane the observer is using, he will see the presence of a dipole, or singularity-free field. Also as he moves the observation plane from A to B he sees that there is an attractive force between the two vortices that attract each other and at one stage they collide with each other and annihilate. But the foliation shows that there is a single vortex that has formed a loop trajectory whereas the different observers have witnessed different

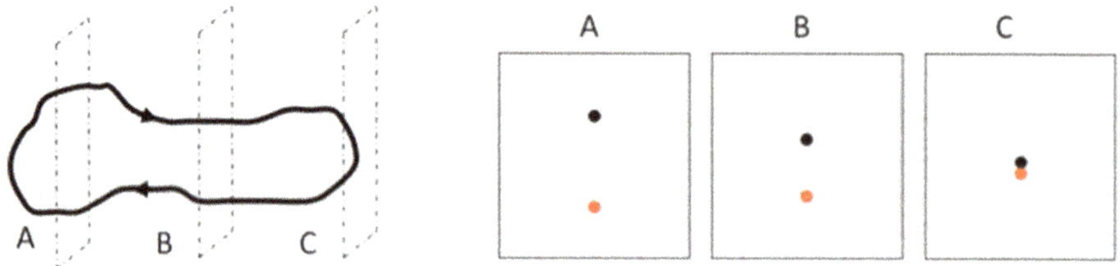


**Figure 2.16.** Berry's paradox. The oriented curve is the vortex trajectory. An observer sees that there is an attractive force between the oppositely charged vortices and there is an attraction between them that draws them closer to each other by observing planes A, B and C sequentially.

events. Also when the foliations are at different (small) angles, the observations will be completely different. First of all the charge of the vortex as defined in equation (2.5) is no longer required, as the same vortex is considered as a dipole with positive and negative vortices separated by a distance. Hence, what is suggested is that instead of charges we can use directed arrows for the vortices. The arrow directed towards the observer is seen as positive and the arrow directed away from the observer is seen as a negative vortex. Then what about the sign rules? They are still there, and there are two conservation laws, namely vortex topological charge conservation and topological index conservation. These indices are shown conserved under changes in foliation [23].

The location of the vortices in a plane and their trajectories in three dimensions are invariants (foliation-independent). The stationary points computed using zeros and two-dimensional gradients are dependent on the coordinate system used and hence their locations and trajectories in three dimensions are foliation dependent.

Here we simply use the term trajectories to refer to these oriented lines, since much of what follows applies to the general complex field independent of whether or not it is embedded within a wave. As stressed by Berry, the essence of the paradox is that what one sees depends on how one looks, so that different observers examining the same wavefield from different perspectives will see different arrangements of point vortices and will infer different reactions for these points. Berry notes that the problems embedded in the paradox arise whenever a three-dimensional field is described in terms of two-dimensional section foliations. Referring to figure 2.17(a), even though the trajectory of the vortex core is a single one, it is interpreted as a positive vortex in one perspective view and negative vortex from another perspective view and even the single vortex is interpreted as a dipole. Vortex creation and annihilation cannot happen anywhere on the curve. For a single vortex moving along a curved trajectory, how can the vortex charge, which is a topological invariant, change along the trajectory? Hence what is suggested by Freund is that the vortex charge can be dispensed with, by using oriented trajectories.

## 2.17 Manifolds and trajectories

Vortex trajectories are the curves obtained [25] by the intersection of surfaces of real zero and imaginary zero. These surfaces generally consist of many disconnected pieces. In figure 2.6 some of the real zero and imaginary zero curves obtained by the intersection of these surfaces with the observation plane are shown disconnected.

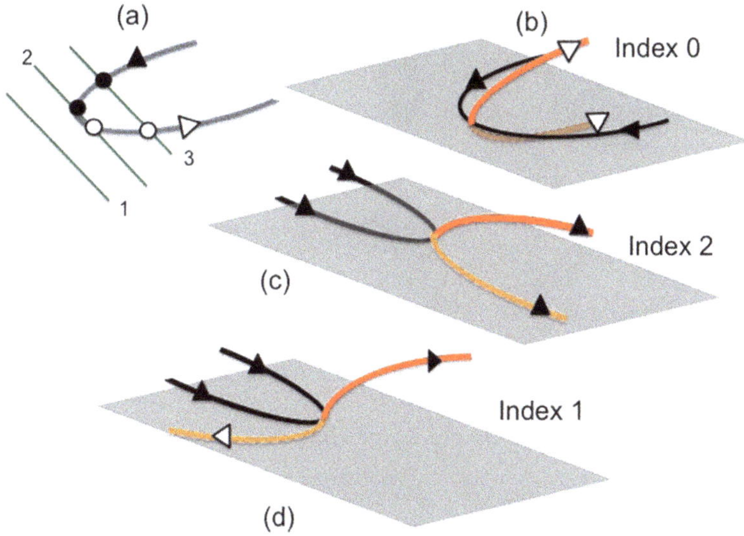


**Figure 2.17.** Birth and death of vortex pairs along with other critical points. Black and white arrowheads represent vortices of positive and negative charge respectively. Black thick trajectories are for vortices and colored trajectories are for other stationary points (saddles and extrema). The trajectory above the plane of the paper (where vortex trajectory lies) is depicted in red and the trajectory below is depicted in orange. Trajectories in (a) are for charges, (b) for indices, (c) for indices, (d) for indices. In (a) an observer moving along the plane 1, 2, 3 sees the creation of vortex pairs (dipole) hence the net charge is always zero. In (a) the index is not conserved in the three leaves of the foliation. To conserve the index, stationary point trajectories can be added to it, and three different possibilities with different topological indices are discussed in figures (b)–(d). In (b), the observer sees the creation of a vortex dipole along with two saddles as he goes from plane 1 through 3, in this case the index as well as the charge are always zero. In (c), the observer sees a maximum and a minimum transforming into a vortex dipole. Here the charge is zero and the index is always two. In (d) a minimum is transforming into a saddle and a vortex dipole. Here in all the leaves of foliation the net charge is zero and the index is one.

Any of these pieces (surface) is referred as a manifold and can be denoted by $R_0$ or $I_0$. On the surface of the manifold (take any one surface of real zero), continuity, differentiability, and single valuedness of the wave function forbids the presence of holes, slits or any sort of discontinuities. Some of the compact manifolds are shown in figure 2.18. Mathematically, a compact manifold is a manifold which is compact. This definition is not helping us in any way. Maybe, compact means finite and the manifold is a surface. A closed manifold is a compact manifold (surface) without a boundary. Some of the closed compact manifolds are disc, sphere, torus and generalized tori [24]. By defining a genus that indicates the number of holes a surface has, a sphere is a genus zero surface, a donut or torus has genus one. A torus is a sphere with one handle. A sphere with two handles has genus two; it is a coffee cup with two handles. The term ‘generalized tori’ used by Freund [24] is a surface with genus greater than one. Discs and spheres do not have a handle, whereas a torus has a handle and generalized tori have many handles and none of them have a boundary. An open manifold is a manifold without boundary and hence not compact. All the open edges of the manifolds are at infinity. The vortex trajectory can be seen as a curved line drawn on the surface of $R_0$ without any reference to

**Figure 2.18.** Different types of compact manifolds. First row—simple and compact manifold and equivalent surfaces, second row—torus and equivalent surfaces, and the third row—genus with two surfaces with multiple handles and equivalent surfaces.

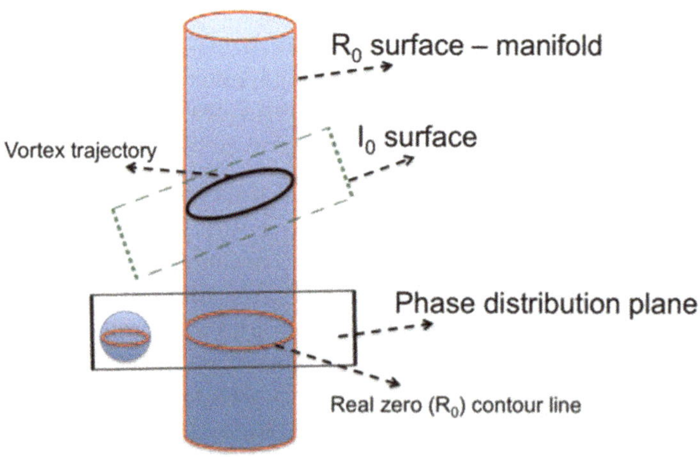


**Figure 2.19.** Manifold construction from phase contours drawn on a two-dimensional observation plane. Let us consider a circle which represents a real zero contour line curve in the two-dimensional phase distribution. The real zero contour surface is a cylindrical tube extending to infinity on both directions. This tubular surface is the manifold. Another manifold which is a compact manifold, for the other real zero contour which is disconnected from the other manifold is shown as a sphere. The intersection of the imaginary zero surface with the manifold (cylinder) gives the vortex trajectory.

imaginary zero surface. We can also use $I_0$ surface as a manifold and vortex trajectories on them without reference to real zero surface. These manifolds generally consist of many disconnected pieces. In figure 2.19 a sphere and a cylinder are two manifolds which are disconnected. But one is compact and the other is not. In the observation plane (phase distribution) these two manifolds, are observed as two different circles. In this example they are shown as two different manifolds. Otherwise, (another possibility) through these two circles, a torus can pass through, and these two circles may correspond to one single closed compact manifold.

We have seen earlier that by adding a constant phase to the wave function, one can move the phase contour lines as well as real and imaginary zero lines in a plane. This exercise was done to make any of the phase contours become either a real zero or an imaginary zero curve that can go through the saddle. But at the vortex point this perturbation does not disturb the crossing of real and imaginary zeros. Hence adding a constant phase results in the change of shape of the surfaces $R_0$ and $I_0$, but the same trajectory looks different on the manifold as the shape has now changed. But as far as the trajectory of the vortex core is concerned, it remains the same in three dimensions. We note that the geometry of a vortex trajectory is unchanged by such a transformation, which is referred to as generalized gauge transformation [23, 26]. The orientation of the trajectory also remains invariant under gauge transformation.

Manifolds may be closed (compact) or open that have boundaries at infinity. Spheres, and other closed shapes are compact. A torus and the surface of a tea cup are also compact and each of them can be considered as a sphere with a handle. A generalized tori is a sphere with an arbitrary number of handles. The handles are like bridges that allow the trajectories to cross over each other—like one segment of the trajectory is over the bridge and the other below the bridge. Hence the crossing in a torus helps to create knots. Handles open a route to exotic trajectories which otherwise would be impossible. The trajectory must be on the surface and no part of it is allowed to leave the surface of its manifold.

One open manifold example is a plane with its boundary at infinity. Tube shaped manifolds are also possible in which case the open ends of the tube are located at infinity. Planes and tubes appear to be the generic manifolds in Gaussian laser beams.

Self-intersecting surfaces such as interpenetrating spheres are non-generic and they are excluded from discussion for the sake of simplicity. Under small perturbations, they decompose into two or more independent pieces.

### Trajectories

On a plane or on an infinitely long cylinder, both closed and open trajectories are possible. Trajectories may be simple closed loops, lines that start at say $(-\infty)$ and end at $(+\infty)$ or lines bent into a U with both ends of the U ending at $(\pm\infty)$. These three types of trajectories are possible on an open manifold.

On compact manifolds only closed trajectories are allowed. On a torus, which is a compact manifold, three types of closed trajectories namely p, r and t are possible [24]. All these three p-, r- and t-type trajectories are depicted in figure 2.20. The p-type is a closed trajectory that can be shrunk to a point and hence is allowed on all

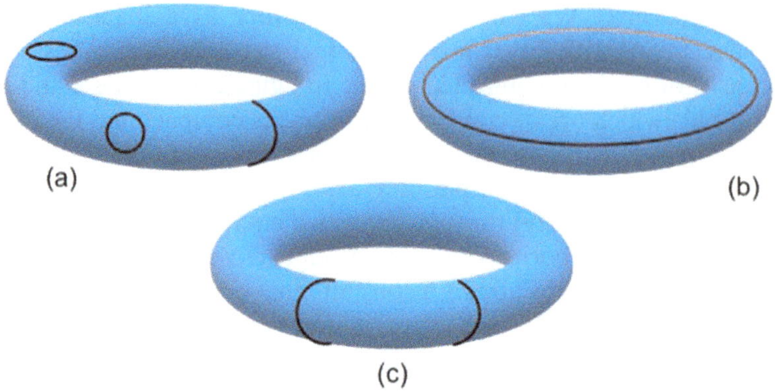


**Figure 2.20.** (a) Two p-trajectories on the left and one r-trajectory on the right are shown on the surface of a torus. (b) One t-trajectory is shown. (c) Two r-trajectories are shown on the surface of a torus.

types of manifold. That is the only possible trajectory on a sphere as in a compact manifold only closed trajectories are allowed. The trajectory labelled r is like a ring on a finger and sits on the handle of the torus. The third trajectory labelled t is a closed loop running along and on the surface of torus, mirroring the shape of the torus. If a torus contains only a single trajectory, this must be a p-trajectory. All these three types of closed trajectories on a torus can be resized, slid along the surface of the manifold, rearranged, provided that during this operation the trajectory does not break or open up, leave the surface of the manifold, or cross another trajectory. These operations facilitate the application of the sign rule.

Having introduced the types of trajectories, now let us concentrate on the relative orientations of the trajectories. If the orientation (arrowhead) of any one trajectory on a given manifold is known, the orientations of all other trajectories on the manifold can be automatically determined. To find the orientations, the trajectories need to be brought into close proximity with each other.

The sign rule [20, 21] will be satisfied if the arrowheads on two immediately adjacent trajectories point in opposite directions.

**p-trajectories** The implication of the sign rule is that on any manifold, sets of independent p-trajectories undivided by any other trajectory type, all must have the same orientation, while nested p-trajectories (one inside the other) must have opposite orientations.

**r-trajectories** For the examination of r-trajectories on a torus or r-trajectories on different handles on a tori, bringing the trajectories in close proximity to each other is difficult as these trajectories cannot be resized or rearranged as done before for p-type trajectories. Hence to bring them closer to each other, we need to make burrows and eventually make a wormhole and connect the trajectories. This process of making a burrow and wormhole is explained in figure 2.21 These burrows are created by pushing and extending the trajectory on the supporting manifold without violating any rule mentioned before and hence are permissible (figure 2.22). This exercise creates two counter-propagating, self cancelling trajectory segments. Then the process of merging the adjacent counter-propagating segment is carried out to form a wormhole. Opening

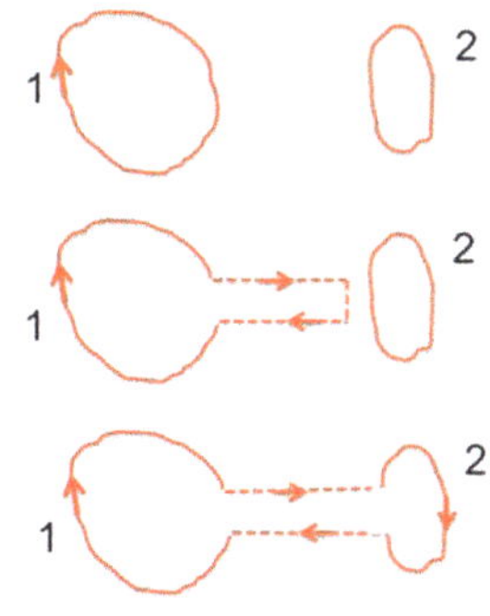


**Figure 2.21.** Process of forming a burrow and wormhole.

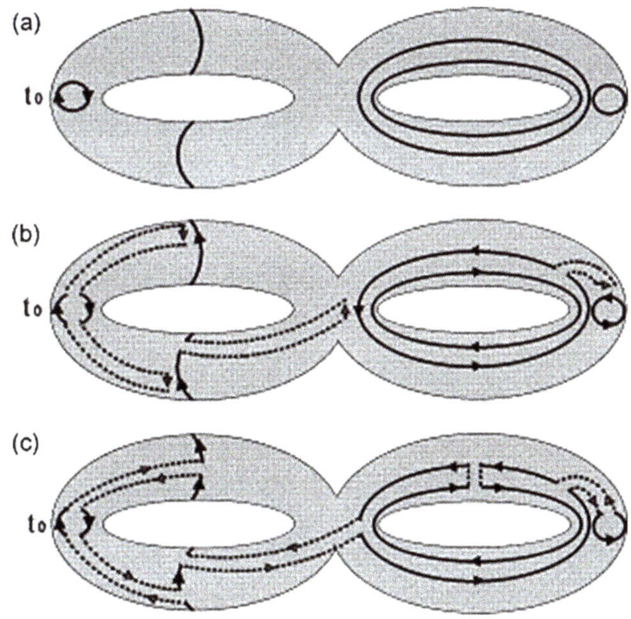


**Figure 2.22.** Process of forming a burrow and wormhole on a genus two torus that hosts p-, r- and t-type trajectories. Although this process decides the direction of legal arrowheads on every closed trajectories, use of wormholes lead to a single closed trajectory at the end. Reprinted from [24]. Copyright (2000), with permission from Elsevier.

or forming a wormhole merges two trajectories into one while closing the wormhole cuts a single trajectory into two. The arrowheads in all these new trajectories must be consistent and any inconsistency means that the sign rule has been violated. Violation of sign rule happens for two r-trajectories (figure 2.23) with same orientation and two t-trajectories (figure 2.24) with same orientation on a torus and hence these are not allowed. In figure 2.25 violation of sign rule for two r-trajectories with same orientation is explained by using t-trajectories and forming burrows and wormholes.

**t-trajectory**: It is a closed loop running along and on the surface of torus, mirroring the shape of the torus. Sign rule does not allow a single t-trajectory. This

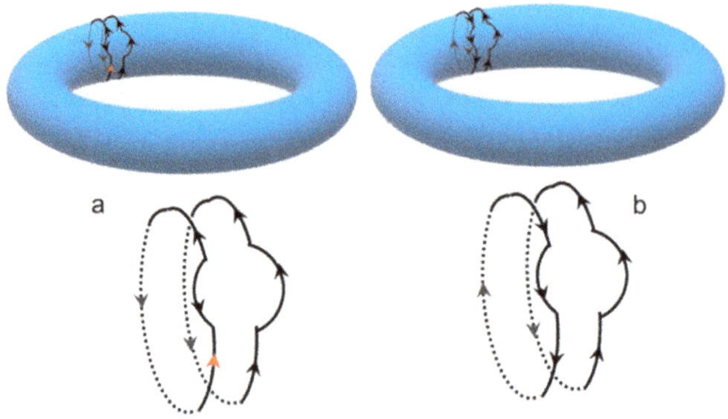


**Figure 2.23.** Two r-trajectories with (a) same orientation is not allowed but (b) forming a null pair is allowed. This has been explained with the help of p-trajectory and by making burrows and wormholes.

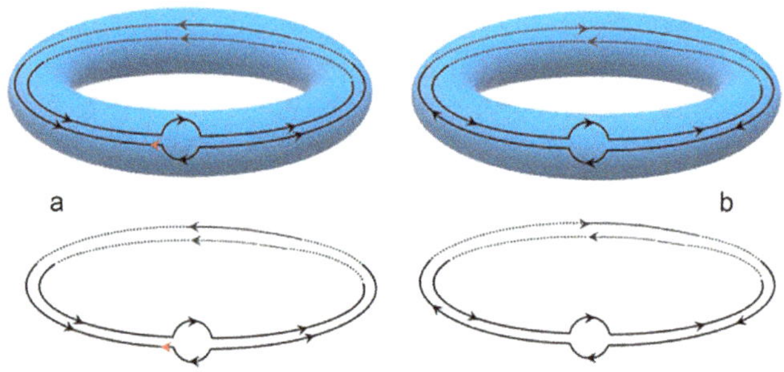


**Figure 2.24.** Two t-trajectories with (a) same orientation is not allowed but (b) forming a null pair is allowed.

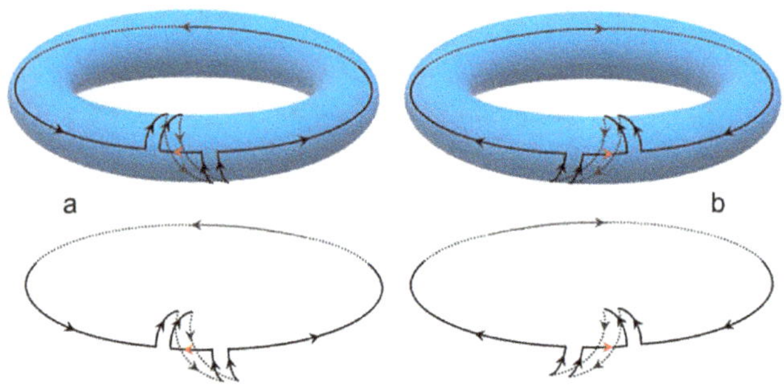


**Figure 2.25.** Two r-trajectories with same orientation is not allowed. This is shown by using a t-trajectory oriented in two different possible directions (a) and (b). This same example can be used to show that single t-trajectory is not allowed on a torus. By connecting different parts of the t-trajectory using burrows and wormholes formed using pair of r-trajectories, this can be seen.

has been depicted in figure 2.25. It should appear in pairs on the surface of the torus. In the pair the two trajectories must have opposite orientation and form a null pair as depicted in figure 2.24.

Connecting different parts of the r-trajectory by a t-trajectory wormhole, leads to a contradiction of the direction of the arrowhead. Similarly connecting different

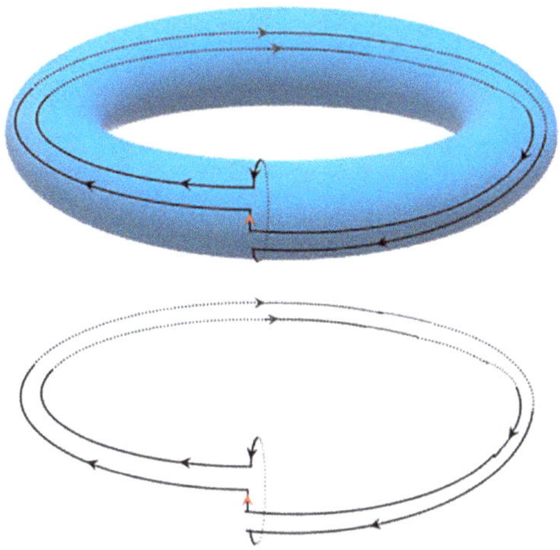

**Figure 2.26.** Single r-trajectory on a torus is not allowed. Using burrows and wormholes made of t-trajectories, it has been explained.

parts of the t-trajectory by r-trajectory wormhole, leads to contradiction on arrowhead direction as shown in figure 2.23. Hence a single r- or a single t-trajectory on a handle violates the sign rule (figures 2.25 and 2.26) and is therefore forbidden [24].

Two counter-propagating trajectories close to one another called null pairs can be inserted anywhere (as long as they do not overlap with other trajectories) on the manifold as they cancel each other and this is tantamount to having no trajectory at all. Higher-order vortices involve self-intersecting manifolds and that analysis is beyond the scope of the book.

In figure 2.22, a genus two torus with two handles is shown. In figure 2.22(a) one p-trajectory is shown on the surface of the left side torus. There are also two r-type trajectories on the same torus. In the second torus which is connected to the first, one p-trajectory and two t-trajectories are shown. From the given orientation of the $t_0$ trajectory one can find the orientations of all other trajectories on the manifold using the process of forming burrows and wormholes as shown in figures 2.22(b) and (c). The orientation of the arrowheads of the trajectories from the knowledge of the initial closed trajectory are unique. This means that there is only one set of correct orientations for the trajectories possible.

## 2.18 Links and knots

Consider now exotic trajectories on the surface of the manifold. They are spirals, links, and knots. Links and knots are just oriented closed curves.

On compact manifolds trajectories are closed curves (loops) and cannot have a beginning or an end. Such closed trajectories form wavefield dislocation loops. The exotic trajectories such as spirals, links and knots are possible on manifolds with suitable handles.

On a torus t-trajectories must always appear as null pairs. Since an unpaired t-trajectory is not allowed on a torus, linked manifolds cannot give rise to unpaired

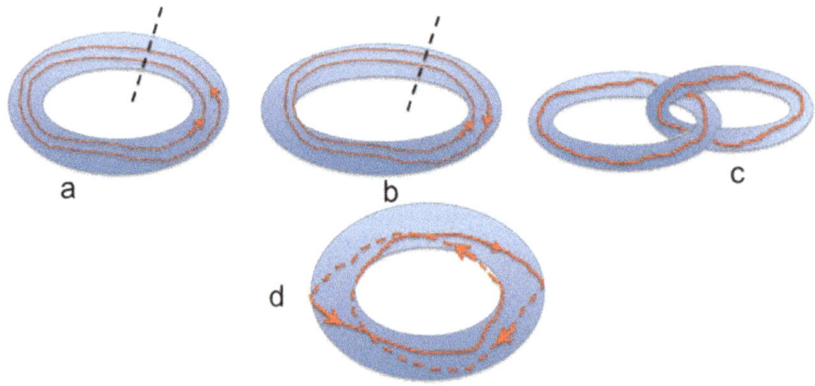


**Figure 2.27.** Link formation. (a) Two co-propagating t-trajectories on a torus is not allowed. This can be seen by inserting an observation plane (marked by the dashed line) one can observe that there is a sign rule violation. (b) Hence two counter-propagating t-trajectories called a null pair is allowed on a torus. (c) Link formation by linked manifold with single t-trajectory in each of the torus is not allowed because a torus supports only null pairs. (d) Link formation on a torus. In (d) the two trajectories appear to be t-trajectories and are indeed two counter-propagating r-trajectories.

linked trajectories (figure 2.27). But a null pair on a single torus can form a link. Although the trajectories in the link appear to be t-trajectories, they are in fact counter-propagating r-trajectories. Torus links require only one manifold (either $R_0$ or $I_0$) to be compact.

Using the sign rule, it is easily seen that a spiral trajectory scribed into a torus handle or other closed manifold is forbidden. A single trajectory is scribed into a manifold that is itself twisted into a spiral. The reason for this exclusion is the same as for the exclusion of single r- and t-trajectories on a torus. Spirals constructed from counter-propagating null pairs, however, are always allowed. Since a spiral may be formed by a line scribed into a twisted sheet open manifold, single spirals can be implanted into Gaussian laser beams.

A minimum of three crossings are required to form a knot. Freund [24] has shown that knotted trajectories are not possible on closed manifolds. Violation of the sign rule prevents knot formation in many cases as shown in figure 2.28. Knots of any order formed from counter-propagating null pairs, however, are, as always, allowed. A single t-trajectory is permitted on a non-compact torus, i.e. a torus whose outer circumference has been slit open and the cut edge extended to infinity. A knotted trajectory on a generalized, non-compact higher-order torus therefore remains a possibility.

A random field formed by the interference of scattered coherent light, is shown to have complicated vortex trajectories as shown in figure 2.29. In fact a speckle field is formed due to the interference of light from many scattering points and is a kind of multiple beam interference. Therefore it suggests that complicated vortex trajectories are possible in multiple beam interference. It has been shown by multiple beam interference (figures 2.30 and 2.31) that many of the complicated trajectories are

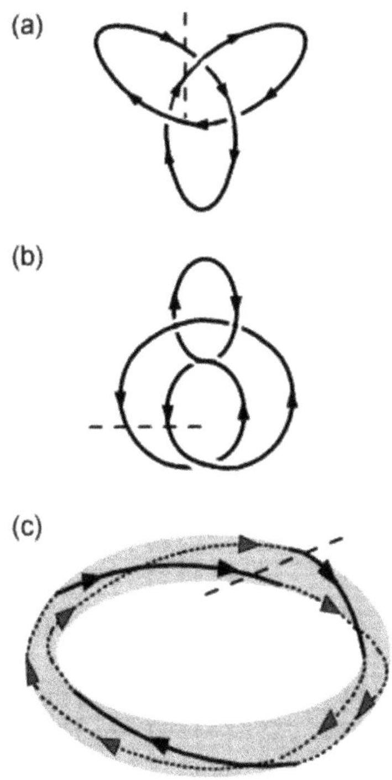


**Figure 2.28.** (a) Trefoil knot. (b) Figure of eight knots are not possible as the insertion of the observation plane (shown as dashed line) reveals that the sign rule is violated. (c) Torus knot $T_{23}$—a higher-order knot is also not allowed. Reprinted from [24]. Copyright (2000), with permission from Elsevier.

possible [27]. Using coaxial superposition of four or more beams the formation of loops, links and knot structures can be achieved [27–31].

## 2.19 Different types of phase defects

### Point, edge and mixed phase defects

An optical vortex is a point phase defect. It is also called a wavefront screw dislocation. The wavefront structure looks like a cork-screw and many of its topological aspects were discussed in this chapter. There are also other types like line and mixed type phase defects in a wavefront [4]. In an edge dislocation the phase ambiguity occurs along a line in contrast to the vortex in which the phase ambiguity occurs at an isolated point. This edge dislocated wavefront can be realized by cutting a part of a wavefront and phase shifting it by $\pi$. The resulting optical field can be represented by two sets of wavefronts, one shifted from the other by $\pi$. The wavefront looks like a step in a staircase. The phase discontinuity here is along a line. The wavefront can be considered to have edge dislocation even when there is phase discontinuity other than $\pi$ (but not integral multiples of $2\pi$) along the line. In a mixed type dislocation, the indeterminate phase occurs along a line but this ambiguity terminates at a point somewhere in the middle of the wavefront. The point at which the defect terminates is like a point dislocation and the line that

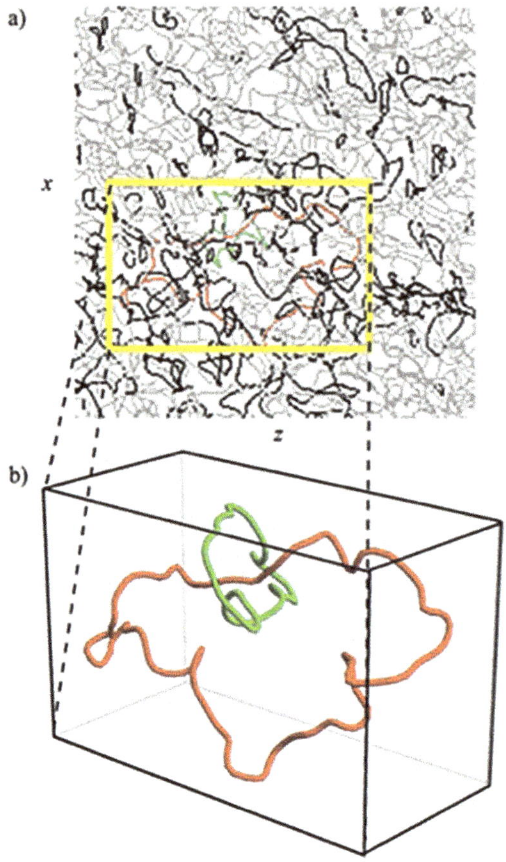


**Figure 2.29.** Complicated structure of the vortex trajectories in random phase distributions. Reprinted with permission from [32]. Copyright (2009) by the American Physical Society.

terminates on the point is like an edge dislocation. Since both types of defects are present, it is called a mixed type dislocation. It can be considered as a vortex with fractional charge. These three types of phase defect [6] are depicted in figure 2.32. A Burgers vector of wave dislocation can be assigned to these phase defects [33] in analogy with the defects in crystals.

### Isotropic and anisotropic vortex

An isotropic vortex is the one in which the azimuthal phase gradient is constant. The phase contour lines that are radial in a vortex are equally spaced azimuthally. In an anisotropic vortex the azimuthal phase gradient is not constant. As a consequence, crowding of phase contour lines at some azimuthal angular positions may happen. For the r-vortex given in equation (2.4) for example, the phase gradient $\frac{d\phi}{d\theta} = 1$. Consider the anisotropic vortex [34] given by

$$V_a(x, y) = x + i\sigma y = r\exp(i\phi_a). \tag{2.9}$$

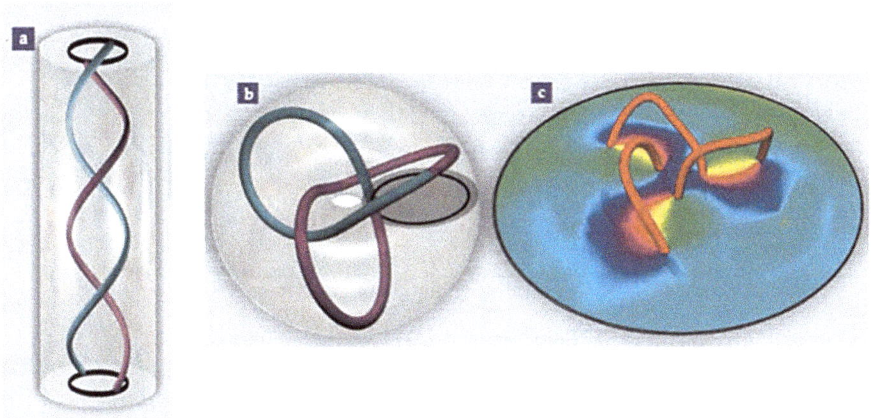


**Figure 2.30.** Constructing an isolated trefoil knot out of vortex lines in an optical field [29]. (a) In cylindrical geometry it is straightforward to devise a complex-valued function that prescribes where vortex lines, drawn here as two strands, should reside as a function of cylinder height as the strands wind around each other. (b) The braid becomes a knotted loop when the cylinder is topologically mapped into a torus, folding the black loops at the top and bottom onto each other. (c) One can experimentally embed the 'knot' function in an optical field by sending a laser beam through a diffractive hologram that shapes the beam's destructive interference pattern. One can then map out the knot in space by measuring where the phases (shown as different colors in this cross section) become singularities (red). Reprinted by permission from [29]. Copyright (2010) by Springer Nature.

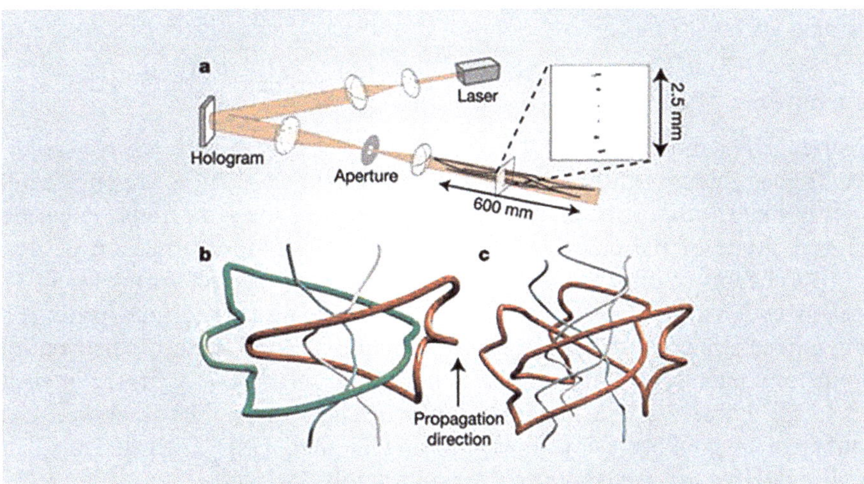


**Figure 2.31.** Experiment to realize knots and links. Knotted lines of darkness. (a) Reconstructed light from the hologram is spatial filtered and a moving camera captures different leaves of foliation. Black lines are threads of darkness that form the link or knot. Inset: over-saturated beam cross-sections at the beam waist of the link with the dark points indicating positions where the optical vortex threads cross the observation plane. (b, c) Three-dimensional representations of measured link (b) and trefoil knot (c) configurations. The link and knot are threaded by further vortices (represented by thc thinner tubes) that follow the axis of the light beam. Reprinted by permission from [27]. Copyright (2004) by Springer Nature.

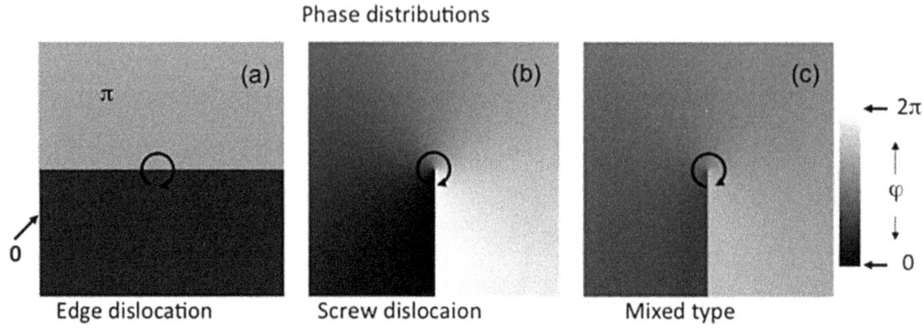


**Figure 2.32.** (a) Edge, (b) point and (c) mixed type dislocations.

The phase variation is given by $\phi_a = \arctan\left(\frac{\sigma y}{x}\right)$ and the azimuthal part of the phase gradient is given by

$$\frac{d\phi_a}{d\theta} = \left\{\frac{\sigma}{\cos^2\theta + \sigma^2\sin^2\theta}\right\} \tag{2.10}$$

The azimuthal phase gradient can be seen varying in an anisotropic vortex. One can also have other types of non-uniform phase gradients for anisotropic vortices.

**Canonical and non-canonical vortex**

A canonical vortex is an isotropic vortex and a non-canonical vortex is an anisotropic vortex.

**Perfect vortex**

In a vortex, the diameter of the dark core increases in size with the charge of the vortex. Higher charge vortices have bigger cores. Perfect vortices are engineered in such a way that the diameter of the vortex core remains the same and is independent of the magnitude of the charge of the vortex. Fourier transformation of Bessel–Gauss (BG) beams of different order is used to generate perfect vortices [35]. These BG beams with vortices in them can be achieved using curved fork gratings [36]. Having a fixed size core allows the beam to maintain good intensity distribution in the donut, and makes it suitable for use in non-linear media [37, 38] as the intensity is linked to the refractive index modulation in such crystals. The fractional charge measurement is possible using phase shifting method [39]. There are also other methods available for perfect vortex generation [40–45].

**Fractional vortex**

In a vortex the accumulated phase change along a closed path around the vortex is $l2\pi$ where $l$ is the topological charge. As a consequence there is no phase discontinuity anywhere in the wavefront except at the location of the singularity. If the accumulated phase change along a closed path is not equal to an integral multiple of $2\pi$ and is fractional, then the vortex is called a fractional vortex. Hence in

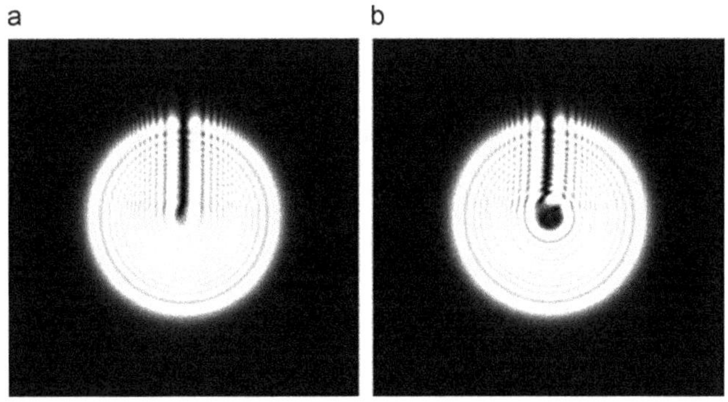


**Figure 2.33.** Fractional vortex—simulated diffraction intensity pattern. A radial intensity cut in the donut structure for a fractional vortex of charge (a) 0.5 and (b) 3.5. Reprinted from [49]. Copyright (2010), with permission from Elsevier.

a fractional vortex there is a phase discontinuity along a line and this line terminates at a point of undefined phase [46–51]. It is a combination of line and point dislocation and is also termed a mixed type phase dislocation. In a pure edge dislocation there is phase discontinuity along the dislocation line, which means that the wavefront is cut along a line and displaced by a distance less than $\lambda$. Hence all along the cut line the phase is singular and this leads to the development of amplitude zero line during propagation. In a mixed type screw dislocation the dark line terminates on the vortex point and hence the donut intensity structure of a integer charged vortex is not there. For a fractional vortex there is a radial intensity cut in the donut structure as shown in figure 2.33.

When multiple fractional charge vortices are present, the dark radial cut pairs with adjacent vortices. This pairing happens irrespective of the polarity of fractional vortex charges. A positive vortex can be connected to another positive vortex or to another negative vortex in the diffraction pattern by dark intensity line as shown in figure 2.34.

Fractional vortices are unstable structures. During propagation along the dark line that is breaking the circular symmetry of the donut structure, a series of integer charged vortices are produced. They are in the form of dipole chains, and these multiple vortices are arranged along the dark line. The intensity distribution in the focal plane of fractional vortex lenses are examined by interference with a plane wave and fringe pattern with multiple forks pointing in opposite directions can be observed as shown in figure 2.35. This confirms the presence of multiple vortices with alternating signs in the radial cut.

## Riemann–Silberstein vortex

Over a century ago Ludwick Silberstein [52] introduced a new vector, which was later known as the Riemann–Silberstein (RS) vector given by

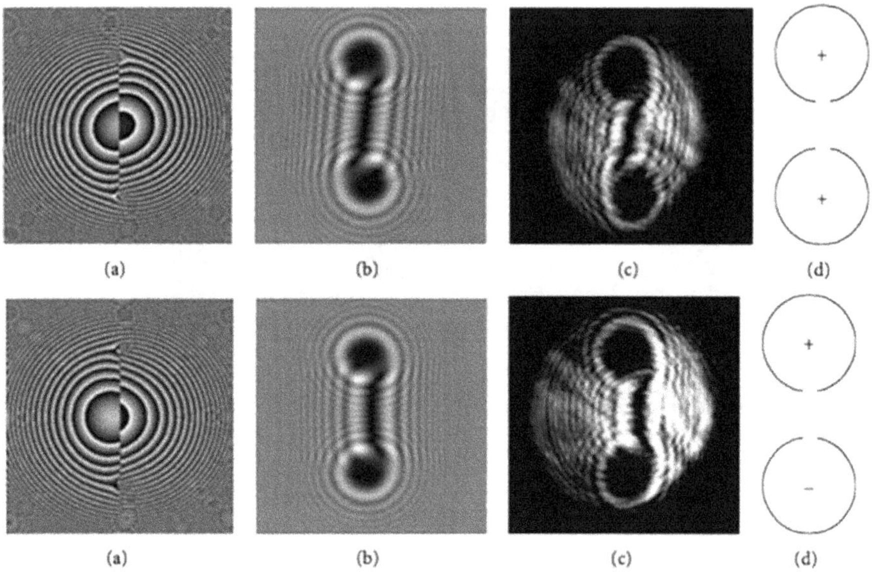


**Figure 2.34.** Top row: Fresnel lens with fractional vortices of the same charge (3.5) embedded at off-axis locations. Bottom row: Fresnel lens with fractional vortices of opposite charge (3.5) embedded at off-axis locations [56]. (a) Phase distribution displayed on SLM. (b) Simulated intensity distribution near the focal plane. (c) Experimentally observed intensity distribution near the focal plane. (d) Distribution of topological charge in the intensity pattern. Reproduced from [56].

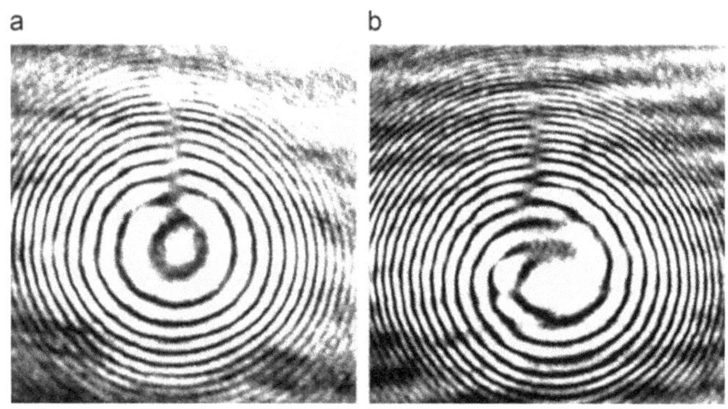


**Figure 2.35.** Fractional vortex–diffraction. Reprinted from [49]. Copyright (2010), with permission from Elsevier.

$$\vec{F} = \sqrt{\frac{\epsilon_0}{2}}(\vec{E} + ic\vec{B}) \tag{2.11}$$

where $\vec{E}$, $\vec{B}$ are electric and magnetic field vectors and $c$ is the velocity of light. This RS vector simplifies many problems in EM theory and is a very useful tool. The circularly polarized plane wave can be written as

$$\vec{F}(r, t) = A\exp\{i(kz - \omega t)\}(\hat{x} + i\hat{y}) \tag{2.12}$$

The total energy is written as the norm of $\vec{F}$

$$E = \|\vec{F}\|^2 = \int d^3r \vec{F}^* \cdot \vec{F} \tag{2.13}$$

The use of the RS vector greatly simplifies the description of the electromagnetic fields [53]. Vortices present in such RS vector fields are called Riemann–Silberstein vortices [54, 55].

## References


[1] Nye J F and Berry M V 1974 Dislocation in wave trains *Proc. R. Soc. Lond. Ser. A* **336** 165–90

[2] Baranova N B, Mamaev A V, Pilipetsky N F, Shkunov V V and Ya Zel'dovich B 1983 Wavefront dislocations: topological limitations for adaptive systems with phase conjugation *J. Opt. Soc. Am.* **73** 525–9

[3] Bazhenov V Y, Vasnetsov M V and Soskin M S 1990 Laser beams with screw dislocation in their wavefronts *JETP Lett.* **52** 429–31

[4] Basisty I V, Soskin M S and Vasnetsov M V 1993 Optics of light beams with screw dislocations *Opt. Commun.* **103** 422–8

[5] Freund I, Shvartsman N and Freilkher V 1993 Optical dislocation network in highly random media *Opt. Commun.* **101** 247–64

[6] Basistiy I V, Soskin M S and Vasnetsov M V 1995 Optical wavefront dislocations and their properties *Opt. Commun.* **119** 604–12

[7] Soskin M S, Gorhkhov V N, Vasnetsov M V, Malos J T and Heckenberg N R 1997 Topological charge and angualr momentum of light beams carrying optical vortices *Phys. Rev. A* **56** 4064–75

[8] Bazhenove V Y U, Soskin M S and Vasnetsov M V 1992 Screw dislocations in light wavefronts *J. Mod. Opt.* **39** 985–90

[9] Soskin M S and Vasnetsov M V 2001 Singular optics *Progress in Optics* **vol 42**

[10] Ghai D P, Senthilkumaran P and Sirohi R S 2008 Shearograms of optical phase singularity *Opt.Commun.* **281** 1315–22

[11] Kowaz M W 1995 Homogeneous and evanescent contributions in scalar near-field diffraction *Appl. Opt.* **34** 3055–63

[12] Berry M V 1994 *Faster than Fourier in Quantum coherence and Reality* (Singapore: World Scientific), 55–65

[13] Lochab P, Senthilkumaran P and Khare K 2016 Near-core structure of a propagating optical vortex *J. Opt. Soc. Am. A* **33** 2485–90

[14] Berry M V 2008 Waves near zeros *Conf. on Coherence and Quantum Optics* (Washington DC: OSA), 37–41

[15] Berry M V 2005 Phase vortex spirals *J. Phys. Math. Gen.* **38** L745–51

[16] Senthilkumaran P and Bahl M 2015 Young's experiment with waves near zeros *Opt. Express* **23** 10968–73

[17] Freund I 1995 Saddles, singularities, and extrema in random phase fields *Phys. Rev. E* **52** 2348–60

[18] Freund I 1999 Saddle point wave fields *Opt. Commun.* **163** 230–42

[19] Freund I 1999 Critical point explosions in two-dimensional wave fields *Opt. Commun.* **159** 99–117

[20] Shvartsman N and Freund I 1994 Vortices in random wave fields: Nearest neighbor anticorrelations *Phys. Rev. Lett.* **72** 1008–11

[21] Freund I and Shvartsman N 1994 Wave-field phase singularities: the sign principle *Phys. Rev. A* **50** 5164–72

[22] Roux F S 2003 Optical vortex density limitation *Opt. Commun.* **223** 31–7

[23] Freund I 2001 Critical foliations *Opt. Lett.* **26** 545–7

[24] Freund I 2000 Optical vortex trajectories *Opt. Commun.* **181** 19–33

[25] Berry M V and Dennis M R 2000 Phase singularities in isotropic random waves *Proc. R. Soc. Lond. Ser. A* **456** 2059–79

[26] Freund I 2001 Critical foliations and Berry's paradox *Opt. Photon. News* **12** 56

[27] Leach J, Dennis M R, Courtial J and Padgett M J 2004 Knotted threads of darkness *Nature* **432** 165

[28] Wilson M 2010 Hologram tie optical vortices in knots *Phys. Today* **63** 18–20

[29] Dennis M R, King R P, Jack B, O'Holleran K and Padgett M J 2010 Isolated optical vortex knots *Nat. Phys.* **6** 118–21

[30] Leach J, Dennis M R, Courtial J and Padgett M J 2005 Vortex knots in light *New J. Phys.* **7** 1–11

[31] Dennis M R 2009 Topological configurations of optical phase singualrities *Topologica* **2** 1–10

[32] Holleran K O, Dennis M R and Padgett M J 2009 Topology of light's darkness *Phys. Rev. Lett.* **102** 143902

[33] Dennis M R 2009 On the burgers vector of a wave dislocation *J. Opt. A: Pure Appl. Opt.* **11** 094002

[34] Kim G-H, Lee H J, Kim J-U and Suk H 2003 Propagation dynamics of optical vortices with anisotropic phase profiles *J. Opt. Soc. Am. B* **20** 351–9

[35] Ostrovsky A S, Rickenstorff-Parrao C and Arrrizon V 2013 Generation of the perfect optical vortex using a liquid-crystal spatial light modulator *Opt. Lett.* **38** 534–6

[36] Karahroudi M K, Parmoon B, Qasemi M, Mobashery A and Saghafifar H 2017 Generation of perfect optical vortices using a Bessel–Gaussian beam diffracted by curved fork grating *Appl. Opt.* **56** 5817–23

[37] Chaitanya N A, Jabir M V and Sharma G K 2016 Efficient nonlinear generation of high power, high order, ultrafast perfect vortices in green *Opt. Lett.* **41** 1348–51

[38] Jabir M V, Chaitanya N A, Aadhi A and Sharma G K 2016 Generation of perfect vortex of variable size and its effects in angular spectrum of the down-converted photons *Sci. Rep.* **6** 1–8

[39] Ma H, Li X, Tai Y, Li H, Wang J, Tang M, Wang Y, Tang J and Nie Z 2017 In situ measurement of the topological charge of a perfect vortex using the phase shift method *Opt. Lett.* **42** 135–8

[40] Garca-Garca J, Rickenstorff-Parrao C, Ramos-Garcia R, Arrizon V and Ostrovsky A S 2014 Simple technique for generating the perfect optical vortex *Opt. Lett.* **39** 5305–8

[41] Oemrawsingh S S R, Eliel E R, Woerdman J P, Verstegen E J K, Kloosterboer J G and 't Hooft G W 2004 Half-integral spiral phase plates for optical wavelengths *J. Opt. A: Pure Appl. Opt.* **6** S288

[42] Kotlyar V V, Kovalev A A and Porfirev A P 2016 Optimal phase element for generating a perfect optical vortex *J. Opt. Soc. Am. A* **33** 2376–84

[43] Chen Y, Fang Z-X, Ren Y-X, Gong L and Lu R-D 2015 Generation and characterization of a perfect vortex beam with a large topological charge through a digital micromirror device *Appl. Opt.* **54** 8030–5
[44] Deng D, Li Y, Han Y, Su X, Ye J, Gao J, Sun Q and Qu S 2016 Perfect vortex in three-dimensional multifocal array *Opt. Express* **24** 28270–8
[45] Fu S, Wang T and Gao C 2016 Perfect optical vortex array with controllable diffraction order and topological charge *J. Opt. Soc. Am. A* **33** 1836–42
[46] Baistiy I V, Pasko V A, Slyusar V V, Soskin M S and Vasnetsov M V 2004 Synthesis and analysis of optical vortices with fractional topological charges *J. Opt. A: Pure Appl. Opt.* **6** S166–9
[47] Berry M V 2004 Optical vortices evolving from helicoidal integer and fractional phase steps *J. Opt. A: Pure Appl. Opt.* **6** 259–68
[48] Lee W M, Yaun X-C and Dholakia K 2004 Experimental observation of optical vortex evolution in a Gaussian beam with an embedded fractional phase step *Opt. Commun.* **293** 129–35
[49] Vyas S, Singh R K and Senthilkumaran P 2010 Fractional vortex lens *Opt. Laser Technol.* **42** 1150–9
[50] Singh B K, Mehta D S and Senthilkumaran P 2013 Visualization of internal energy flows in optical fields carrying a pair of fractional vortices *J. Mod. Opt.* **60** 1027–36
[51] Sharma M K, Joseph J and Senthilkumaran P 2014 Fractional vortex dipole phase filter *Appl. Phys. B* **117** 325–32
[52] Silberstein L 1907 Grundgleichungen in bivectorieller behandlung *Ann. Phys.* **22** 579
[53] Bialynicki-Birula I and Bialynicka-Birula Z 2013 The role of the Riemann-Silberstein vector in classical and quantum theories of electromagnetism *J. Phys. A: Math. Theor.* **46** 053001
[54] Berry M V 2004 Riemann-Silberstein vortices for paraxial waves *J. Opt. A. Pure Appl. Opt.* **6** S175–7
[55] Kaiser G 2003 Helicity, polarization and Riemann-Silberstein vortices *J. Opt. A. Pure Appl. Opt.* **6** S243–5
[56] Vyas S, Singh R K, Ghai D P and Senthilkumaran P 2012 Fresnel lens with embedded vortices *Int. J. Opt.* **2012** 249695

**IOP** Publishing

Singularities in Physics and Engineering
Properties, methods, and applications
**Paramasivam Senthilkumaran**

# Chapter 3

# Generation and detection methods

## 3.1 Introduction

At the dark regions in speckle fields optical vortices do occur [1–5]. Their generation is due to the interference of scattered light. But there are many methods reported for the generation of vortices, in the form of isolated singularities and in the form of networks of vortices in random and lattice fields. To harness the unique properties of the optical phase singularities for various applications, it is important to have methods of controlled generation and manipulation of speckles. Detection methods are also equally important. In this chapter vortex generation and detection methods are presented.

## 3.2 Generation

There are various ways that an optical vortex can be generated. These methods involve the use of specially designed phase elements such as spiral phase plates, wedges, and diffractive elements such as fork gratings, zone plates and spatial light modulators. Interference methods are also used in generating optical vortices. There is also intra-cavity manipulation of optical fields and laser mode converters used for vortex generation. Use of a mirror with a defect inside a laser resonator, or the use of a specially designed helical mirror for vortex generation have been successfully demonstrated. Some methods offer real time manipulation of the charge of the vortex, making them suitable for specific applications.

### 3.2.1 Spiral phase plate

A spiral phase plate (SPP) is an optical element [6] that generates a phase singularity by an azimuthal dependent retardation on the optical field. It is a phase element in which the phase retardation is achieved by controlling the thickness of the plate spatially during fabrication. The thickness variation is proportional to the azimuthal angle

doi:10.1088/978-0-7503-1698-9ch3

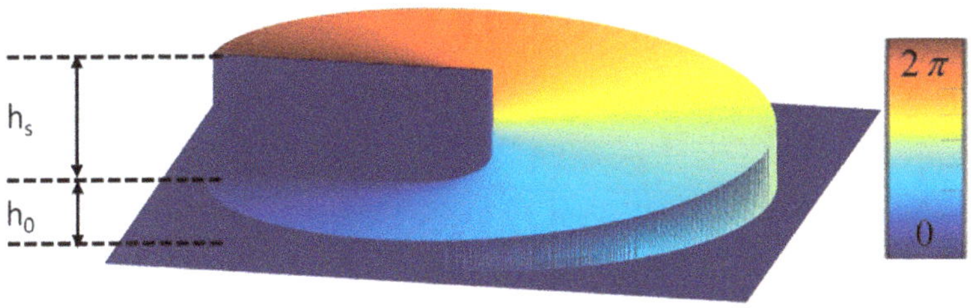


**Figure 3.1.** A spiral phase plate.

$$h(\theta) = h_s \frac{\theta}{2\pi} + h_0 \tag{3.1}$$

where $h(\theta)$ is the height at the angular position $\theta$, $h_0$ is the base thickness of SPP, $\theta$ is the azimuthal angle and $h_s$ is the maximum step height. Figure 3.1 shows such a spiral phase plate. When this plate is introduced in the path of a uniform plane wavefront or in the waist region of a Gaussian beam, it results in an azimuthal dependent optical phase delay given by $\phi(\theta, \lambda) = k(\mu - \mu_0)h(\theta) + k\mu_0 h_0$, where $\mu$ is the refractive index of the SPP and $\mu_0$ is the refraction index of the surrounding medium [7]. The azimuthal phase delay imprints a helical profile on the input field, resulting in the formation of a phase singular beam.

The topological charge of the phase singular beam, thus generated is given by $m = \frac{\phi(\theta, \lambda)}{2\pi}$. Thus the charge of the vortex beam is a function of step height, wavelength of the input beam and the difference in refractive index of the SPP and that of the surrounding. Any mismatch would result in the generation of a singular beam with fractional charge. For generation of phase singular beams of small topological charge, the step height of the SPP should be comparable to the wavelength of light, which is easily achievable for millimeter waves. However, for lasers, particularly at visible wavelengths, generating continuous azimuthal surface height variations in the sub-micron range exhibits practical limitations. A possible solution to the problem is to generate SPP with its refractive index matched to that of the surroundings. This is achieved by immersing the spiral phase plate in a liquid bath having a refractive index comparable to that of the SPP. The index matching allows the use of a higher value of step height (in the mm range) of the SPP. However, density variation and temperature dependence of the refractive index of the liquid used for index matching affect the performance of the device. A better technique, which is most commonly used for the fabrication of SPP, is to have a transparent plate of constant thickness and to coat the plate such that the thickness of the coating is a function of azimuthal angle [6]. In that case $\mu$ and $\mu_0$ would represent refractive indices of the coating and that of the substrate respectively. Since it is difficult to have a controlled variation of surface height along the azimuthal direction, the entire SPP is divided into a number of segments, and incremental step height is provided to successive segments so that the total step height, for one round trip, is equal to an integral multiple of the wavelength of operation. Such a staircase-like structure can act like a spiral phase plate. Recently new techniques of generating

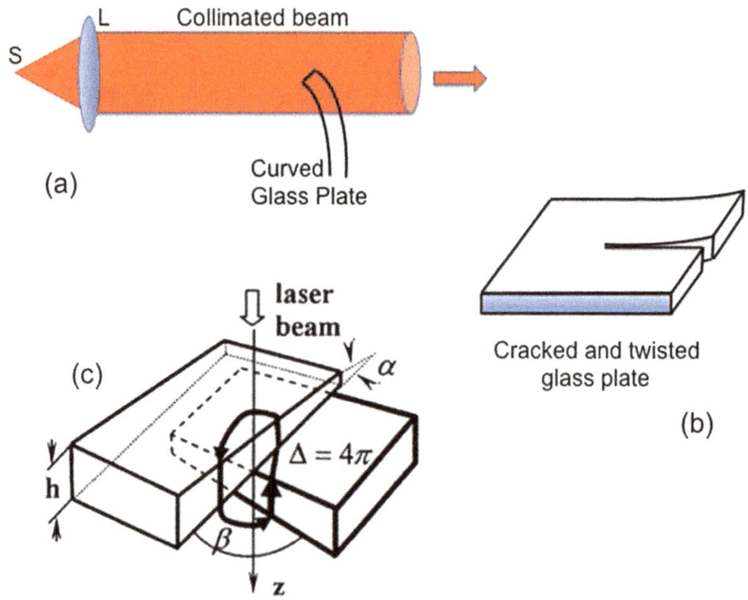


**Figure 3.2.** (a) Vortex generation by introducing a curved glass plate in half of the beam, (b) cracked and twisted transparent plate, (c) stack of wedge plates (reproduced with permission from [19]. Copyright (2005) by OSA).

SPP, which use photo polymerization and micro machining, have been introduced [7, 8]. It is also possible to generate vortices using non-spirals [9], where half of the beam cross section passes through a glass plate which is curved and adjustable phase plates [10] in which the amount of twist given to the plate can yield higher charges. Some of them are presented in figure 3.2. Basically, in these cases it is the tilt in different parts of the wavefront that plays an important role in vortex generation. These plates can be used to generate vortices in different wavelengths also, as these are not wavelength-specific devices.

One of the limitations of the use of SPP is that phase singular beams generated in this way are not pure LG modes. Moreover, the method generates phase singular beams of fixed topological charge and for a specific wavelength. An SPP made of glass material is able to take high power laser illumination with a low failure rate.

### 3.2.2 Fork grating

Fork gratings [11] are linear amplitude gratings in which the fringe period in the upper and lower halves differ by the magnitude of the topological charge of the vortex that it produces. The number of diffraction orders produced by this grating depends on its transmittance profile. Let us first consider a diffraction grating with sinusoidal amplitude profile. The transmittance function of the grating $t_g(x, y)$ is given by

$$t_g(x, y) = A + B\cos\{Gx - m\theta(x, y)\} \tag{3.2}$$

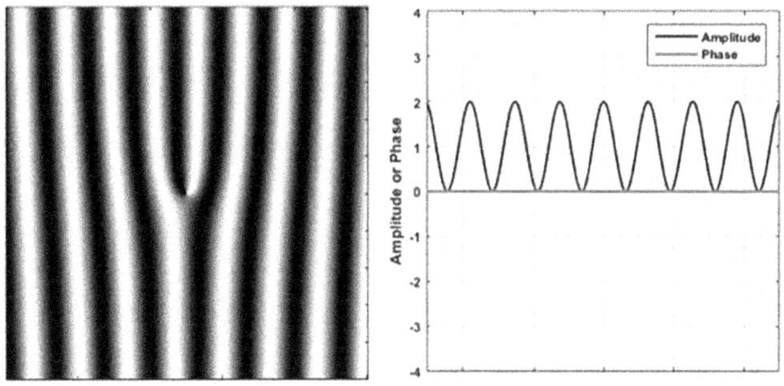


(a) (b)

**Figure 3.3.** (a) Sinusoidal amplitude fork grating and (b) the amplitude transmittance profile.

where $\theta(x, y) = \text{Arctan}\left(\frac{y}{x}\right)$ is the azimuthal angle. The grating vector is $G = \frac{2\pi}{d}$ where $d$ is the grating period. $A$ and $B$ are constants and $m$ is an integer. A typical sinusoidal amplitude fork grating is shown in figure 3.3.

To explain the working of the fork grating in vortex generation, let us first look at the concept of detour phase. When diffraction gratings are made using a diamond tool, the spacing between the rulings are supposed to be constant throughout the grating. But when the machine is switched off and then on, due to the backlash the spacing between the rulings at one location becomes different to the correct period $d$. This results in two identical gratings with one wrong period of the grating between the two pieces of grating. This grating under plane wave illumination produces plane waves in each of the diffraction orders it produces. Let us consider the first diffraction order. Due to the defect in the grating, the first order will have a dislocated wavefront (two sets of wavefronts traveling at angle $\theta$ but with a phase delay between them). This is explained in figure 3.4 in which the grating on the left is a perfect grating and the plane waves going along the first order are depicted in green. In figure 3.4(b), the grating period at one location is different and is given by $d + \Delta$. As a result the first order diffraction that occurs at an angle (say $\theta$) under plane wave illumination produces two sets of plane waves with a phase delay of $k(d + \Delta)\sin(\theta) = 2\pi + \delta$ where $\delta = k\Delta \sin(\theta)$ is the detour phase. In a similar way in a fork grating around the point where a grating line terminates, the period of the grating is gradually changing from one value to another and this gives rise to an additional phase arising due to the detour explained above. In figure 3.5 the change of period along the horizontal direction as we go from line scans numbered 1 to 5 shows conspicuous and continuous period change. In fact for a vortex phase variation, the grating period is continuously changing from one value to another at all locations in the grating and hence the combined effect of detour phase from each location results in the generation of a helical wavefront in the first diffraction order. Note that in a vortex, the phase gradient field is azimuthal and is non-zero everywhere across the cross-section of the beam, and this indicates that the grating period is changing all across the wavefront. Any radial scan through the center of the

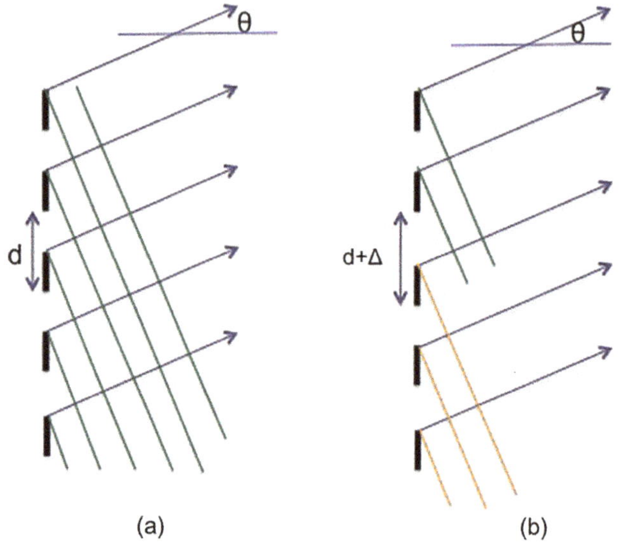


**Figure 3.4.** Concept of detour phase. (a) depicts a normal grating producing a plane wave in the first order under plane wave illumination. The period of the grating is *d*, the plane wavefronts are shown in green and the direction of propagation is shown by blue rays. (b) A defect in the grating with a different grating period $d + \Delta$ is shown. This defect is only at this place (local) and the grating period in the regions above and below is *d* as depicted. This grating also produces plane waves in the first order in the same direction as that shown in (a), but due to the defect, one set of plane waves shown in green and the other in orange have a phase delay between them. This is the detour phase and is proportional to $\Delta$.

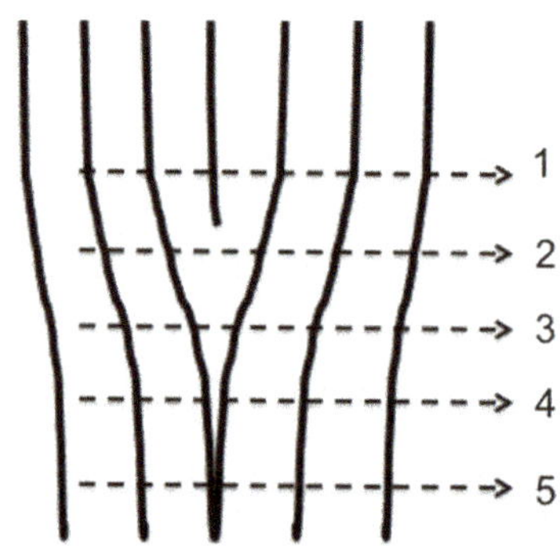


**Figure 3.5.** Schematic of a fork grating. Horizontal scans numbered 1 to 5 see different grating element spacings. The grating period changes gradually and hence the detour phase also changes gradually. The dominant periodicity of the grating (carrier frequency) is along the horizontal direction and hence the overall beam deflection is in the horizontal going towards first order. The small varying detour phase occurring due to change of periodicity throughout the grating leads to the generation of helical waves.

fork grating sees maximum phase detour that develops the helical wave in the first order.

The number of diffraction orders and amount of light diffracted into each order depends on the transmittance profile of the grating. For the sinusoidal amplitude grating, there are only three orders, namely $-1$, 0 and $+1$ orders and hence for a fork

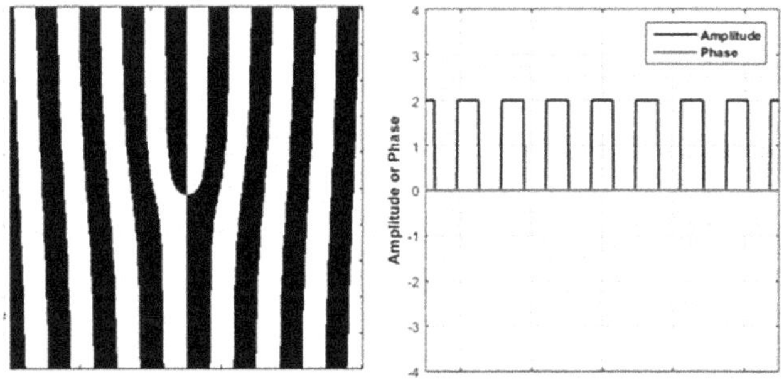


(a) (b)

**Figure 3.6.** Binary amplitude fork grating and the amplitude/phase transmittance profiles are shown on the right.

grating, only $-m$, 0 and $+m$ charged vortices can be produced. Any variation in the amplitude modulation that results from fluctuations in the experimental recording conditions, will not affect the charge of the vortex. Only the efficiency in different orders undergoes changes. This is the positive aspect of using a grating over the spiral phase plate. Fork gratings work for different wavelengths whereas a spiral phase plate cannot be used in different wavelengths other than the one it is designed for. In fork gratings, by going from a sinusoidal amplitude profile to a binary amplitude profile or sinusoidal/binary phase profile, multiple orders can be realized and in each case the efficiency of light realizable in the first order differs. For example, gratings with different transmittance functions can be given by

$$\begin{aligned} t_1 &= A + B\,\mathrm{sgn}\,[\cos(Gx - m\theta(x, y))] \\ t_2 &= \exp\{i[A + B\cos(Gx - m\theta(x, y))]\} \\ t_3 &= \exp\{i[A + B\,\mathrm{sgn}\,[\cos(Gx - m\theta(x, y))]]\} \end{aligned} \tag{3.3}$$

where $t_1$ is the transmittance function of a binary amplitude grating, $t_2$ is the sinusoidal phase grating and $t_3$ is the binary phase grating.

A signum function or Heaviside function is defined as

$$\begin{aligned} f(t) &= -1 \quad \text{for} \quad t \leqslant 0 \\ &= +1 \quad \text{for} \quad t > 0 \end{aligned} \tag{3.4}$$

The amplitude and phase transmittance profiles for $t_1$ is shown in figure 3.6, whereas the transmittance functions ($t_2$ and $t_3$) of the phase gratings are shown in figure 3.7. The constants $A$, $B$ are such that $A \geqslant B$ and $(A + B) = 1$. By changing their values, the efficiency in different diffraction orders can be manipulated. Further other grating profiles such as blazed amplitude or phase profiles can be employed to achieve selectively diffracted light into different orders. Countless modifications based on computer generated holography techniques can be employed to generate pure and efficient vortex beam generation. When the fork grating produces multiple diffraction orders, as in the case of binary amplitude or phase fork grating, the topological charge of the diffracted beams depends on the order of diffraction.

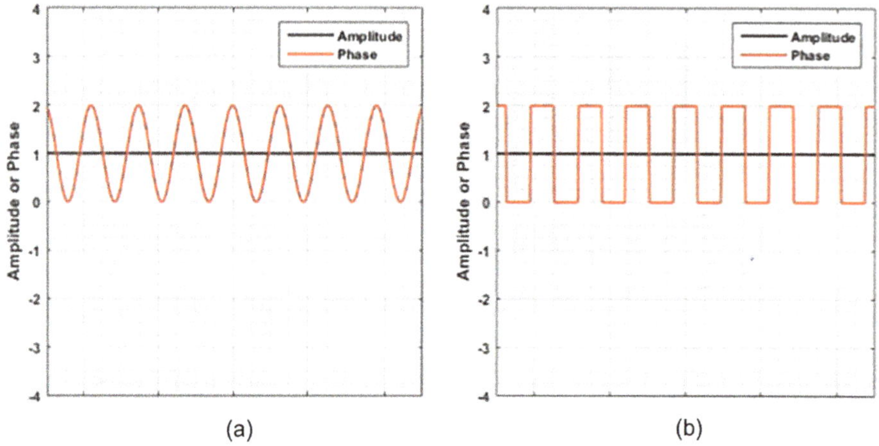


**Figure 3.7.** The amplitude and phase transmittance profiles of $t_2$ and $t_3$ corresponding to phase gratings.

A fork grating designed to produce a vortex of charge $m$ will produce multiple vortex beams of order $m$, $2\,m$, $3\,m$ ... in the first, second, third orders and so on. Similarly negatively charged vortices will appear in the negative diffraction orders. Normally the number of order in a diffraction grating is limited to $n < \frac{d}{\lambda}$, and hence the number of vortices produced are also limited.

### 3.2.3 Spiral zone plate

Spiral zone plates [12] can be used to produce singular beams. The fringe width in a normal zone plate is large at the center of the zone plate and keeps decreasing as we move radially outward. A binary amplitude zone plate has a transmittance function given by

$$t(r) = \frac{1}{2} + \frac{1}{2}\text{sgn}\left\{\cos\left(\frac{\pi r^2}{\lambda f}\right)\right\} \tag{3.5}$$

where $f$ is the principal focal length of the zone plate. The transmittance function of a normal zone plate is a function of $r$ whereas a spiral zone plate transmittance is a function of both $r$ and $\theta$ and is given by

$$t(r, \theta) = \frac{1}{2} + \frac{1}{2}\,\text{sgn}\left\{\cos\left(m\theta - \frac{\pi r^2}{\lambda f}\right)\right\} \tag{3.6}$$

This can be thought of as a binary amplitude computer generated hologram in which the object and reference beams are a singular beam and a spherical beam [13] respectively. Since both the beams are on-axis, during reconstruction, the sinusoidal amplitude spiral zone plate, upon illumination by a plane wave will produce a converging vortex beam of charge $m$ and a diverging vortex beam of charge $-m$. Figure 3.8 shows the amplitude transmittance of a sinusoidal amplitude spiral zone

**Figure 3.8.** Sinusoidal amplitude spiral zone plate that can produce a vortex of charge 1. Along with this vortex a plane wave and a diverging beam with vortex of opposite charge will also be reconstructed under plane wave illumination.

plate. Spiral zone plates can be made to have various transmittance profiles [14]. One of them corresponds to the real image and the other corresponds to the virtual image. Further, the binary amplitude profile leads to multiple diffraction orders, which means multiple real and virtual images are produced and all have the nominal propagation direction along the optical axis of the system. One can introduce carrier frequency to the plane reference wave to separate the orders angularly to avoid overlap. Such an angle multiplexed spiral zone plate has the transmittance function given by

$$t(r, \theta) = \frac{1}{2} + \frac{1}{2}\,\mathrm{sgn}\left\{\cos\left(2\pi f_x x + m\theta - \frac{\pi r^2}{\lambda f}\right)\right\} \tag{3.7}$$

The linear phase variation with carrier frequency $f_x$ ensures that diverging vortex beams are on one side and the converging vortex beams are on the other side of zeroth order diffraction.

A diffractive optical element (DOE) that can produce an intensity null at the focal plane was suggested by Roux [15] in which the transmittance function of the element has the features of a zone plate and a fork grating. Actually that element can be realized by the interference of a spherical wave with considerable tilt and a vortex beam.

### 3.2.4 Tilts

Generally on a wavefront which is a continuous and differentiable surface, a phase difference of $\pi$ between adjacent points does not occur. Circulating phase gradients also do not occur on wavefronts with smooth surfaces. One way to have adjacent points having a phase difference of $\pi$ is to introduce tilts in two portions of the same

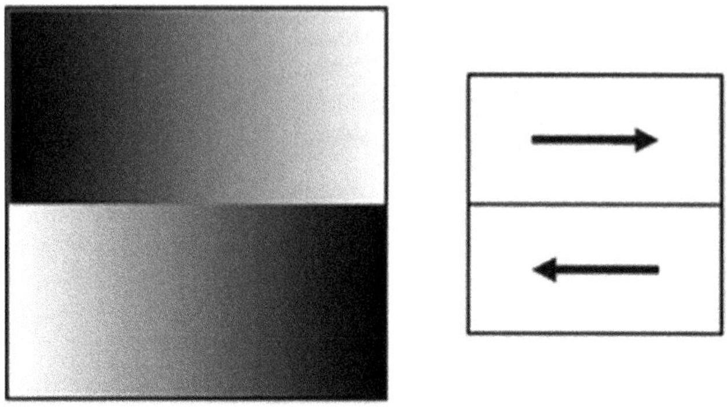

**Figure 3.9.** The phase ramp structures in the upper and lower part of the wavefront has a phase discontinuity along the horizontal line. The phase gradients in the two regions are oppositely oriented.

wavefront as shown in figure 3.9. It is two oppositely oriented linear phase ramps juxtaposed. There is a phase discontinuity along a line that separates these two regions. Since the tilts on either side of this line are opposite, there exist possibilities that some points on either side of this line are out of phase. Consider a typical complex field distribution in which two regions of the wavefront have opposite tilts. During propagation of this complex field, vortices evolve from discrete points that lie on the line of phase discontinuity that separates these two regions [16, 17]. The complex field is given by

$$\begin{aligned} \psi(x, y) &= \exp\{i2\pi\alpha x\} \quad &\text{for} \quad y > 0 \\ &= \exp\{i2\pi\alpha(1 - x)\} \quad &\text{for} \quad y < 0 \end{aligned} \tag{3.8}$$

where $\alpha$ is a constant which decides the slope of the phase variation across the wavefront. If the wavefront is tilted by an angle $\theta$, then $\alpha = \frac{\sin\theta}{\lambda}$. There is a line of phase discontinuity at $y = 0$ as shown in figure 3.9.

Consider two points with coordinates $\left(\frac{1}{4\alpha(1+2\alpha)}, 0\right)$ and $\left(\frac{1}{4\alpha(1-2\alpha)}, 0\right)$. Any two points along $x_1 = \left(\frac{1}{4\alpha(1+2\alpha)}\right)$ and on either side of $y = 0$ are out of phase. Similarly any two points along $x_2 = \left(\frac{1}{4\alpha(1-2\alpha)}\right)$ on either side of $y = 0$ are also out of phase. For example the phase difference between points $(x_1, y_1)$ and $(x_1, -y_1)$ is given by $2\pi\alpha x_1 - 2\pi\alpha(1 - x_1) = \pi$. This is also true for other diametrically opposite points about $(x_1,0)$ with coordinates $(x_1 + \Delta x, y)$ and $(x_1 - \Delta x, -y)$. Further, by decomposing the phase gradient at its neighborhood into radial and azimuthal components one can see the circulating phase gradient around this point (figure 3.10). These circulating phase gradients and the presence of $\pi$ phase difference between diametrically opposite points about $(x_1,0)$ and $(x_2,0)$ suggest that vortices can be induced in the field distribution during propagation. The out of phase condition is not met for all points on the line of phase discontinuity. It happens to only these two points and these two points are found to develop into two vortices of same polarity upon propagation. Propagation plays an important role in the generation of vortices. In

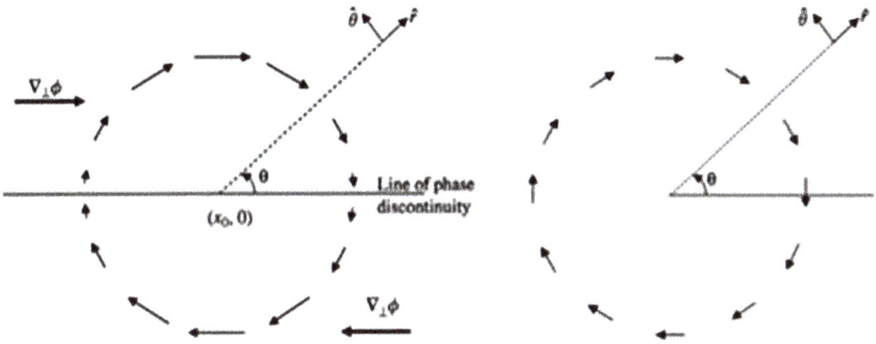


**Figure 3.10.** The azimuthal component of the phase gradient due to double wedge is shown in (a) and the circulating phase gradient of a vortex is shown in (b) for comparison. Note the radial component is not shown in (a). Reprinted from [17]. Copyright (2010), with permission from Elsevier.

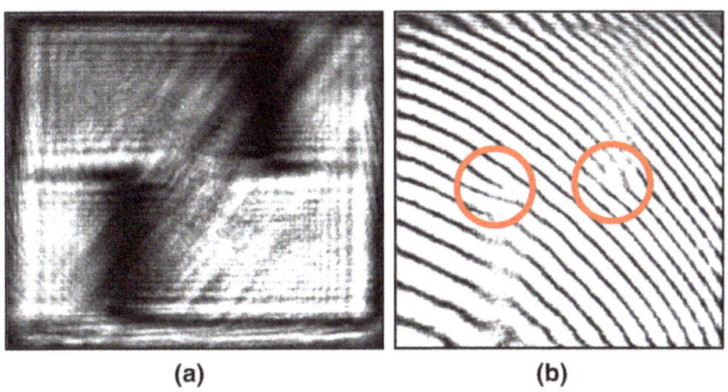


**Figure 3.11.** Observed intensity and the fork fringe patterns of the diffracted field from a double wedge configuration. Reprinted from [16]. Copyright (2010), with permission from Elsevier.

figure 3.11(a) the azimuthal part of the phase gradient around any of the two points is shown. The tilt in the wavefront is decomposed into the radial part and the azimuthal part of the gradient in which the radial part is ignored and the azimuthal part is compared with the actual phase gradient that exists around a phase vortex as shown in figure 3.11(b). Like the azimuthal gradient, the radial part also is spatially varying around a circle about the point and its magnitude is also considerable, but is ignored for intuitive comparison of the azimuthal part with a vortex gradient.

According to charge conservation, only pairs of vortices in the form of dipoles develop in the field [18]. But here, the pair of vortices that are generated have the same sign. This is due to the fact that the wavefront is subjected to abrupt phase changes along the line joining the two regions of the wavefront. Because of the abrupt change there is no natural evolution, the conservation rule no longer holds good here. The new topological feature seen here is the presence of a line phase discontinuity in the wavefront. This line phase discontinuity is the boundary that separates regions of different tilts and allows the emergence of many vortices of the same topological charge. The phase gradient in the neighborhood of two points (that develop into vortices) have similar gradient field distribution. Vortices produced in this way using tilt are of unit charge only and generation of higher-order vortices is not possible using this method.

In a similar way a linear phase gradient in one half of the beam and the rest of the beam having constant phase can also produce a vortex [17]. This has been demonstrated by passing the fundamental Gaussian beam through the edge formed by the row of the dielectric wedge so that one half of the beam propagates through the wedge while the other spreads through the free space [19]. The vortex position at the wedge is restricted by only one condition: the contour over the mask surface inside the beam must give an integral multiple of $2\pi$ excess phase. Thus, the slight deformation of the phase mask does not deform the vortex as a whole. It is shown that two wedge plates arranged in a crossed position create a large circulating phase gradient and can result in the generation of higher charge vortices. Ya *et al* [19] have showed that this stack of wedges generates single beams with higher-order vortices. The topological charge of the vortex is equal to the number of wedges in the stack. In contrast to the spiral phase plate, a wedge plate does not produce vortices with a fractional topological charge. Somewhere along the wedge edge, the phase difference on either side of the edge is $\pi$ that can induce a vortex. Using annular illumination and optical wedge [20] higher-order vortex generation was demonstrated. Yuan *et al* [21] extended the idea and developed a systematic approach to the wedge method for forming optical vortices and extended the method to consider wavelength scalability. Parallel alignment of wedges of different wedge angles [22] and parallel alignment of opposite wedges also can work for vortex generation. Vortex lattice generation consisting of line of vortices of the same charge or interlaced with lines of oppositely charged vortices and other combinations were also realized using many phase wedges aligned parallel/anti-parallel [16].

### 3.2.5 Interference methods

In general, when three or more scalar waves interfere in space, complete destructive interference occurs on lines called nodal lines, and on points called phase singularities, wave dislocations, or optical vortices (OVs). Hence, three plane waves can be made to interfere at small angles [23, 24]. The interference due to three uniform amplitude plane waves can generate vortex arrays if the tilt between the three interfering beams can be suitably changed.

Let us first consider the case of orthogonal tilt given to two of the three wavefronts. The complex fields for the three plane waves are given by $e^{ikz}$, $e^{i(k_x x+k_z z)}$ and $e^{i(k_y y+k_z z)}$. At the observation plane the interference of these three plane waves leads to the generation of a lattice of vortices. The resultant field at $z = 0$ plane is given by $1 + e^{ik_x x} + e^{ik_y y}$ and the resultant intensity is given by

$$I(x, y) = 3 + 2[\cos(k_x x) + \cos(k_y y) + \cos(k_x x - k_y y)] \tag{3.9}$$

Vortices are known to occur at intensity nulls and at the zero crossings of real and imaginary parts of the resultant field distribution at the $z = 0$ plane (observation plane). Hence we can write

$$\begin{aligned}\cos(k_x x) + \cos(k_y y) + \cos(k_x x - k_y y) &= -\frac{3}{2}\\ 1 + \cos(k_x x) + \cos(k_y y) &= 0\\ \sin(k_x x) + \sin(k_y y) &= 0\end{aligned} \tag{3.10}$$

Solving these three equations, we see that vortex formation is possible when

$$\begin{aligned}k_x x &= -k_y y\\ \cos(k_x x) &= -\frac{1}{2}\\ \cos(k_x x - k_y y) = \cos(2k_x x) &= -\frac{1}{2}\end{aligned} \tag{3.11}$$

The locations $(x, y)$ where vortices can occur can be deduced from equation (3.11). It can be seen that due to the periodic nature of cos function, a lattice of vortices results in the interference pattern.

As far as uniform amplitude plane waves are concerned, it has been observed that a minimum of three plane waves are required to form the vortex lattice structure. The resulting vortex lattice structure consists of an array of vortex dipoles, whose spatial period of repetition can be controlled by controlling the tilt between the interfering plane waves. It is also possible to realize higher charge vortex lattices, in a quasi-periodic lattice structure. This can be done by interfering multiples of three beams [25, 26], with inter-beam phase engineering. In the earlier case the interfering plane waves had tilt between each of them but there was no phase shift (difference) between them. By introducing phase differences between the interfering plane waves apart from the tilts, structurally diverse patterns can be realized [27]. For forming a vortex of charge $p$ at the center of the interference pattern, we need $q$ plane waves such that $q = p \times 3$ whose propagation vectors could be arranged in an umbrella geometry as shown in figure 3.12(a). The tip of these propagation vectors, as projected on an $k_x$, $k_y$ plane appear as dots on a ring as shown in figure 3.12(b). The amount of phase shift given to each of these plane waves can be understood by first considering the basic unit of three plane waves. For these three plane waves the phase shifts are $\phi_m = m \times \frac{2\pi}{3}$ where $m = 1$–3. This basic combination is repeated $p$ times to realize $q$ plane waves, which are phase shifted. The resulting interference pattern of these $q$ plane waves results in the generation of higher-order vortex of

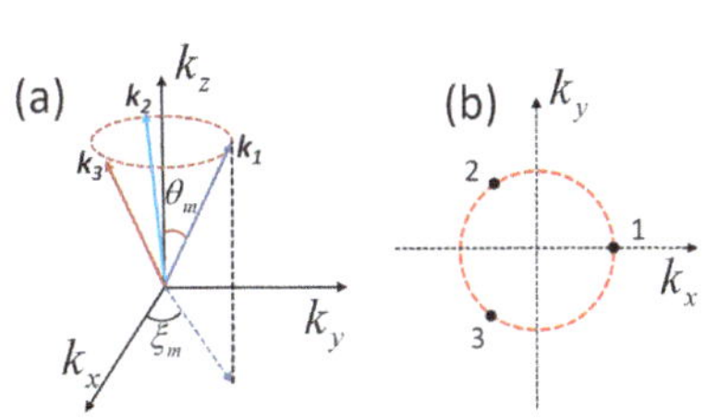


**Figure 3.12.** The propagation vectors of the interfering beams in k-space. (a) Three-dimensional view. (b) Top view—$(k_x k_y)$ plane view.

charge $p$ at the center of the lattice. This phase engineering is important, otherwise the resultant interference pattern will have a maximum at the center instead of intensity null, when more than three plane waves are used. Spiraling of these lattice structures in three dimensions happens when an on-axis plane wave is introduced in the interference [25, 28].

Three beam interference (three plane waves with tilt as shown in umbrella geometry) (a) without phase engineering, (b) with phase engineering and four-beam interference, (c) without phase engineering and (d) with phase engineering are shown in figure 3.13. Similarly multiple beam interferences of six and nine beams are

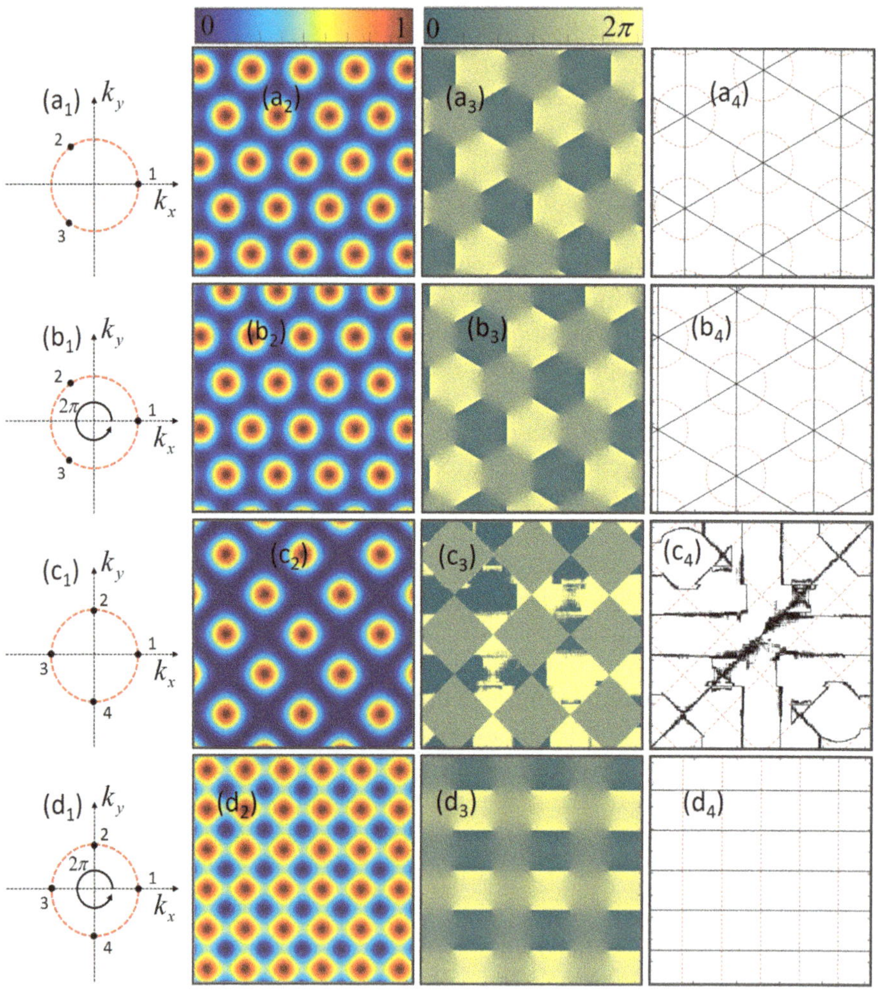


**Figure 3.13.** Column I: arrangement of plane waves in k-space, column-II: intensity distribution in the interference pattern, column III: the phase distribution in the interference field, column IV: zero crossings of the real and imaginary part of the field. The vortices are located at the zero crossings. (a) Three-beam interference. (b) Three-beam phase engineered plane waves interference. (c) Four-beam interference and (d) four-beam phase engineered plane waves interference.

shown in figures 3.14 and 3.15 respectively. In both the six- and nine-beam interferences, in (a) beams are not phase engineered, in (b) beams are phase engineered and in (c) an additional central beam is introduced in the phase engineered beam interference. In figure 3.14(b) the lattice consists of vortices of charge +2 and charge −1. The central perturbing beam makes these double charge vortices disintegrate into single charge vortices as shown in figure 3.14(c). When the number of interfering beams is more the lattice loses translational symmetry and become quasi-periodic. This is evident from figure 3.15. When the nine beams are not phase engineered a bright spot is produced at the center as all the beams constructively interfere at the central spot. The phase engineering results in the replacement of the bright spot by a dark spot with charge three vortex in it. This is because the total phase variation for all the beams is $6\pi$ and since it is equal to $3 \times 2\pi$ the charge at the center is three. This can be verified from the zero crossings of real and imaginary zeros of the resultant complex field due to interference at the center. The maximum charge that is realizable at the center here is three as the number of beams involved in the interference is 9 and we need three beams per vortex charge, which means 9/3 = 3. To generate a charge of eight at the center we need minimum $8 \times 3 = 24$ beam interference and each of the plane wave must be phase engineered

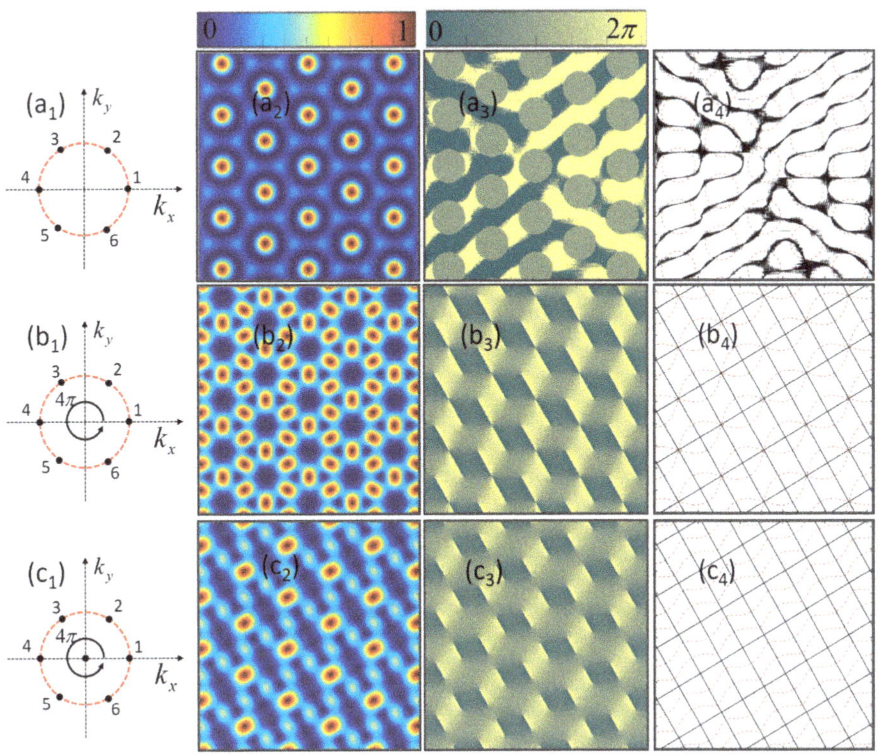


**Figure 3.14.** Columns I–IV: same as in figure 3.39. (a) Six-beam interference, (b) phase engineered six-beam interference and (c) phase engineered six-beam interference with an on-axis perturbing beam added.

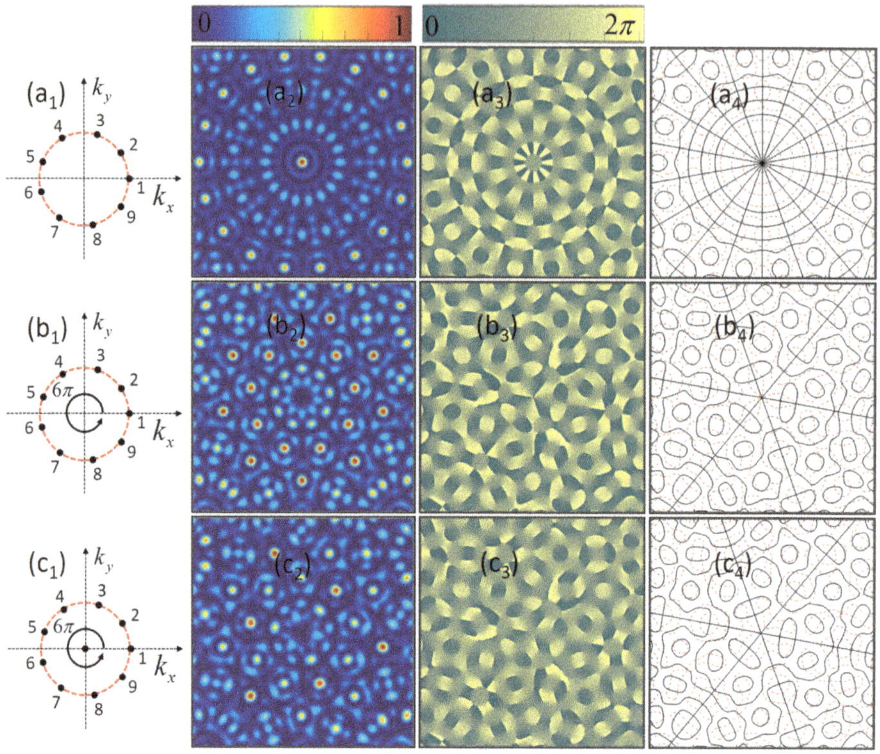


**Figure 3.15.** Column I–IV: same as in figure 3.39. (a) Nine-beam interference, (b) phase engineered nine-beam interference and (c) phase engineered nine-beam interference with an on-axis perturbing beam added.

so that the total phase variation is $8 \times 2\pi = 16\pi$. In figure 3.15(c) the charge three vortex is seen disintegrated due to the addition of perturbing central beam.

To realize interference of plane waves, interferometers can be used. When multiple beams are involved in interference, the interferometer also becomes bulky and cumbersome to use. To avoid using such bulky interferometers, for multiple beam interference we use the setup shown in figure 3.16. Each plane wave is represented by a point in the k-space. Hence the number of plane waves that are needed are first generated as a number of points in the k-space. Since the orientation of each of the plane waves is different, each point in k-space corresponding to these plane waves are different. To realize this distribution, the wavefront that has the spatial frequency variation corresponding to the complex field resulting from the multiple beam interference is synthesized using a computer. This resulting phase distribution of multiple beam interference is then displayed on a reflective spatial light modulator. A plane wave reflected from the spatial light modulator will have a wavefront shape in such a way that its Fourier transform will produce spot distribution in the Fourier plane. The number of spots is equal to the number of beams involved in the multiple beam interference. For phase engineering, the phase shifts to each of the plane waves is taken into account while computing the phase distribution of the resultant complex amplitude of the interference. Due to

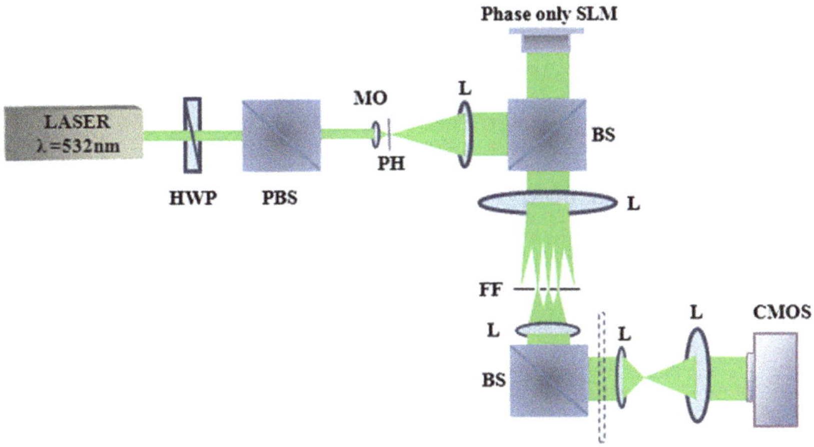


**Figure 3.16.** Experimental setup to realize multiple beam interference. HWP: half wave plate, PBS: polarizing beam splitter, MO: microscope objective, PH: pin hole, L: lenses, BS: beam splitter, FF: Fourier filter, SLM: spatial light modulator, CMOS: camera. HWP–PBS is to select the appropriate plane polarized light as the SLM is sensitive to only one polarization. The beam from the collimating lens to phase SLM is a plane wave whereas the beam reflected from SLM is the field due to interference of multiple plane waves. This beam is further purified (by discarding unwanted spatial frequency components) at the FF plane filter. Multiple plane waves are transformed into point sources and again transformed into multiple beams between L–FF–L path. Multiple beam interference pattern is captured using CMOS camera. Reproduced with permission from [26]. Copyright (2012) by OSA.

phase engineering the phase distribution that is to be displayed on the SLM also differs. All these spots are transformed into plane waves by the use of a collimating lens. These plane waves interfere to produce the vortex lattice structures.

The experimental setup [26] used to generate a lattice of vortices is shown in figure 3.16. A vertically polarized laser of wavelength 532 nm is spatially filtered and collimated by spatial filter assembly (SF) and lens respectively. The collimated light is allowed to pass through a cube beam splitter (BS) and the reflected light coming out of the beam splitter falls on a programmable reflective phase-only spatial light modulator (SLM). The SLM is used to modify the wavefronts of a beam falling on it. The reflected wave from the SLM carries the phase variation corresponding to the multiple beam interference of scalar waves. The modified beam is focused to the Fourier plane (FF plane) of the lens where undesired Fourier components are filtered out using an amplitude filter. The filtered spots are point sources which when placed at the focal plane of a collimating lens produce multiple overlapping plane waves. Spiraling vortex structures (figure 3.17) in three dimension can be created in a phase engineered multiple beam interference with a central perturbing beam.

### 3.2.6 Speckles

When a coherent beam passes through a random medium, speckles are produced. This can happen by sending a laser light through a turbulent medium or by scattering the light by reflection at rough surface or transmission through a diffuser. The randomly scattered waves interfere with each other forming speckles. Speckles are known to contain vortices [1, 4, 29–32]. On average, each dark region in a

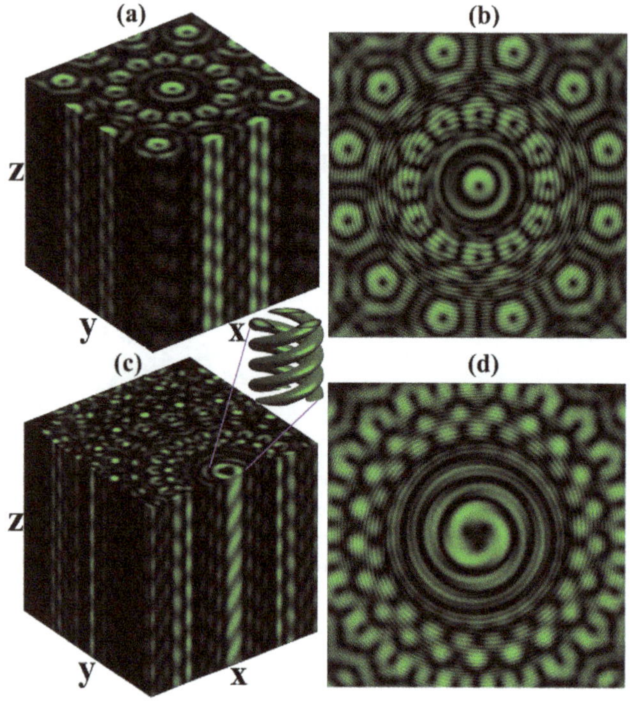


**Figure 3.17.** Tunable vortex spirals generated by multiple beam interference. (a) and (b) 12 beam interference with an additional central perturbing beam. (c) and (d) 21 beam interference with central perturbing beam. On-axis perturbing beam is required to create the spirals. Reproduced with permission from [25]. Copyright (2011) by OSA.

speckle pattern has a vortex. Speckle generation, therefore, is an example of the natural way of generation of vortices, but with less or no control. Formation of speckles is indicative of the coherence of the light. Highly coherent light beams tend to produce high contrast speckles like the fringes in interferometry. One way of avoiding the speckle effect is to use low coherent sources of light. The apparent randomness in the speckle field also has some structures. The vortices in a speckle field obey the sign rule and by knowing the sign of any one of the vortices, the signs of all the other vortices can be found.

### 3.2.7 Spatial light modulator

A spatial light modulator (SLM) can be used for phase singularity generation [32]. An SLM consists of liquid crystal molecules whose birefringence can be tuned by the application of an electric field. This change in the birefringence can be used to realize the required phase shift in one of the polarization components of the light. Liquid crystals are filled between transparent pixelated electrodes and polarization selection is made by using polarizing elements along with the liquid crystal device. The amount of phase shift or amplitude modulation at a specific pixel can be controlled electronically or optically and accordingly the SLMs are called electronically addressed SLM or optically addressed SLM.

The phase variation corresponding to an optical vortex can be realized on an SLM operating on phase mode. To realize a vortex, the amount of maximum phase retardation that an SLM is required to produce is $2\pi$ at a specific wavelength. An SLM that does not produce the required maximum phase retardation at a specific wavelength, therefore produces a vortex of fractional charge instead of an integer charge vortex. To avoid such a mismatch between what is electronically addressed to the SLM and what is realized, such SLMs can be used as computer generated holograms and the light can be diffracted through these holograms to produce a vortex. The required complex field (vortex field) is realized as a reconstructed field from the hologram. The transmittance of the phase hologram can be computer generated. For example, the interference intensity between a suitable reference wave and the complex field is computed and the phase distribution that is to be displayed on the SLM is made to vary as a function of interference intensity distribution. When the SLM is used in the amplitude mode, the amplitude transmittance of the SLM can be a function of computed interference intensity distribution. In this way the virtual image term produced by the hologram in the first order diffraction is extracted for use. By this method, the SLM that is not capable of producing a phase shift of $2\pi$ can be made to produce $2\pi$ phase shifts. This limited ability of the SLM is only translated into reduced diffraction efficiency of the hologram. Moreover, note that to realize a phase hologram with maximum diffraction efficiency, the required phase modulation is not $2\pi$, but is less than that. Since the field of computer generated holography is well developed, there are many techniques available to realize complex fields with any amplitude and phase variation by using the SLM as a hologram.

### 3.2.8 Dammann vortex grating

A Dammann grating is a binary phase grating that can produce equal intensity in all its diffraction orders [33]. The diffraction efficiency in a normal grating is not the same in all orders. In Dammann gratings this is achieved by using a special grating structure design using computation. In a binary phase grating, which does not produce equal intensity light spots in its diffraction orders, the transition points from one phase value to another, at the grating plane is modified. Hence in one period of the normal binary phase grating instead of having one transition point, multiple transition points at which the phase value changes from one level to another are made to produce a Dammann grating. The same concept is used in the making of a Dammann vortex grating (DVG).

A DVG is a kind of vortex grating into which the Dammann phase-encoding method is introduced, and this DVG can produce a vortex with equal light energy and different topological charge for each diffraction order [34]. The Dammann vortex grating [35] has transmittance function given by

$$T_{\mathrm{DVG}}(\rho, \varphi) = \sum_{m=-\infty}^{\infty} c_m \exp\left\{im\left(\frac{2\pi}{\Lambda}\rho\cos\varphi + l\varphi\right)\right\} \tag{3.12}$$

The coefficient $c_m$ of the $m$ order is given by

$$\begin{aligned} c_m &= -\frac{i\Lambda}{2m\pi}\left[1 + 2\sum_{n=1}^{N-1}(-1)^n \exp(-i2\pi m x_n) + (-1)^N \exp(-i2\pi m x_N)\right] \quad \text{for} \quad m \neq 0 \\ &= \Lambda\left[2\sum_{n=1}^{N-1}(-1)^n x_n + (-1)^N (x_N)\right] \quad \text{for} \quad m = 0 \end{aligned} \tag{3.13}$$

where $\{x_n\}$ are normalized phase transition points in one period with boundary values of $x_0 = 0$ and $x_N = 1$. The total number of these transition points is $N$ and $\Lambda$ is the grating period.

It is shown that this formula is the same in form as that of a conventional vortex grating, and the difference is only caused by the coefficients, which are determined by the phase transition points for binary pure-phase 0; $\pi$ structures. By optimizing the values of these transition points, the light energy can be redistributed into several desired orders with good uniformity and high efficiency, and this is the essence of the Dammann phase-encoding method. Theoretically, the light energy distribution among different diffraction orders can be arbitrarily controlled by modulating these normalized phase transition points. Generation of dipole vortex arrays is also possible using spiral Dammann zone plates [36].

### 3.2.9 Mode conversion methods

A mode converter comprises a pair of cylindrical lenses separated by a distance. It transforms a Hermite–Gaussian (HG) optical mode to its equivalent Laguerre–Gaussian (LG) mode or vice versa [37, 38]. Both HG modes and LG modes are solutions of the paraxial wave equation. The transverse field distribution of laser beams that exhibit rectangular symmetry, can be described in terms of superposition of HG modes. The field amplitude of given HG mode is described by

$$\begin{aligned} U_{nm}^{\mathrm{HG}} &= C_{nm}^{\mathrm{HG}} H_n(x\sqrt{2}/w) H_m(y\sqrt{2}/w)\frac{1}{w}\exp\left[-\frac{ik}{2R}(x^2 + y^2)\right] \\ &\times \exp\left[-\frac{(x^2 + y^2)}{w^2}\right]\exp[-i(n + m + 1)\psi] \end{aligned} \tag{3.14}$$

where $C_{nm}^{\mathrm{HG}}$ is the normalization constant, $H_n(x\sqrt{2}/w)$ and $H_m(y\sqrt{2}/w)$ are Hermite polynomials, $w(z)$ is the beam waist. The field amplitude, $U_{nm}^{\mathrm{LG}}$, of an LG mode is given by

$$\begin{aligned} U_{nm}^{\mathrm{LG}} &= C_{nm}^{\mathrm{LG}} L_{\min(n,\,m)}^{n-m}(2r^2/w^2)\frac{1}{w}\exp\left[-\frac{ik}{2R}(r^2)\right]\exp\left[-\frac{(r^2)}{w^2}\right] \\ &\times \exp[-i(n + m + 1)\psi]\exp[-i(n - m)\phi](-1)^{\min(n,\,m)}(r\sqrt{2}/w)^{n-m} \end{aligned} \tag{3.15}$$

where $C_{nm}^{LG}$ is the normalization constant and $L_{min(n,m)}^{n-m}(2r^2/w^2)$ is the generalized Laguerre polynomial. In both the equations

$$\frac{1}{R(z)} = \frac{z}{(z_R^2 + z^2)} \tag{3.16}$$

$$w(z)^2 = \frac{2(z_R^2 + z^2)}{kz_R} \tag{3.17}$$

$$\psi(z) = \arctan\left(\frac{z_R}{z}\right) \tag{3.18}$$

The Rayleigh distance $z_R = \frac{\pi w_0^2}{\lambda}$ is the $z$ distance at which the beam has maximum curvature, (i.e. distance between minimum (zero) curvature plane ($z = 0$) and maximum curvature plane). $k$ is the propagation constant. The azimuthal phase term, $\exp[-i(n - m)\phi]$, gives rise to interwined helical wavefronts and is the origin of the orbital angular momentum within the beam. In both HG and LG beams apart from the amplitude terms described by their respective polynomials, the quadratic phase terms and Gaussian amplitude terms are common. Hence one can guess that the conversion between LG and HG modes involves phase modulation.

Since, both HG and LG modes form a complete basis set, linear superposition of one set can be used to describe the other, i.e. any LG mode can be expressed in terms of HG modes with appropriate weights and phase factors and vice versa (figure 3.18). Fundamental modes in both HG and LG basis sets have Gaussian amplitude distribution, centered on the beam axis. It is therefore the higher-order HG modes which are transformed to LG modes. Since the gain of the laser is large for the fundamental mode, there is a need to suppress this mode for generation of pure higher-order HG modes. This can be achieved by inserting a cross-wire in the laser cavity with wires aligned with the nodes of the desired mode.

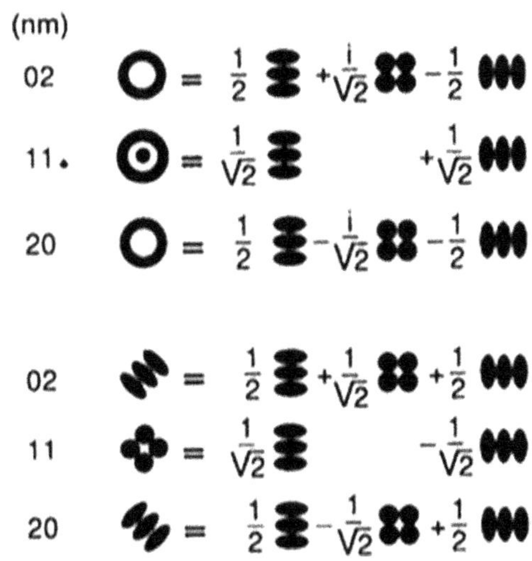


**Figure 3.18.** Decomposition of $HG_{nm}$ and $LG_{nm}$ modes of order 2. Reprinted from [37]. Copyright (1993), with permission from Elsevier.

In order to convert a given HG mode to corresponding LG mode, it is necessary to introduce a phase shift of $\frac{\pi}{2}$ between successive mode components of the HG mode. This is achieved by manipulating the Gouy phase shift, using cylindrical mode converters [37, 38]. The scheme used for the conversion of an HG mode to its equivalent LG mode is depicted in figure 3.19. When an HG beam is focused by a cylindrical lens, the resulting Gouy phase shift depends both on the mode indices and the degree of astigmatism. The astigmatism can be characterized in term of different Rayleigh ranges $z_{Rx}$ and $z_{Ry}$ governing the divergence of the beam in the $x$–$z$ and $y$–$z$ planes. The resulting Gouy phase factor [37] is given by:

$$\psi = \exp\left\{-i\left[\left(n+\frac{1}{2}\right)\arctan\left(\frac{z}{z_{Rx}}\right)+\left(m+\frac{1}{2}\right)\arctan\left(\frac{z}{z_{Ry}}\right)\right]\right\} \tag{3.19}$$

If two identical cylindrical lenses of focal length $f$ are separated by $\sqrt{2}f$, then the transmitted constituent HG modes of a beam with a Rayleigh range of $(1 + 1/\sqrt{2})f$ undergo a relative phase shift due to the Guoy phase of $(n - m)\frac{\pi}{4}$. As $(n-m)$ differs by two for each of the constituent modes, an incident $HG_{10}$ mode rotated at 45° is converted to a pure $LG_0^{\,1}$ mode. The device is also called a $\frac{\pi}{2}$ mode converter as it introduces a phase difference of $\frac{\pi}{2}$ between the constituent modes. When the separation between the lenses is $2f$, the phase difference between the constituent HG modes becomes $\pi$. Such an arrangement, also called $\pi$ mode converter, reverses the handedness of any transmitted LG mode. Figures 3.19 and 3.20 show the operation of $\frac{\pi}{2}$ and $\pi$ mode converters respectively.

### 3.2.10 Intra-cavity methods

In a He–Ne laser cavity, by introducing a spot defect in one of the resonator mirrors, direct generation of a vector donut mode has been demonstrated [39–41]. A low reflectivity circular-shaped spot defect of 30 μm in diameter is created on the inner surface of one of the cavity mirrors by laser ablation. The presence of this low reflectivity spot suppresses the central maximum of a Gaussian beam during oscillations inside the resonator. This means that the oscillation of a fundamental Gaussian mode ($TEM_{00}$ mode) is suppressed and the resonator allows the oscillation of higher-order transverse modes instead. If cylindrical symmetry of the cavity is maintained, ring-shaped transverse modes, including donut modes, oscillate.

In lasers with cylindrical cavity resonators, the $TEM_{01}^*$ donut mode, is composed of two $TEM_{01}$ modes in phase and space quadrature and the cophasal surface has helical form [42]. In other words, the $TEM_{01}^*$ mode is a superposition of a $TEM_{01}$ mode and 90° rotated version of it. This mode is known to contain a vortex. It has been demonstrated that a $TEM_{01}^*$ donut mode with a helical wavefront can be created by frequency locking $TEM_{01}$ and $TEM_{10}$ modes in phase quadrature.

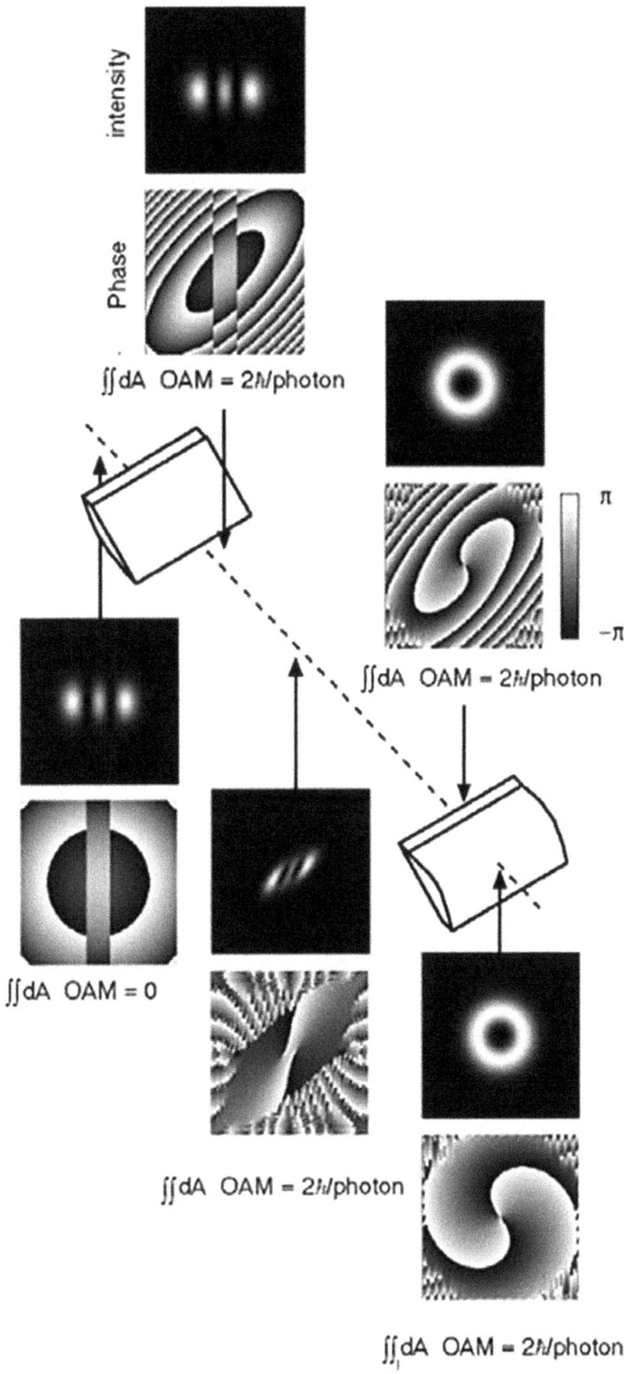


**Figure 3.19.** A $\frac{\pi}{2}$ mode converter. Two cylindrical lenses each of focal length $f$ are separated by a distance $\sqrt{2}f$. An input HG mode is converted into an output LG mode. Intensity and phase distributions at different planes are shown. Reproduced with permission from the corresponding author [38]. 

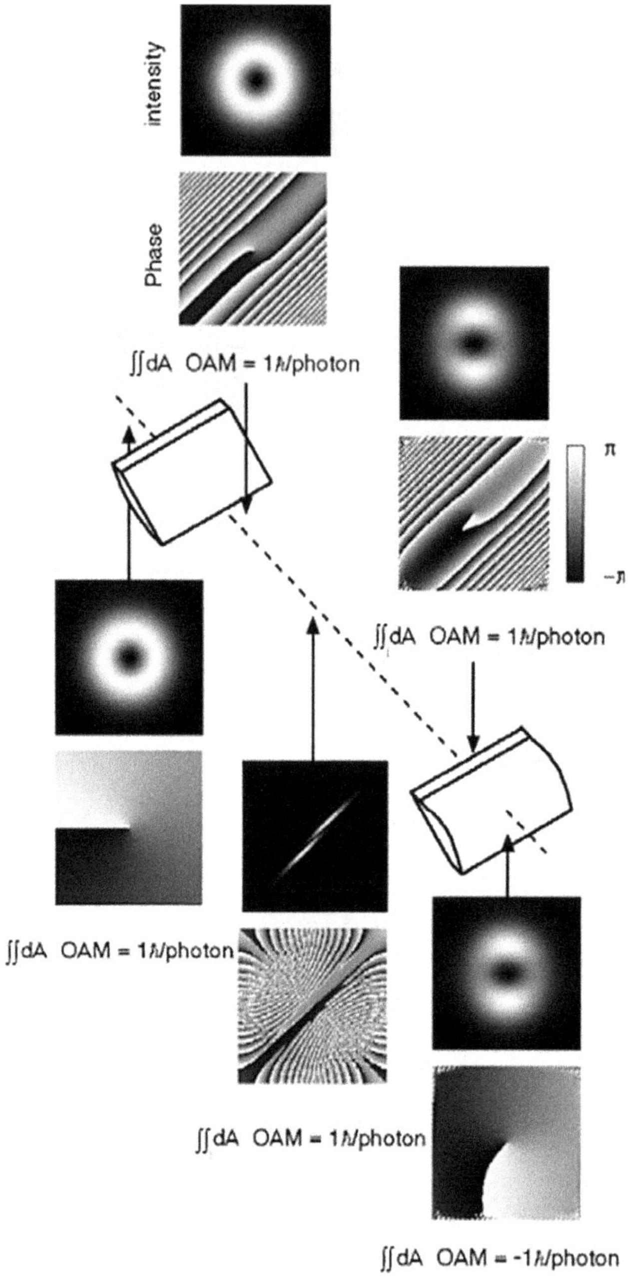


**Figure 3.20.** A $\pi$ mode converter comprising two cylindrical lenses of focal length $f$ separated by $2f$. A collimated LG mode with azimuthal index $l$ is converted to a LG with index $-l$. Reproduced with permission from the corresponding author [38]. 

### 3.2.11 Adaptive helical mirror

An adaptive helical mirror (AHM) can by used for vortex generation [43, 44]. Bending a mirror continuously into a helical shape is possible using a thin flat mirror with a radial cut, actuated by a PZT [Pb(ZrTi)$O_3$] actuator fabricated in a tubular form. The azimuthal phase variation in the AHM is continuous all around, except near the cut. The optical vortex of any topological charge can be generated using this device (shown in figure 3.21) as long as the piezo-ceramic and the thin mirror with a radial cut are used within their breakdown limits. It is possible to vary the charge of the vortex in real time by varying the piezo device excitation voltage. This capability of changing the charge in real time is useful in adaptive optics, optical switching and communication. This makes this device attractive from the point of view of adaptive optics, quantum computing, optical switching, and communication. Vortices of fractional charge can also be generated using this device.

A PZT tubular actuator is specially prepared with a spiral-shaped electrode on its outer surface. The PZT tube is also given a cut along its length, so that the expansion along the axial direction of the tube is not hampered during actuation. It is then glued to the back of the mirror in such a way that the axis of the PZT tube passes through the center of the mirror. Upon actuation by applying a voltage to the piezo actuator, the longitudinal expansion of the tube is different and is a function of azimuth angle. This results in bending the mirror in the form of a single-turn helix.

### 3.2.12 Vortex generation in optical fibers

In this section we describe three different ways of producing optical vortices using fibers. The first method uses stressed optical fibers [45], the second method uses coupling of phonons–photons in optical fibers [46], and the third method is based on using asymmetric fiber couplers [47] for the generation of optical vortices.

**Figure 3.21.** Adaptive helical mirror that can be electronically controlled to generate vortex beams of different order. Reproduced with permission from [43]. Copyright (2008) by OSA.

Alternative methods for generating OVs efficiently use optical fibers, such as, a hollow-core optical fiber [48] and a holey fiber for generating hollow beams [49].

A short length of stressed, near-single-mode fiber-optic waveguide is used to transform a linearly polarized $HG_{10}$ mode into a circularly symmetric annular beam with a well-defined azimuthal phase [45]. This is similar to the cylindrical lens mode converters [37]. The input mode is expressed as two orthogonal $HG_{10}$ modes which are in-phase and aligned with the stress axis. These HG modes couple into $LP_{11}$ modes. The stress changes the propagation constant that governs the phase velocity of the modes. The two orthogonal modes experience different phase velocities as the stress is applied along a specific direction. With appropriate adjustment of the stress, the two modes appear at the end of the fiber with 90° out of phase, and the output is a circularly symmetric annular beam. The mode purity is, however, not very high.

However, the azimuthal phase dependence is well defined, resulting in a well-defined orbital angular momentum. The stress on the fiber also introduces birefringence into the fiber core that interacts with the polarization state of the transmitted light. A circularly polarized beam launched into this setup becomes elliptically polarized. Since circularly polarized light is associated with spin angular momentum of $\pm\hbar$ per photon, to avoid any uncertainty between spin and orbital angular momentum that is created, the output from the fiber is polarized linearly with a simple polarizer. Figure 3.22 shows the experimental setup that converts the HG mode to the donut beam with a well-defined azimuthal phase term.

The He–Ne laser with an output power of approximately 2 mW at 632 nm is forced to oscillate in a $HG_{10}$ mode by the inclusion of an intra-cavity cross-wire. The output beam is coupled into a single-mode (1.06 μm) fiber-optic waveguide. The

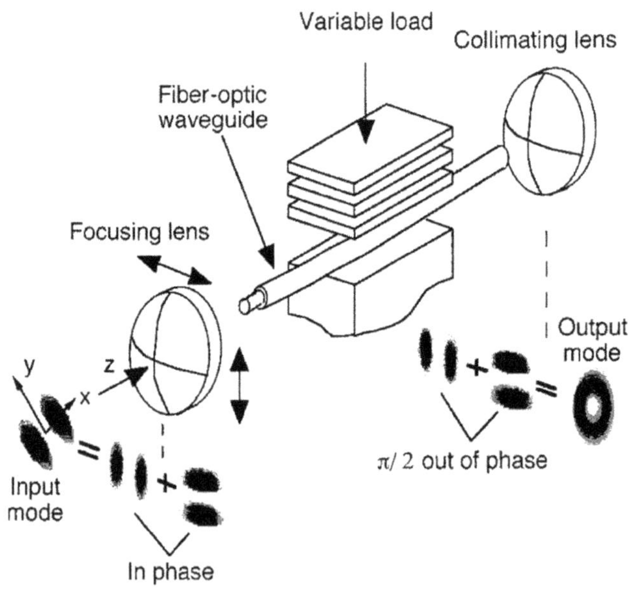


**Figure 3.22.** Experimental arrangement, based on a stressed near-single-mode optical fiber, for producing annular beams with orbital angular momentum. Reproduced with permission from [45]. Copyright (1998) by OSA.

optical fiber is 300 mm in length and is stressed over a 100 mm region by the application of rectangular lead weights with a total mass of approximately 2 kg. A 30 mm focal length lens is used to collimate the output of the fiber, and the resulting beam can be recorded with a CCD array. The overall efficiency of the fiber-based mode converter is limited by the coupling efficiency of the input beam into the fiber and is measured to be approximately 10%.

In the second method, the interaction in an optical fiber can lead to transfer of OAM from an acoustic vortex to an optical vortex [46] and that spin and orbital angular momentum are conserved independently among the interacting photons and phonon. This acousto-optic interaction can be used to generate pure and stable optical vortices in the fiber medium starting from its fundamental mode.

The interaction between acoustic waves and optical waves in a fiber can be engineered for optical vortex generation. In an acousto-optic interaction, a refractive index grating is created using acoustic waves through the elasto-optic effect. This leads to coupling between optical modes. The acoustic waves are phonons and their interaction with photons results in a scattering process. The total energy and linear and angular momenta (both SAM and OAM) of the photon–phonon particles are conserved in the scattering. Phonons do not have intrinsic spin angular momentum, but can carry orbital angular momentum by forming vortices in the medium in which they propagate. In an optical fiber their interaction can lead to transfer of OAM from an acoustic vortex to an optical vortex.

The acousto-optic interaction can be viewed from the photon–phonon picture. When optical and acoustic waves propagate in the same direction, the fundamental mode photon emits a phonon identical to the phonons of the interacting acoustic mode (stimulated emission) and becomes a higher-order mode photon. During this emission process, the photon loses a portion of its linear momentum and energy to the phonon, and therefore, its momentum is reduced to that of a higher-order mode and its frequency is down-shifted by the acoustic frequency. If the initial photon carries a SAM of $\hbar$ and the interacting acoustic mode is a vortex mode with an OAM of $-\hbar$ per phonon, meaning that the emitted phonon has an angular momentum of $-\hbar$, then conservation of angular momentum requires that the generated higher-order mode photon carry a total angular momentum of $2\hbar$. Since spin can account for only one $\hbar$ per photon, the remaining $\hbar$ must be carried by the photon's OAM. Thus, the higher-order mode photon manifests itself as a vortex mode bearing an OAM of $\hbar$ and a SAM of $\hbar$. A closer examination of the acousto-optic coupling coefficient based on vortex modes shows that spin and orbital angular momenta are conserved independently.

In the fiber-optic systems discussed [45, 46] specially designed optical fibers and complicated experimental setups for generating optical vortices are used. In the next method, the use of a fiber coupler in the generation of vortex beams is discussed. There are reports [50–52] on the phenomenon of generating two modes using fiber couplers: one fundamental mode and the other higher-order mode. Most of the fiber couplers are fabricated for communication wavelengths and when they are operated at shorter wavelengths (632.8 nm), they act as multimode fibers. In the third method mentioned here [47], a fused taper fiber coupler which is a standard coupler made for optical communication wavelength, but operated at shorter wavelength (632.8 nm)

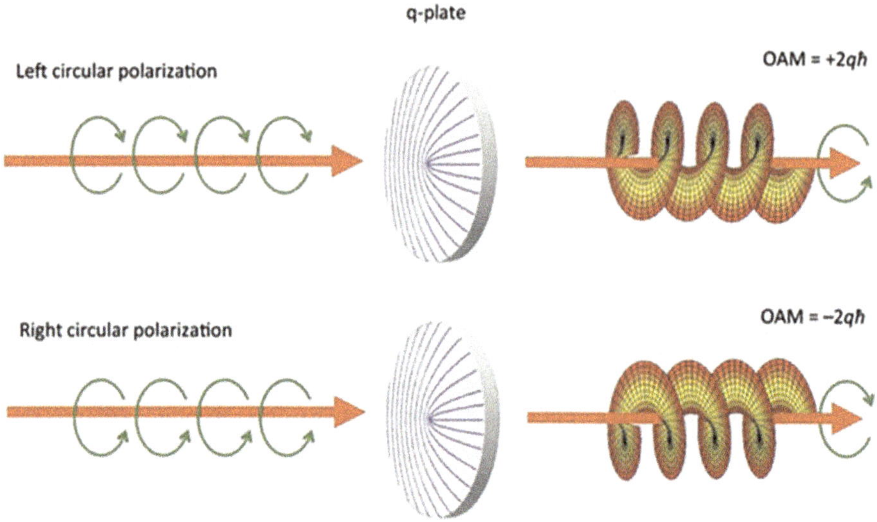


**Figure 3.23.** *q*-wave plates of different charges can be used for phase singularity generation. Reprinted from [53], with permission from SPIE.

is used. The fiber coupler is similar to the all fiber mode converter reported in [50, 51]. The circular waist supports several polarization modes. When an input light beam ($TEM_{00}$) is launched into the fiber which is equivalent to $LP_{01}$ mode of the fiber, the same mode is coupled out from the un-pre-tapered fiber which is a fundamental mode and the $LP_{11}$ mode is coupled out from the fused pre-tapered fiber. Thus, one of the fibers of the Y-coupler was used to generate a spherical beam and the second a donut beam, respectively.

### 3.2.13 *q*-wave plates for vortex generation

*q*-wave plates are spatially varying half wave plates [53–55]. The principle axis of the HWP is spatially varying in these plates. These plates are used to produce polarization singularities, a subject that will be discussed in the last chapter. These polarization singularities are superpositions of phase singularities in circular polarization basis. The charge $q$ of the *q*-wave plate is equal to half the charge of the optical vortex that it can produce. These plates can also be used to produce phase singular beams. Figure 3.23 show the generation of phase singular beams when circularly polarized light is made to incident on the *q*-wave plate. The polarity of the phase singularity can be changed by changing the handedness of the circular polarization from left to right or vice versa depending on the charge of the *q*-wave plate.

## 3.3 Detection

The focal plane field distribution of a lens is the typical Airy pattern for a non-singular beam. Singular beams are known to produce donut shaped focal spots whose central dark core size is dependent on the charge of the vortex. Although this fact can be used as a technique for vortex detection, there are many other ways a vortex can be detected. The helical phase variation can be detected using various techniques. In this section, we discuss various interference, diffraction based experimental techniques used for vortex detection. In computation optics the fields are handled in mathematical form and various methods of detecting vortices in such fields are also presented at the end of this section.

### 3.3.1 Interference methods

Vortices can be detected using phase detection techniques such as interferometric methods. The presence of vortices in the test wavefront can be found by interfering it with a reference beam and analyzing the interference pattern. The shape of the interference fringes depends on the type of the reference beam used.

*Interference with plane waves*

First we deal with on-axis plane waves. On-axis plane waves and a vortex beam interference results in a radial fringe pattern [56]. Since the interference fringes represent phase difference contour lines, and in a vortex phase variation, phase contours are radial, we get radial fringes. The number of fringes in the interference pattern is equal to the magnitude of the charge $m$ of the vortex. A new fringe starting from the center of the interference pattern is the signature of the optical vortex. The fringe starts from the vortex core and runs along one of the phase contours of the vortex, where the phase difference between the vortex beam and the plane reference beam is $m2\pi$. The resultant amplitude in the interference can be given by

$$A = \exp\{i\phi_0\} + \exp\{im\phi(x, y)\} \tag{3.20}$$

Here $\phi_0$ is a constant, $\phi$ is the azimuthal angle. The bright fringe formation condition is given by

$$\phi_0 - m\phi(x, y) = 2n\pi \tag{3.21}$$

Here $n$ is the order number and $m$ is the charge of the vortex. In this equation the interference maximum order number is equal to the magnitude of the topological charge of the vortex.

*Interference with tilted plane waves*

A tilted plane wave interfering with vortex beams leads to the resultant field given by

$$A_{res} = \exp\{i2\pi(f_x x + f_y y)\} + \exp\{im\phi(x, y)\} \tag{3.22}$$

The fringes form fork-like patterns as shown in figure 3.24 for $f_y = 0$. Otherwise the fork grating is rotated and the nearly parallel fringes enclose an angle with the $x$ axis. Tilt decides the constants $f_x$ and $f_y$ and $m$ is the charge of the vortex. The number of new fringes forming the fork at the vortex site decides the topological charge of the vortex. Depending on the sign of the vortex and the tilt in the plane reference wave, the forks are pointing up or down (say). To elucidate this, consider the four cases given below:

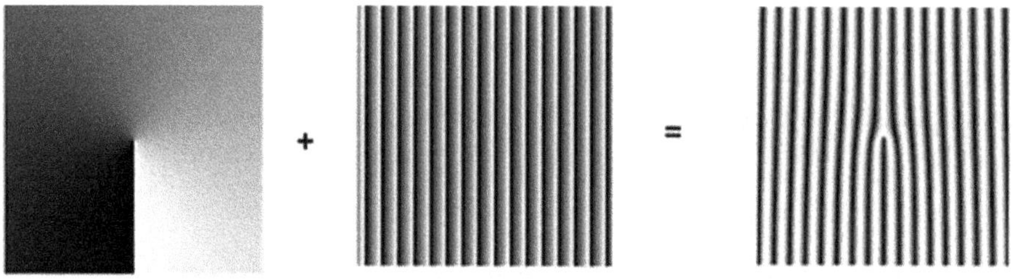

**Figure 3.24.** Interference of a plane wave with a tilt and a helical wave produces fork interference pattern as vortex signature. From left to right: phase distributions of vortex, plane beam and the resulting interference pattern. Note in this depiction the first two frames are phase distributions and the third is an intensity distribution.

$$\begin{aligned} A_{1R} &= \exp\{-i2\pi f_x x\} + \exp\{im\phi\} \\ A_{2R} &= \exp\{-i2\pi f_x x\} + \exp\{-im\phi\} \\ A_{3R} &= \exp\{i2\pi f_x x\} + \exp\{im\phi\} \\ A_{4R} &= \exp\{i2\pi f_x x\} + \exp\{-im\phi\} \end{aligned} \tag{3.23}$$

The interference patterns for fields $A_{1R}$ and $A_{4R}$ are the same as either the direction of the tilt or the sign of the charge in each of them is inverted. Similarly, interference patterns for the fields $A_{2R}$ and $A_{3R}$ are the same as both the direction of tilt and the charge are inverted simultaneously in them. Recall that using a fork grating upside down flips the sign of charge of the vortex produced by the fork grating in any order. In a dipole vortex lattice, the forks can be seen pointing in opposite directions as shown in figure 3.25. The types of fringes obtained when a vortex beam interferes with another vortex beam, and with plane waves are shown in figure 3.26.

*Interference with spherical waves*

When the reference wave is a spherical wave, interference with the vortex beam produces spiral fringes. The fringes spiral from the vortex core. The spiral fringes are formed only when the phase extremum point of the spherical wave and the vortex core coincide. The number of new fringes starting from the core equals the charge of the vortex. For a positive and for a negative charge vortex, the senses of spiraling are opposite to each other. Like before, the sense of spiraling also depends on the curvature of the reference spherical wave. The resultant amplitude in the interference is given by

$$\begin{aligned} A_{1R} &= \exp\left\{\frac{-ik}{2R}(x^2 + y^2)\right\} + \exp\{im\phi\} \\ A_{2R} &= \exp\left\{\frac{-ik}{2R}(x^2 + y^2)\right\} + \exp\{-im\phi\} \\ A_{3R} &= \exp\left\{\frac{ik}{2R}(x^2 + y^2)\right\} + \exp\{im\phi\} \\ A_{4R} &= \exp\left\{\frac{ik}{2R}(x^2 + y^2)\right\} + \exp\{-im\phi\} \end{aligned} \tag{3.24}$$

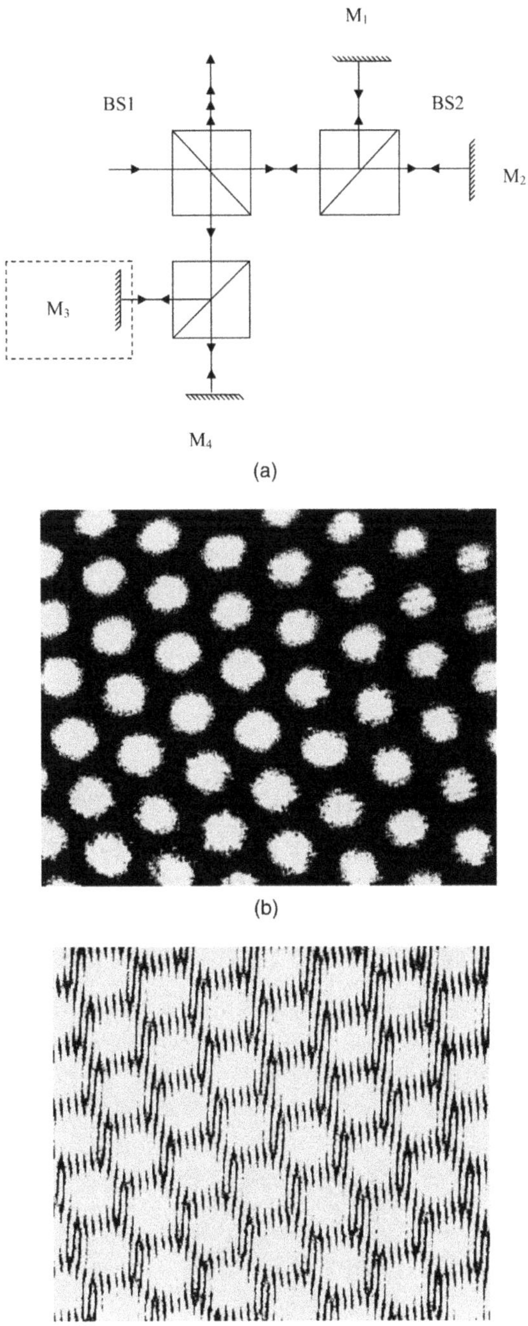


**Figure 3.25.** (a) Interferometer to generate vortex lattice. (b) Intensity distribution in the interference pattern. (c) Fork fringes in a vortex lattice. This is obtained by interfering a vortex lattice field with a tilted plane wave. Reproduced with permission from [24]. Copyright (2007) by OSA.

In the interference, a spherical beam with positive curvature and a positive charge vortex will produce fringes that spiral the same way, if both the curvature of the spherical beam and the sign of the charge are reversed simultaneously. Therefore the sense of spiraling in the fringes corresponding to fields $A_{1R}$ and $A_{4R}$ are same. Likewise $A_{2R}$ and $A_{3R}$ will also produce same type of fringes. These fringes are known as spiral fringes (figure 3.27).

There are some interesting features one can observe in the interference patterns. For a vortex charge of unit magnitude, there is only one fringe starting from the center of the fringe pattern and spiraling out. Hence the fringe order number is the same for the whole pattern and is equal to the topological charge of the vortex. Even though these fringes are Fizeau fringes, how can it be possible to have the same fringe order? In optical testing, it is well known that the quadratic phase variation of the spherical wave normally produces fringes similar to Newton's rings. Since the fringe order is the same, does it mean that the thickness (or phase difference, or air-gap—as experimentalists put it) is the same? The problem lies with the use of a singular beam in interference.

When the vortex core location does not coincide with the phase extremum, the fringe formation is governed by the resultant field given by

$$A_R = \exp\{im\phi'(x, y)\} + \exp\left\{\frac{-jk}{2R}(x^2 + y^2)\right\} \tag{3.25}$$

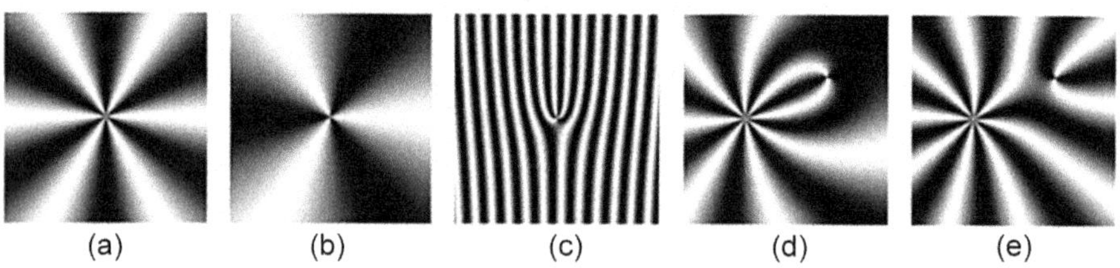


**Figure 3.26.** Fringes obtained due to interference between a vortex of charge +3 with (a) another beam with charge −3, (b) an on-axis plane wave, (c) an off-axis plane wave. Interferograms obtained when both the interfering beams have off centered vortices of (d) opposite unequal charges, (e) same signed unequal charges. Reprinted from [13].

**Figure 3.27.** Interference of a spherical wave with a helical wave produces a spiral interferogram. From left to right: phase distributions of vortex, spherical beam and the interference pattern. Note in this depiction the first two frames are phase distributions and the third is intensity distribution.

where $\phi' = \text{Arctan}\left[\frac{y - y_1}{x - x_1}\right]$ is the phase variation due to a vortex at $(x_1, y_1)$. Further different types of fringes produced by the vortex beam interfering with spherical and conical beams are shown in figure 3.28.

*Lateral shear interferometer*

In a lateral shear interferometer, the reference wave is derived from the test wavefront itself and hence it is a self-referencing interferometer. The fringes obtained here are called shearograms [57] and they differ from the fringes discussed in the earlier subsection. The fringes here represent phase gradient contours. The schematic of the shear interferometer is shown in figure 3.29. Consider a test wavefront described by the path function $W(x, y)$, which describes the path difference between the test wavefront and a plane wavefront (taken as a reference wavefront). A sheared (or displaced) version of the wavefront with path function $W(x + \Delta x, y)$ is produced in a shear interferometer and is made to interfere with the original wavefront. The complex amplitudes of the two waves that interfere in a shear interferometer is given by $\exp\{ikW(x, y)\}$ and $\exp\{ikW(x + \Delta x, y)\}$. The two waves are assumed to have the same amplitude. The interference intensity is given by

$$I(x, y) = I_0\{1 + \cos[k(W(x, y) - W(x + \Delta x, y))]\} \tag{3.26}$$

For beams containing phase vortices, both the test wavefront and the sheared wavefront have a vortex in them. Hence the shearogram obtained will contain two new fringes starting from the vortex locations. In a shear interferometer plane waves

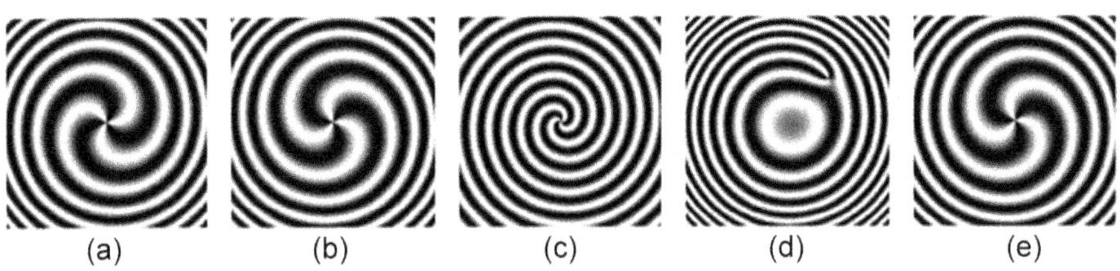


**Figure 3.28.** Fringes obtained due to interference between a vortex of charge three with (a) spherical beam of positive curvature, (b) spherical beam of negative curvature, (c) conical beam. (d) Interference between an off centered vortex and a spherical wave. (e) Interference between negatively charged vortex and a spherical beam with positive curvature. Reprinted from [13].

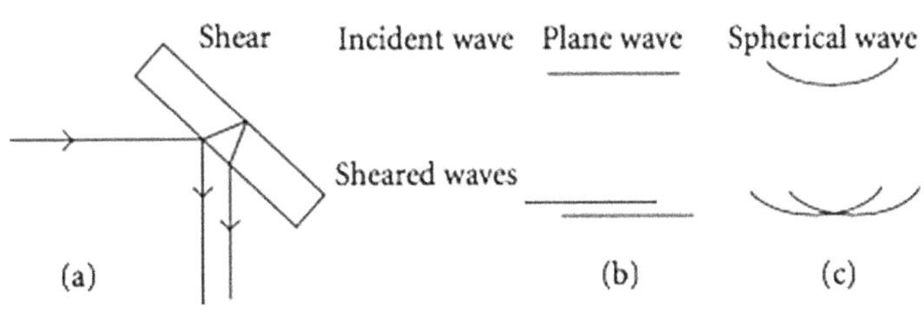


**Figure 3.29.** (a) Plane parallel plate acting as a lateral shear interferometer in which the incident wave is split into two and sheared. (b) Schematic showing that the phase difference between sheared plane waves is constant. (c) Schematic showing the phase difference between sheared spherical waves is not constant and varies linearly in the direction of shear. Reprinted from [13].

produce fringe-free shearograms, but when there is curvature in the test beam, the shearogram will have straight line fringes. A spherical beam in a normal interferometer produces circular fringes. Hence a vortex beam (unit charge) in a shear interferometer will produce only two fringes radiating from the vortex centers. If there is a tilt introduced in the interferometer, the resulting shearogram has fork fringes with two forks aligned opposite to each other. Inherent tilt between the sheared wavefronts can be introduced using a wedge plate lateral shear interferometer (figure 3.30).

Lateral shearing can be realized using a plane parallel plate (PPP). The test beam incident on this PPP at an oblique angle produces two reflected beams, one reflected from the front surface and the other from the back surface of the PPP. Because of the finite thickness of the PPP and oblique angle of incidence the two reflected beams are laterally shifted [13]. The amplitudes of these two beams are nearly the same and hence good contrast shearograms can be observed. Since we look for the birth of two new fringes for every vortex present in the beam and not the shape of the fringes as required in optical testing, use of an ordinary glass plate in place of PPP will suffice [58–60]. Hence the simplest and most elegant technique for vortex detection is to insert an ordinary glass plate in the test beam and probe the shearogram obtained in reflection. Additional tilts between sheared wavefronts can be introduced by using a wedge plate instead of PPP and depending upon the direction of the tilt and shear the fringe spacing or orientation can be controlled. A vortex embedded in a spherical beam produces fringes as shown in figure 3.31. When multiple vortices are present in a beam, like in a speckle pattern, the shearogram obtained is shown in figure 3.32.

### 3.3.2 Diffraction methods

*Diffraction through a single slit*

When a singular beam is incident on a narrow slit, the field just behind the slit is given by

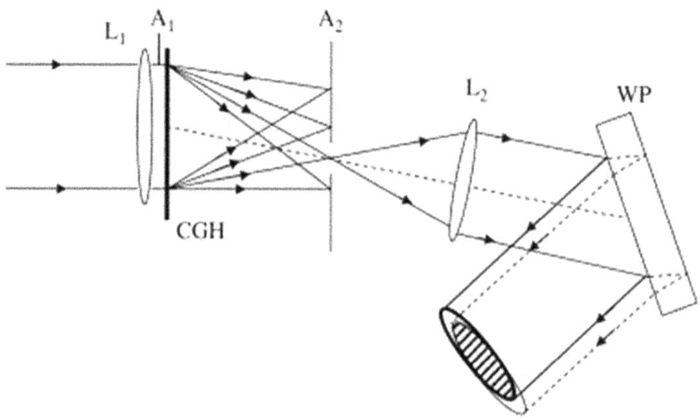


**Figure 3.30.** Wedge plate lateral shear interferometer for vortex detection. A wedge plate usually provides tilt to one of the sheared wavefronts in addition to shear. For pure shear, a plane parallel plate can be used. Reprinted from [60]. Copyright (2008), with permission from Elsevier.

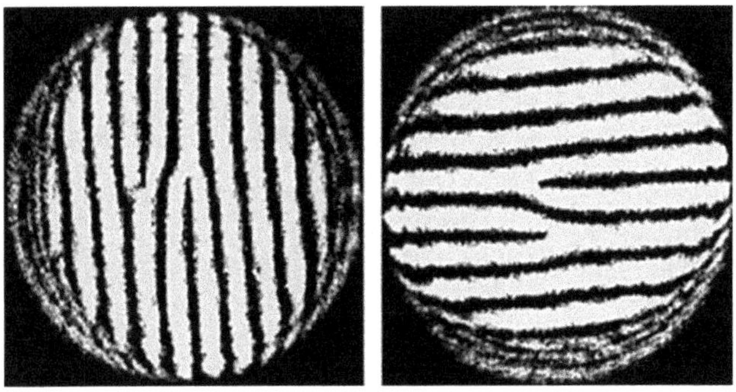

**Figure 3.31.** Lateral shear interferograms called shearograms of vortex beams. Shearograms obtained when the shear is in the horizontal and in the vertical directions are shown on the left and right sides respectively. Reprinted from [58]. Copyright (2008), with permission from Elsevier.

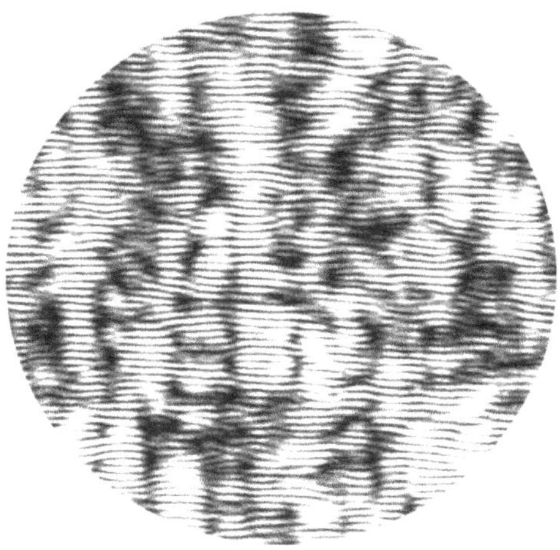

**Figure 3.32.** Lateral shear interferogram of a speckle pattern. A number of fork fringes can be observed with forks pointing in both directions. Use of a glass plate is one of the simplest and easiest ways to test for the presence of a phase singularity in a beam. Reprinted from [58]. Copyright (2008), with permission from Elsevier.

$$U(x, y) = rect\left(\frac{x}{b}\right)\exp(im\phi) \tag{3.27}$$

where $b$ is the slit width, $\phi$ is the azimuthal angle and $m$ is the topological charge of the singular beam. For a plane wave, i.e. $m = 0$, the diffraction pattern has intensity distribution [61], given by

$$I(x, y) = b^2\frac{\sin^2\beta}{\beta^2} \tag{3.28}$$

where $\beta = \frac{1}{2}kb\sin(\theta)$ and $\theta$ is the diffraction angle. The quantity $\beta$ represents the phase difference between the waves reaching the point of observation from two points in the slit, separated by half the slit width. For a uniform on-axis plane wave, the complex field inside the slit is the same, and all along the length of the slit. For a

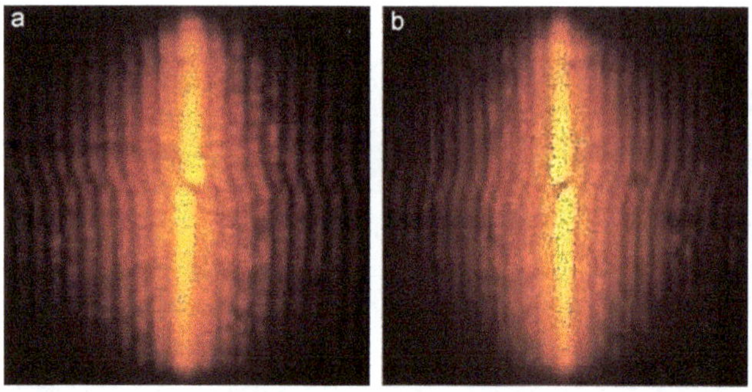


**Figure 3.33.** (a) Single-slit diffraction pattern of singular beam [62] with charge +1 and (b) with charge −1. Reprinted from [62]. Copyright (2009), with permission from Elsevier.

singular beam illuminating the slit such that the core of the vortex is at the mid-way between the slit width, this $\beta$ is a function of position coordinates [62]. This is because the phase difference between the right and left edge of the slit is a function of position coordinates. As a result, the fringes obtained due to single-slit diffraction undergoes bending as shown in figure 3.33. This bending can be used to figure out the sign and magnitude of optical vortex.

*Diffraction through a double slit*

Due to the azimuthal phase dependence of singular beams the double slit interference pattern is affected [63]. The changes in the Young's fringes can be used to study the helical phase structure. Two slits, each of width $b$ and with separation distance $d$ under plane wave illumination produces the intensity distribution given by

$$I(x) \propto \cos^2 \frac{\delta}{2} \tag{3.29}$$

where $\delta = \frac{1}{2}kd \sin \theta$ is half the phase difference between the lights reaching an observation point from the two slits. But when a singular beam is illuminating the double slit setup this phase difference is modified so that the intensity distribution in the interference pattern is now given by

$$I(x, y) \propto \cos^2 \left[ \frac{\delta}{2} + \frac{\Delta\phi(y)}{2} \right] \tag{3.30}$$

where $\Delta\phi(y) = \phi_1(y) - \phi_2(y)$ is the phase difference between the points at the two slits that arise due to the helical wavefront. Because of this extra term the Young's fringes undergo a bending as shown in figure 3.34. This bending of fringes is an indication of the presence of vortices in the beam.

*Double slit diffraction from waves near zeros*

The helical phase variation of singular beams is responsible for the bending of fringes in the single and double slit experiments. The phase variation has azimuthal dependence. But near the vortex core, where the phase gradients are very high, the phase variation is found to have radial as well as azimuthal dependence. This has been predicted theoretically [64, 65]. In the only experimental evidence reported [66], the spacing of the Young's fringes is shown to defy the usual dependence of pin-hole separation and is governed by the radial part of the phase gradient. Experimental observations (figure 3.35) show that the number of fringes in the pattern depends on the topological charge of the incident beam. The larger the charge of the vortex, the larger the radial component of the phase gradient and the higher the number of fringes observed.

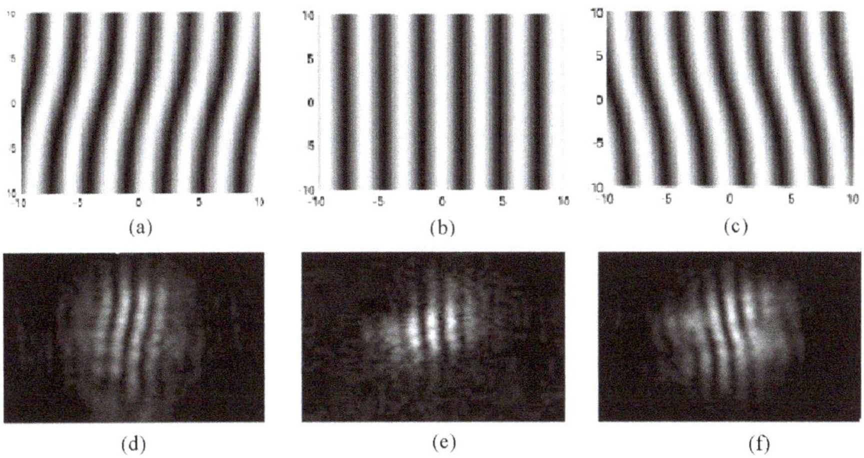


**Figure 3.34.** Interference from a LG wavefront incident on two slits. Simulated interference patterns when the charge of the incident beam is (a) $l = +1$, (b) $l = 0$ and (c) $l = -1$. Experimentally obtained far-field interference patterns are shown for (d) the $l = +1$, (e) $l = 0$, and (f) $l = -1$ when the beam is passing through two slits. Reproduced with permission from [63]. Copyright (2006) by OSA.

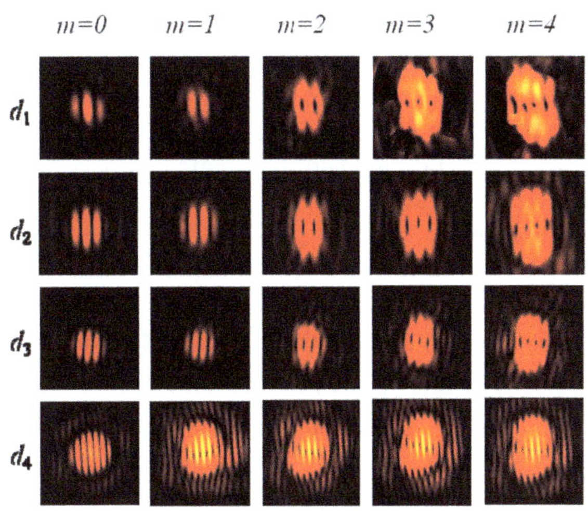


**Figure 3.35.** Young's double slit experimental patterns from waves near zero of a vortex. Anomalous behavior of Young's fringes, where the fringe width defies the slit separation distance relation. Reproduced with permission from [66]. Copyright (2015) by OSA.

*Diffraction through multiple slits—diffraction grating*

Diffraction gratings that produce multiple diffraction orders can be used for vortex detection [67]. To understand how this method works let us consider a diffraction grating with binary transmittance function. The transmittance function of such a grating can be given by $V_l(\phi)G_\gamma(x)$, where $V_l(\phi)$ is a vortex phase function of charge $l$ and $G_\gamma(x)$ is linear phase variation that forms the grating structure with vertical lines. The resulting phase variation is a grating with fork structure. $\gamma = \frac{2\pi}{d}$ and $d$ is the grating period. Here $x$ is the horizontal spatial coordinate. The grating is then made binary phase-only with two phase transmittance values 0 and $\phi$. This new grating can be expressed in Fourier series as

$$G_{l,\,\gamma}(x,\, y) = \sum_{n=-\infty}^{+\infty} c_n \exp\{iln\phi\}\exp\{in\gamma x\} \tag{3.31}$$

where the magnitude of the Fourier coefficients $|c_n|$ are given by

$$|c_n| = \frac{[\sin(n\phi wd)]}{[n\phi]} \tag{3.32}$$

where $w$ is the width of the grating. This fork grating is illuminated with a vortex beam of charge $m$. Now the field behind the grating is given by

$$\exp\{im\phi\}G_{l,\gamma}(x,\, y) = \sum_{n=-\infty}^{+\infty} c_n \exp\{i(m + ln)\phi\}\exp\{in\gamma x\} \tag{3.33}$$

In the diffraction pattern in each of the diffraction orders there is a vortex having a charge $ln + m$ where $n$ is the diffraction order number. The coefficients $c_n$ still decide the intensity at each spot. If the incident field has a vortex of charge $m = +1$, an intensity peak corresponding to the delta function appears in the first order ($n = 1$). Further, each focus spot at a location $p = n\gamma$ is characterized by a vortex having a charge of $ln + m$. So the existence of the delta function on the $n$th diffraction order indicates the illumination by a vortex beam of charge $m = n$. This way vortices can be detected in a beam. Figure 3.36 shows the intensity distribution in each of the diffraction orders of a fork grating when illuminated by a vortex beam.

*Variable frequency grating*

Novel diffraction gratings with variable frequencies can be used for vortex detection [68]. By defining functions $C_1(x) = \cos[2\pi x/(T_0 + n_1 x)]$ and $C_2(x, y) = \cos[2\pi x/(T_0 + n_1 y)]$, two binary amplitude gratings with transmittance functions

$$\begin{aligned} t_1(x, y) &= 1 \quad \text{if} \quad C_1(x) \geqslant 0 \\ &= 0 \quad \text{if} \quad C_1(x) < 0 \end{aligned} \tag{3.34}$$

and

$$\begin{aligned} t_2(x, y) &= 1 \quad \text{if} \quad C_2(x, y) \geqslant 0 \\ &= 0 \quad \text{if} \quad C_2(x, y) < 0 \end{aligned} \tag{3.35}$$

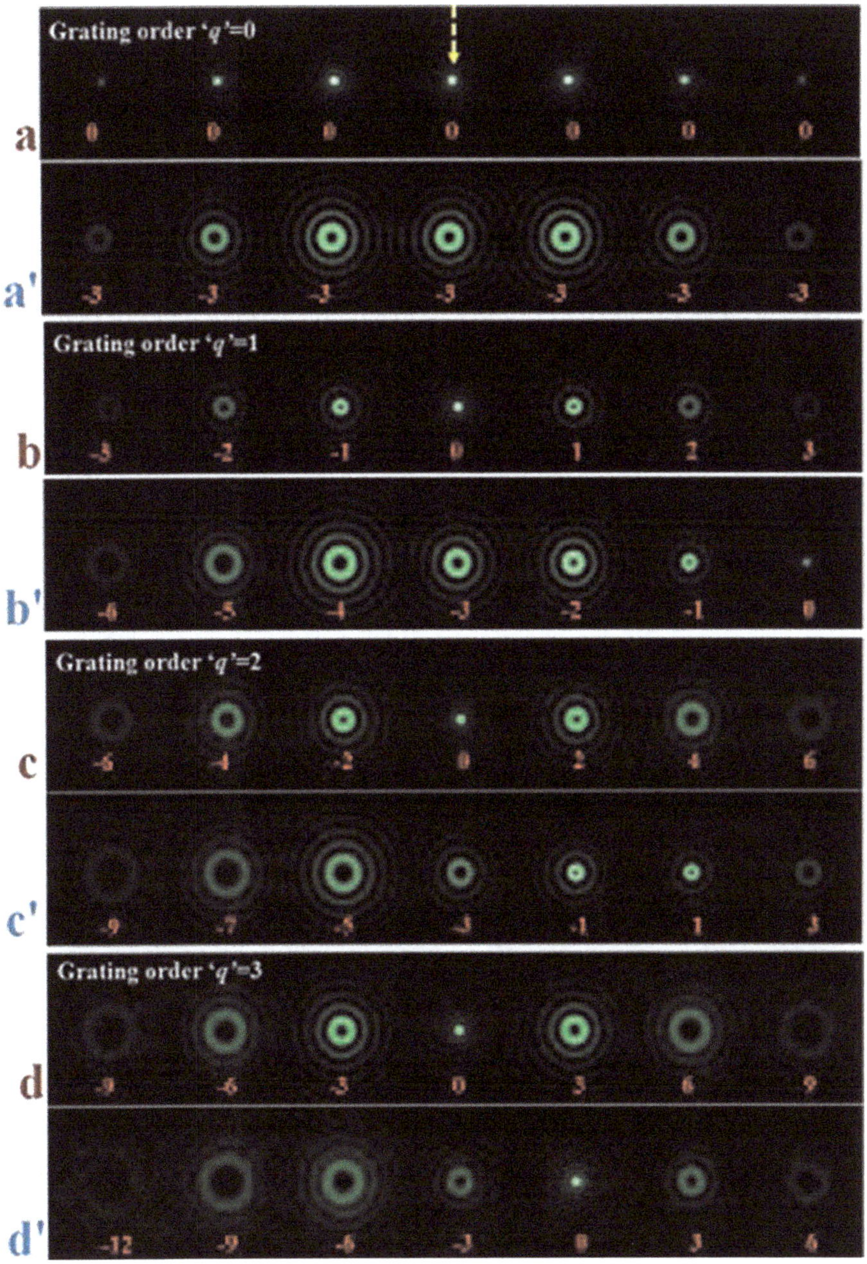


**Figure 3.36.** Diffraction orders from vortex sensing (fork) gratings. The zero diffraction order is at the center marked by an arrow in the first figure. The charge of the fork grating in a, b, c and d is 0, 1, 2, 3 respectively. The number below each of the diffraction order indicates the vortex charge obtained. The diffraction orders under plane wave illumination are shown in (a)–(d) and when illuminated by a vortex beam of charge −3 are shown in (a')–(d') respectively.

can be considered. When a vortex beam is incident on this grating, the far-field diffraction pattern splits the charge of the incident vortex into elementary charges in the diffraction order which is spread over a region because of the spatially varying angle of diffraction from the grating. When the grating has the spatially varying period in the $x$ direction, the spread is in the $xy$ plane. When the period varies in both $x$ and $y$ directions, the spread occurs in the $y$ direction in the positive diffraction order and in the $x$ direction for the negative diffraction order. The diffraction gratings and their diffraction patterns under singular beam illumination are shown in figure 3.37. One of the positive aspects of this method is that the test beam can be made to be incident on any part of the grating and there is no center of element and vortex core alignment requirement as demanded by many other methods.

*Dammann grating*

A fork grating, which is useful in the generation of vortex beam can also be used as a vortex sensing element. In a similar way a Dammann grating that is useful in the generation of vortex beams can also be used as a vortex sensing element.

Dammann gratings are binary phase gratings that are used to create equal intensity light spots. In a sinusoidal amplitude grating the number of diffraction orders are restricted to $-1$, 0 and $+1$, whereas in a binary amplitude grating, multiple diffraction orders occur. The central beam or the zero order diffraction has

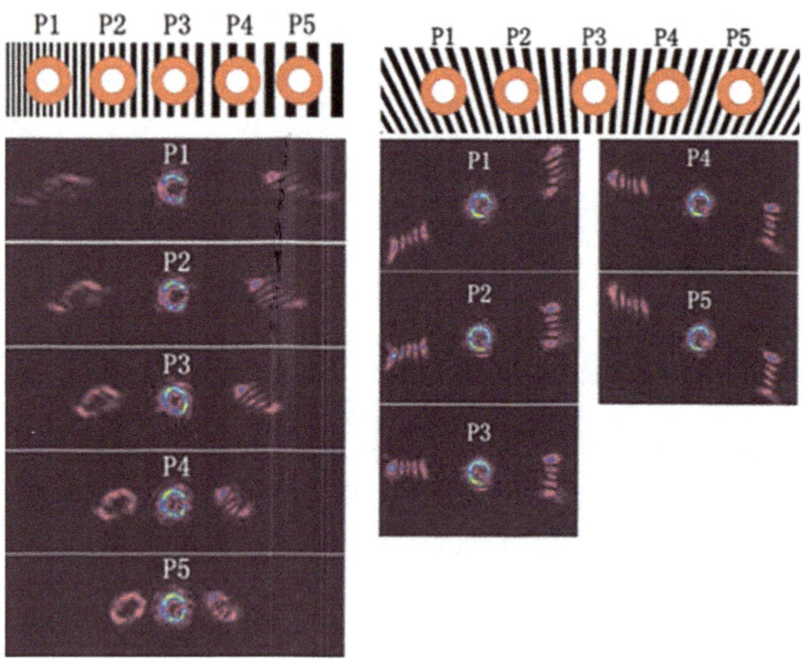


**Figure 3.37.** Variable frequency grating for charge determination. The period of the grating is gradually changing (left panel) and the period and the orientation change simultaneously for the grating (right panel). This results in separation (disintegration) of higher charge vortex into lower charge vortices. Reproduced with permission from [68]. Copyright (2015) by OSA.

maximum intensity, whereas only 6.25% of light intensity is diffracted into the first order of the sinusoidal amplitude grating and 10.3% of light is diffracted into the first order of the binary amplitude grating [69, 70]. Also the multiple orders are not of the same intensity. In a sinusoidal phase grating, multiple diffraction orders are possible and the maximum diffraction efficiency achievable is 33% in the first orders and the remainingf 34% (as ±1 orders have 33% + 33% = 66%) of light energy is distributed over all the other orders including the central undiffracted light. In this chapter, it has been shown that a multiple order producing diffraction grating can be used as a vortex sensing grating [67]. This vortex sensing grating is a fork grating. If the fork grating is made for a vortex charge of one, then a singular beam with charge one incident on this grating produces a bright spot at the −1 order as the charge of the incident beam gets cancelled at this order. Hence a bright spot is produced in −1 order and the rest of the orders have a donut intensity profile. Hence, to check a vortex beam of charge five using a fork grating with single fork, the −5 diffraction order of the grating produces a bright spot indicating that the vortex charge of the incident beam is +5.

But the problem lies with the fact that the diffraction orders in a grating do not have equal intensity. Hence to check higher-order charges a grating that can produce equal intensity spots at higher-order is desired. A Dammann grating is one such grating which is capable of producing different diffraction orders arranged in a single line or in a two-dimensional grid of points. Reference [34] reports the generation of a new kind of composite phase diffractive grating, by integrating a $5 \times 5$, two-dimensional Dammann vortex grating with a 12th or −12th order spiral phase plate. This grating is demonstrated to have a vortex charge detection capacity ranging from +24 to −24. Methods to increase the range of detection are also discussed in [71].

*Annular grating*

The vortex beams with different topological charges (i.e. OAM states), as shown in figure 3.38(a), illuminate the annular gratings at an offset position from the center of the annular gratings (red ring in figure 3.38(b)) [72]. The far-field diffraction intensity patterns are recorded after passing through the annular gratings. The magnitude of the topological charge can be distinguished by observing the number of dark fringes after diffraction. Moreover, the sign of the topological charge can be determined by observing the orientation of the diffraction pattern.

This annular grating has equal fringe width but the orientation of the grating is changing continuously. Over the illuminating area, the grating vector continuously changes direction, and this suggests that the effect of the variable frequency grating shown in figure 3.37 in disintegrating a higher charge vortex comes into play.

*Diffraction through triangular aperture*

In this section the diffraction pattern of singular beams through an equilateral triangle aperture [73], annular triangular aperture [74] and an isosceles right triangular aperture [75] are discussed. These diffraction patterns can be used for the detection of the charge of a vortex.

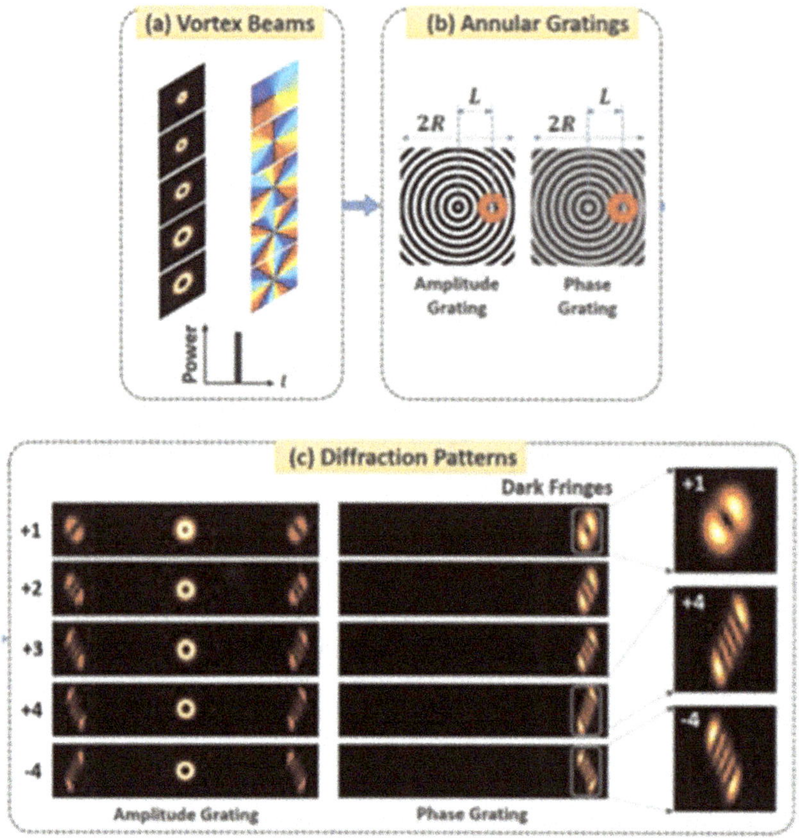


**Figure 3.38.** Vortex diffraction patterns when different parts of annular grating are illuminated. (a) Vortex beams with different charge, (b) region of illumination indicated by red ring and (c) diffraction patterns observed. Reprinted with permission from [72]. Copyright (2017) by Springer Nature.

The far-field diffraction of helical beams by triangular apertures produces intensity lattice patterns that are correlated to the charge of the helical beam. This provides a direct way to measure the charge and sign of the vortex by diffraction. Vortices at laser wavelength of 514 nm produced using computer generated hologram are diffracted through an equilateral triangular aperture with side length 1.75 mm. A 30 cm focal length lens immediately after the aperture produces the far-field diffraction pattern at the focal plane.

The phase singular beam is made to fall at the center of the aperture with its edges well illuminated [73]. The interference of edge waves from the sides of the triangle contribute to the diffraction pattern. By taking each edge separately, simulation is carried out and it is shown that the diffraction pattern generated matches with the experimentally observed lattice pattern. The charge of the vortex is equal to the number of spots on one side of the triangular lattice pattern minus one and the orientation of the triangular lattice pattern decides the sign of the vortex.

A similar experiment in which an annular triangular aperture is used to diffract the helical wave produces patterns that can be used for vortex detection [74]. In the annular aperture the ratio of the side of the inner triangle to the outer triangle is varied from 1 to 0 and when this ratio approaches 0, the diffraction pattern is seen to match exactly with the pattern obtained using a triangular aperture. These diffraction patterns and the schematic of the annular triangular aperture are shown in figure 3.39.

Instead of equilateral triangular aperture, an isosceles right triangular aperture produces truncated straight line fringes instead of spots [75]. The truncated straight fringes in the diffraction pattern from the isosceles right triangular aperture are perpendicular to the hypotenuse of the triangular aperture and the number of dark fringes directly reveals the topological charge or the OAM state of the incident beam. The formation of straight line fringes indicates that the diffraction is similar to that of a two-aperture system. Because of the geometry of the triangle here, the circular cross-section of the beam does not illuminate the aperture symmetrically as compared to the equilateral triangular aperture (figure 3.40). The donut structure of the beam leaves two large bright areas of the donut passing through the aperture that may be responsible for the fringe formation instead of bright spots. Experimentally observed fringe patterns in the diffraction of helical beams through right angle isosceles triangular apertures are shown in figure 3.41.

The diffracted intensity profiles obtained by irradiating the aperture with oppositely charged vortex beams are seen oriented with a flip. From this flip in the pattern, the polarity of the charge can be ascertained.

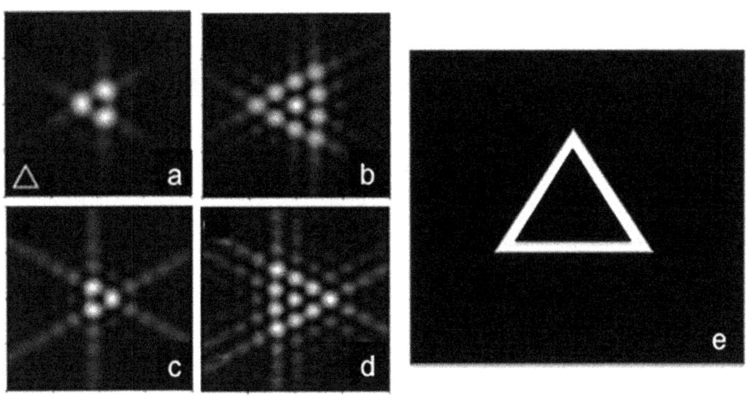


**Figure 3.39.** Fraunhofer diffraction patterns for vortex beams with different topological charges diffracted by an annular triangle aperture. (a) For charge one and (b) for charge three vortex diffracting through an annular aperture when the ratio of the inner to the outer side of the annular triangle aperture, is 0.4; but when this ratio is increased to 0.8, the diffraction patterns produced are shown in (c) for charge one and (d) for charge three vortex beam. (e) shows the schematic of the annular triangular aperture [74]. Note that the diffraction patterns shown in (a) and (c) are closer to triangular aperture diffraction patterns of a singular beam [73]. Reprinted from [74]. Copyright (2011), with permission from Elsevier.

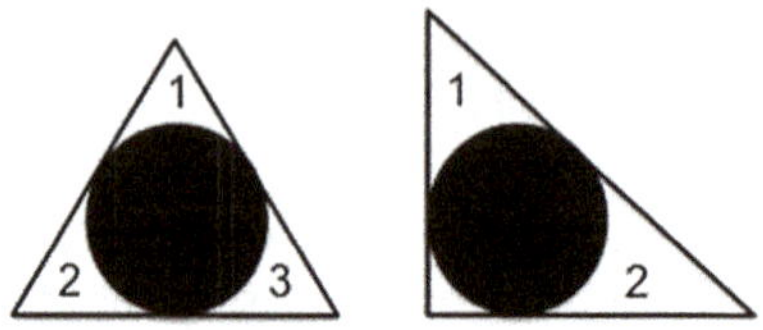


**Figure 3.40.** In an equilateral triangular aperture the dark core of the incident vortex beam divides the aperture into three (numbered in the figure) equal bright regions, whereas in a right angle isosceles triangular aperture the dark core of the vortex divides the beam into two equal bright regions. Reprinted with permission from [75]. Copyright (2015) by the American Physical Society.

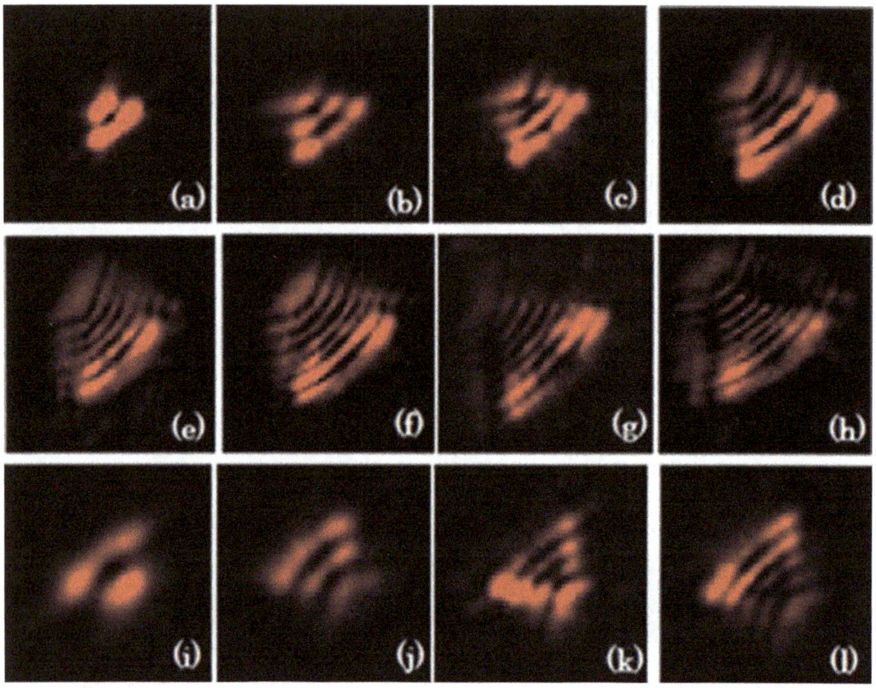


**Figure 3.41.** Experimentally obtained diffracted intensity profiles through a right triangular aperture for singular beams for (a–h) $m = 1$ to $m = 8$ and (i–l) $m = -1$ to $m = -4$. Reprinted with permission from [75]. Copyright (2015) by the American Physical Society.

*Diffraction through other types of apertures*

The diffraction patterns of other types of apertures such as circular [76], diamond shaped [77], annular and angular apertures [78, 79], hexagonal shaped aperture [80] and regular polygon [81] also have shown interesting features dependent on the charge of the beam. The singular beam diffraction is different from the Airy diffraction pattern of plane wave diffraction by a circular aperture [76]. When the aperture size is smaller compared to the incident beam spot size, the concentric rings in the diffraction pattern is given by

$$I \propto \frac{J_{|l|+1}(k\rho a/f)}{k\rho a/f}. \tag{3.36}$$

The location of zeros of the Bessel function $J_{l+1}$ determines the charge of the vortex. Annular circular apertures (ring apertures) [78] and off-axis diffraction through a circular aperture [82] were also reported for vortex detection.

The diffraction of the vortex beam by a diamond shaped aperture splits a higher-order vortex of charge $|l|$ into $l$ vortices with unit charge. It is related with the topological charge $l$ of the incident optical vortex beam. The phenomenon of splitting of higher charged vortex by diffraction is used for charge detection in these methods.

*Angular diffraction*

The topological charge of a vortex beam can be detected using diffraction at a sectorial screen [83]. The diffraction patterns are interesting and the number of zeros and bright spots in the pattern is correlated to the charge of the vortex. The magnitude of the charge can be found by counting the number of bright spots that appear on a broken ring like structure in the diffraction pattern as shown in figure 3.42. Positive and negative charged vortices produce diffraction patterns that are flipped by 180° and the setup has tolerance to misalignment.

### 3.3.3 Detection using lens aberrations

Lenses suffering from some of the Seidal aberrations can be used to detect the topological charge of a vortex. Many of the aberrations are known to split a higher charged vortex into multiple vortices—each of unit charge. This phenomenon can be

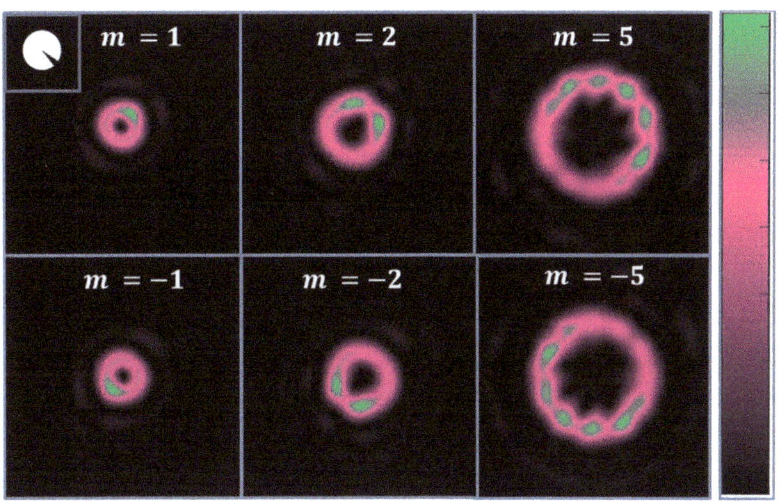


**Figure 3.42.** Simulated far-field diffraction patterns after vortex beams pass through the sectorial screen shown as an inset in the first frame. The topological charge of the vortex beam illuminating the sectorial screen in each case is mentioned in the figure.

used for vortex detection. Lenses suffering from circularly non-symmetric aberrations are found to be very effective for this purpose. The symmetry in the optical vortex field is broken by an aberrated lens.

**Astigmatism**: A lens suffering from the astigmatic aberration can be used for vortex detection. The aberration function for astigmatism is given by $W(\rho, \theta) = A_a\rho^2 \cos^2\theta$. Here $\rho$, $\theta$ are polar coordinates in the exit pupil plane of the lens and $A_a$ is the aberration coefficient. A donut beam focused using a lens suffering from aberration, splits the charge of the vortex into unit charge vortices and they are arranged in a slanted line in the focal plane [84–86]. Figure 3.43(a) shows the focal plane intensity distributions of one such case. Astigmatism is usually referred to as cylinder in aberration studies. Hence the performance of a cylindrical lens, will produce similar splitting [87–89] and it has been demonstrated. A vortex beam of charge $\pm l$ is found to split into $l$ separated dark spots.

*Coma*

The aberration function defining coma given by $W(\rho, \theta) = A_c\rho^3 \cos\theta$ is not circularly symmetric. $A_c$ is the aberration coefficient. Hence splitting of a vortex into multiple unit charged vortices is possible by focusing a vortex beam by a lens suffering from coma [90–93]. Figure 3.43(b) shows the focal plane intensity distribution of a vortex beam focused using a lens suffering from coma aberration.

Hence for a lens that is not properly mounted and used with a tilt, the aberrations in the optical system can be exploited in vortex detection. It has been reported [89, 94] that a tilted lens can be effectively used to detect the charge of a vortex. The propagation dynamics of optical vortices through a tilted spherical lens is observed. At a certain propagated distance past the lens, a vortex beam of charge $|m|$ breaks into $|m| + 1$ intensity spots as shown in figure 3.44. The orientation of the array of spots gives the sign of the topological charge. Vortices with higher charges upto 14 are detected clearly using this simple and effective method.

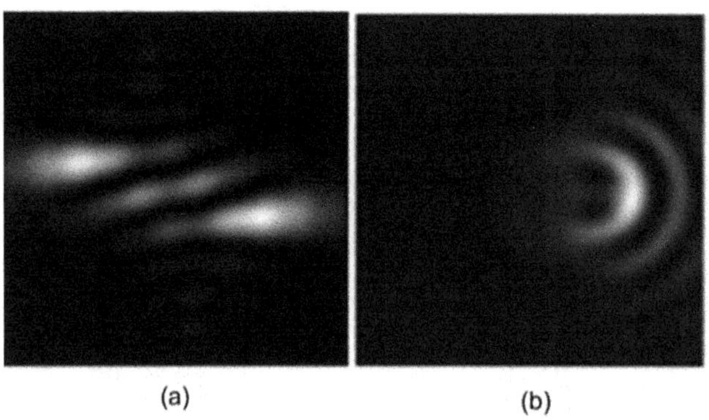


**Figure 3.43.** Intensity distributions at the focal plane of (a) astigmatism afflicted and (b) coma afflicted lenses, in singular beam focusing. Part (a) reprinted from [85]. Copyright (2007), with permission from Elsevier. Part (b) reprinted from [92]. Copyright (2008), with permission from Elsevier.

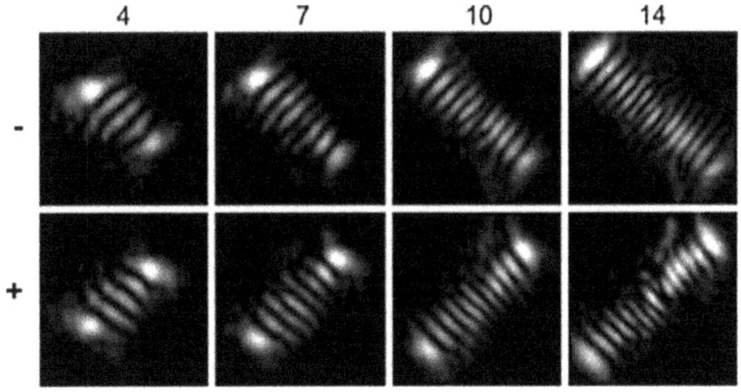


**Figure 3.44.** Experimentally observed intensity distributions at $z = 60$ cm for vortices of different topological charges and signs. The magnitude, and sign of the charges are indicated in the figure. Reprinted with permission from [94]. Copyright (2013) with permission from Elsevier.

### 3.3.4 Detection of vortices in computational optics

Iterative algorithms in computational techniques for wave optical engineering are used to synthesize complex field distributions. These computed fields are used in the fabrication of diffractive optical elements or displayed onto an SLM. Image or phase recovery from experimentally captured data also involves computation techniques. During the iterative procedures the phase distribution can develop optical vortices. To identify them in the complex fields there are many methods [59, 95, 96] reported in the literature.

*Line integral method*

The accumulated phase $\Delta\phi$ around each vortex in a closed loop enclosing a vortex is non-zero. Hence, to identify the presence of a vortex in a field, at every pixel location, the integral given by

$$\Delta\phi(x, y) = \oint \nabla\phi(x, y) \cdot dl \tag{3.37}$$

is evaluated, and if the accumulated phase is non-zero, then that point (pixel) under scrutiny is a vortex point. The closed loop is taken to be as small as possible, usually made up of eight neighboring pixels. This integral only predicts the presence of a vortex inside the closed loop considered. The vortex can be located in between pixels also and hence finding the exact location inside the loop is difficult. Further, another drawback is that the integral only gives the total charges enclosed by the loop and hence the presence of dipoles inside the loop goes undetected. Normally, dipoles with such small separation distances will tend to annihilate each other and do not cause any problem. Scanning the whole field may be time consuming and hence the line integral method can be carried out only at zero amplitude points in the field distribution.

*Zero crossing method*

The zero crossing method is based on the fact that the wave function of the vortex field is comprised of real and imaginary parts, both of which vanish simultaneously at the vortex point. Consider the complex field

$$U(x, y) = a(x, y)\exp\{i\phi(x, y)\} \tag{3.38}$$

The real zero, Re(0), and imaginary zero, Im(0), contours of $U(x, y)$ are given by

$$\begin{aligned} a(x, y)\cos\phi(x, y) = 0 \\ a(x, y)\sin\phi(x, y) = 0 \end{aligned} \tag{3.39}$$

At the vortex point the zero crossing of Re(0) and Im(0) means

$$a(x, y)\cos\phi(x, y) = a(x, y)\sin\phi(x, y) = 0 \tag{3.40}$$

By plotting Re(0) and Im(0) lines, the presence of vortices can be found by locating the intersection of these lines.

*Detection using phase contours*

By definition, contour is a surface of constant value of a parameter. Phase contour is a constant phase surface in a volume. For different values of phase, different contour surfaces can be drawn, and it is useful to draw many contour surfaces, each representing a different phase value. These values are normally at equal intervals. In two dimensions these are curves drawn on a phase distribution. When there are rapid phase variations, crowding of phase contour lines happens and an extrema can be identified by looking for closed contours. Since the phase distribution is a single valued function with a definite phase at each point, two different phase contours can never intersect, i.e. cross each other as at the intersection point, the field cannot represent two different phase values. In a non-vortex field, phase contour lines form closed loops enclosing extrema of phase and two contours touch each other at saddle points. Phase contour lines for non-vortex fields never intersect each other.

Intersection of phase contours occurs only when vortices are present. By plotting the phase contour lines in the field distribution, one can identify the location of the vortices. Vortices with higher topological charges normally do not occur and if they occur they can be identified by noticing that the number of phase contour lines terminating (or originating from) is more at a vortex. The phase contour map in a typical random field is shown in figure 3.45. Vortices are present at those locations where the contour lines intersect. This is because, at the immediate neighborhood of a vortex all phase values ranging from 0 to $m2\pi$ are present. In a random field the phase distribution, phase contours and zero crossings are drawn and shown in figure 3.45. In this figure, vortices can be located by the intersection points of phase contours or by zero crossings.

Vortex detection by the zero crossing method is a special case of the phase contour method. In a complex field Re(0) represents phase values of $\pi/2$, $3\pi/2$ ... and Im(0) represents phase values of 0, $\pi$,.... At a zero crossing on one side of the crossing, the Re(0) line is a phase contour line of phase value $\pi/2$ and on the other side of crossing Re(0) is a phase contour line of phase value $3\pi/2$. Similarly for a Im(0) line, the phase contours for phase values 0 and $\pi$ are on either side of the crossing.

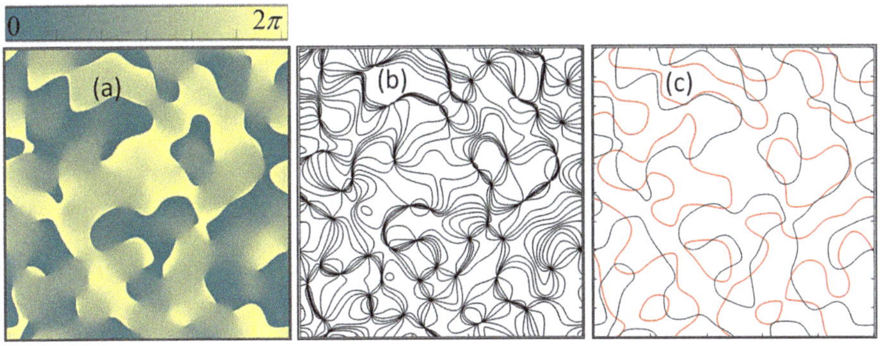


**Figure 3.45.** Presence of vortices in a random field. (a) Random phase distribution (wrapped). (b) Phase contour map. (c) Real and imaginary zero contours. In the phase contour map, vortex locations can be identified by the points at which the contours merge and in (c) vortices are located at points where the zeros of real (red) and imaginary (black) parts of the wave function cross each other.

*Helmholtz–Hodge decomposition method*

From the phase distribution construct a phase gradient field which is a vector field. Helmholtz–Hodge decomposition (HHD) allows a vector field that is defined in a region $\mathbb{R}^3$ on a bounded domain and is twice continuously differentiable to be separated into the divergence-free part (solenoidal) and the curl-free part (irrotational). In a vector field $F$ that is on a bounded domain $V$ in $R^3$ and is twice continuously differentiable and whose divergence $\nabla \cdot F = b(r)$ and curl $\nabla \times F = c(r)$ are known, can be segregated into components $f_1$ and $f_2$ determined by

$$F = f_1 + f_2 \Rightarrow \nabla\varphi + \nabla \times A \tag{3.41}$$

where $\varphi(r)$ and $A(r)$ are scalar and vector potentials, respectively, that can be obtained from Poisson's equations.

$$\varphi(r) = \frac{1}{4\pi}\int_V \frac{b(r)}{r}dv' \tag{3.42}$$

$$A(r) = \frac{1}{4\pi}\int_V \frac{c(r)}{r}dv' \tag{3.43}$$

These potentials allow the field $F$ to be segregated into the curl-free and divergence-free components. This method has been applied to scalar optical fields [97]. Since the phase distribution is a scalar field distribution, construction of vector field distribution namely the phase gradient field distribution is important in this method. Note that the vector field is a phase gradient field and should not be confused with the vector fields (waves) in optics where polarization aspect is also included to EM waves. A typical HHD field is shown in figure 3.46. The solenoidal part of the field represent circulating phase gradients, indicating the presence of optical vortices in the field distribution.

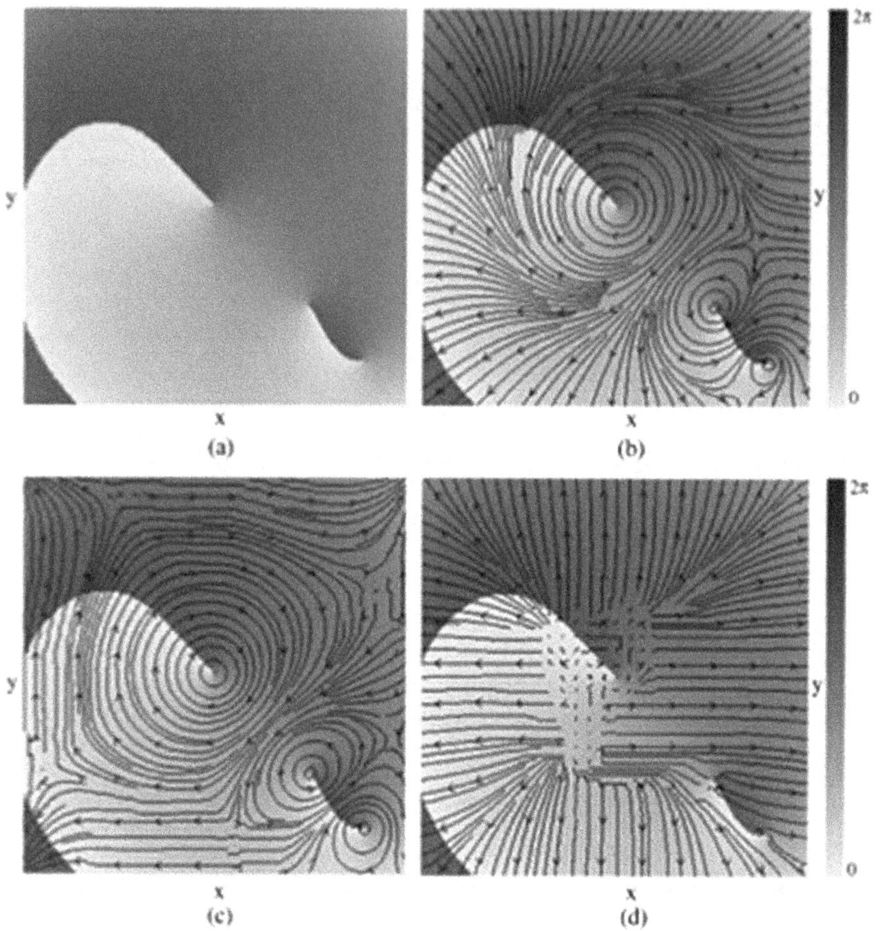


**Figure 3.46.** HHD of a random vortex field added to a positively diverging spherical beam. (a) Transverse phase profile of a random vortex field added to a spherical beam with a positive curvature. (b) Phase gradient field lines of the beam superimposed on the phase profile. (c) Flow lines of the solenoidal component of the Hodge decomposed field. The field lines circulate about the vortex centers. (d) Irrotational component with diverging field lines. This is the vortex free field. Reproduced with permission from [97]. Copyright (2012) by OSA.

# References

[1] Baranova N B, Zel'dovich B Y, Mamaev A V, Pilipetski N F and Shkunov V V 1981 Dislocations of the wavefront of a speckle-inhomogeneous field *JETP Lett.* **33** 195–9

[2] Freund I, Shvartsman N and Freilkher V 1993 Optical dislocation network in highly random media *Opt. Commun.* **101** 247–64

[3] Freund I 1995 Saddles, singularities, and extrema in random phase fields *Phys. Rev.* E **52** 2348–60

[4] Staliunas K, Berrzanskis A and Jarutis V 1995 Vortex statistics in optical speckle fields *Opt. Commun.* **120** 23–8

[5] Aagedal H, Schmid M, Teiwes S, Beth T and Wyrowski F 1996 Theory of speckles in diffractive optics and its application to beam shaping *J. Mod. Opt.* **43** 1409–21

[6] Beijersbergen M W, Coerwinkel R P C, Kristensen M and Woerdman J P 1994 Helical wavefront laser beams produced with a spiral phase plate *Opt. Commun.* **112** 321–2

[7] Oemrawsingh S S R, van Houwelingen J A W, Eliel E R, Woerdman J P, Verstegen E J K, Kloosterboer J G and Hooft G W 2004 Production and characterization of spiral phase plates for optical wavelengths *Appl. Opt.* **43** 688–94

[8] Sueda K, Miyaji G and Nakatsuka M 2004 Laguerre–Gaussian beam generated with a multilevel spiral phase plate for high intensity laser pulses *Opt. Express* **12** 3584–9

[9] Kim G H, Jeon J H, Ko K H, Moon H J, Lee J H and Chang J S 1997 Optical vortices produced witha a non-spiral phase plate *Appl. Opt.* **36** 8614–21

[10] Rotschild C and Zommer S 2004 Adjustable spiral phase plate *Appl. Opt.* **43** 721–7

[11] Basisty I V, Soskin M S and Vasnetsov M V 1993 Optics of light beams with screw dislocations *Opt. Commun.* **103** 422–8

[12] Heckenberg N R, McDuff R, Smith C P and White A G 1992 Generation of optical phase singularities by computer generated hologram *Opt. Lett.* **17** 221–3

[13] Senthilkumaran P, Masajada J and Sato S 2012 Interferometry with vortices Int *J. Opt.* **2012** 517591

[14] Sharma M K, Singh R K, Joseph J and Senthilkumaran P 2013 Fourier spectrum analysis of spiral zone plates *Opt. Commun.* **304** 43–8

[15] Roux F S 1993 Diffractive lens with a null in the centre of its focal point *Appl. Opt.* **32** 4191–2

[16] Vyas S and Senthilkumaran P 2010 Two dimensional vortex lattices from pure wavefront tilts *Opt. Commun.* **283** 2767–71

[17] Vyas S and Senthilkumaran P 2010 Vortices from wavefront tilts *Opt. Lasers Eng.* **48** 834–40

[18] Freund I and Shvartsman N 1994 Wave-field phase singularities: the sign principle *Phys. Rev.* A **50** 5164–72

[19] Ya I, Shvedov V and Volyer A 2005 Generation of higher-order optical vortices by a dielectric wedge *Opt. Lett.* **30** 2472–4

[20] Lin J, Yuan X-C, Bu J, Ahluwalia B P S, Sun Y Y and Burge R E 2007 Selective generation of high-order optical vortices from a single phase wedge *Opt. Lett.* **32** 2927–9

[21] Yuan X C, Ahluwalia B P S, Tao S H, Cheong W C, Zhang L S, Lin J, Bu J and Burge R E 2007 wave length scalable micro-fabricated wedge for generation of optical vortex beam in optical manipulation *Appl. Phys.* B **86** 209–13

[22] Lin J, Yaun X-C, Bu J, Chen H L, Sun Y Y and Burge R E 2007 Generalised model for orbital angular momentum states generated by parallel aligned phase wedges *Opt. Lett.* **32** 2170–2

[23] Masajada J and Dubik B 2001 Optical vortex generation by three plane wave interference *Opt. Commun.* **198** 21–7

[24] Vyas S and Senthilkumaran P 2007 Interferometric optical vortex array generator *Appl. Opt.* **46** 2893–8

[25] Xavier J, Vyas S, Senthilkumaran P, Denz C and Joseph J 2011 Sculptured 3D twister superlattices embedded with tunable vortex spirals *Opt. Lett.* **36** 3512–4

[26] Xavier J, Vyas S, Senthilkumaran P and Joseph J 2012 Tailored complex 3D vortex lattice structures by perturbed multiples of three-plane waves *Appl. Opt.* **51** 1872–8

[27] Boguslawski M, Rose P and Denz C 2011 Increasing the structural variety of discrete nondiffracting wave fields *Phys. Rev.* A **84** 013832

[28] Xavier J, Dasgupta R, Ahlawat S, Joseph J and Gupta P K 2012 Three dimensional optical twisters-driven helically stacked multi-layered micromotors *Appl. Phys. Lett.* **100** 121101

[29] Huntley J M and Buckland J R 1995 Characterization of $2\pi$ phase discontinuity in speckle interferograms *J. Opt. Soc. Am.* A **12** 1990–6

[30] Wang W, Ishii N, Hanson S G, Miyamoto Y and Takeda M 2005 Phase singularities in analytic signal of white-light speckle patern with application to micro-displacement measurement *Opt. Commun.* **248** 59–68

[31] Aksenov V P and Tikhomirova O V 2002 Theory of singular phase reconstruction for an optical speckle field in the turbulent atmosphere *J. Opt. Soc. Am.* A **19** 345–55

[32] Wang W, Yokozeki T, Ishijima R, Takeda M and Hanson S G 2006 Optical vortex metrology based on the core structure of phase singularities in Laguerre–Gauss transform of a speckle pattern *Opt. Express* **14** 10195–206

[33] Dammann H and Gortler K 1971 High-efficiency in-line multiple imaging by means of multiple phase holograms *Opt. Commun.* **3** 312–5

[34] Fu S, Wang T, Zang S and Gao C 2016 Integrating 5x5 Dammann gratings to detect orbital angular momentum states of beams with the range of -24 to .24 *Appl. Opt.* **55** 1514–7

[35] Yu J, Zhou C, Jia W, Hu A, Cao W, Wu J and Wang S 2012 Three-dimensional dammann vortex array with tunable topological charge *Appl. Opt.* **51** 2485–90

[36] Yu J, Zhou C, Jia W, Hu A, Cao W, Wu J and Wang S 2012 Generation of dipole vortex array using spiral Dammann zone plates *Appl. Opt.* **51** 6799–804

[37] Beijersbergen M W, Allen L, van der Veen H E L O and Woerdman J P 1993 Astigmatic laser mode converters and transfer of orbital angular momentum *Opt. Commun.* **96** 123–32

[38] Padgett M J and Allen L 2002 Orbital angular momentum exchange in cylindrical-lens mode converters *J. Opt. B: Quantum Semiclass. Opt* **4** S17–9

[39] Ito A, Kozawa Y and Sato S 2010 Generation of hollow scalar and vector beams using a spot-defect mirror *J. Opt. Soc. Am.* A **27** 2072–7

[40] Vyas S, Kozawa Y and Sato S 2014 Generation of a vector doughnut beam from an internal mirror He-Ne laser *Opt. Lett.* **39** 2080–2

[41] Kano K, Kozawa Y and Sato S 2012 Generation of a purely single transverse mode vortex beam from a He-Ne laser cavity with a spot-defect mirror *Int. J. Opt.* **2012** 359141

[42] Vaughan J M V and Willetts D V 1983 Temporal and interference fringe analysis of excimer $TEM_{01}$ laser *J. Opt. Soc. Am.* **73** 1018–21

[43] Ghai D P, Senthilkumaran P and Sirohi R S 2008 Adaptive helical mirror for generation of optical phase singularity *Appl. Opt.* **47** 1378–83

[44] Ghai D P 2011 Generation of optical vortices with an adaptive helical mirror *Appl. Opt.* **50** 1374–81

[45] McGloin D, Simpson N B and Padgett M J 1998 Transfer of orbital angular momentum from a stressed fiber-optic waveguide to a light beam *Appl. Opt.* **37** 469–72

[46] Dashti P Z, Alhassen F and Lee H P 2006 Observation of orbital angular momentum transfer between acoustic and optical vortices in optical fiber *Phys. Rev. Lett.* **96** 043604

[47] Kumar R, Mehta D S, Sachdeva A, Garg A and Senthilkumaran P 2008 Generation and detection of optical vortices using all fiber-optic system *Opt. Commun.* **281** 3414–20

[48] Won C, Yoo S H, Oh K, Paek U-C and Jhe W 1999 Near-field diffraction by a hollow-core optical fiber *Opt. Commun.* **161** 25–30

[49] Hu M-L, Wang C-Y, Song Y-J, Li Y-F, Chai L, Serebryannikov E E and Zheltikov A M 2006 A hollow beam from a holey fiber *Opt. Express* **14** 4128–34

[50] Witkowska A, Leon-Saval S G, Pham A and Birks T A 2008 All-fiber $LP_{11}$ mode convertors *Opt. Lett.* **33** 306–8

[51] Birks T A, Culverhouse D O, Farwell S G and Russell P St J 1995 All-fiber polarizer based on a null taper coupler *Opt. Lett.* **20** 1371–3

[52] Volpe G and Petrov D 2004 Generation of cylindrical vector beams with few-mode fibers excited by Laguerre–Gaussian beams *Opt. Commun.* **237** 89–95

[53] Marrucci L 2013 The q-plate and its future *J. Nanophoton* **7** 1–4

[54] Marucci L, Manzo C and Paparo D 2006 Optical spin-to-orbital angular momentum conversion in inhomogeneous anisotropic media *Phys. Rev. Lett.* **96** 1639905

[55] Slussarenko S, Murauski A, Du T, Chigrinov V, Marrucci L and Santamato E 2011 Tunable liquid crystal q-plates with arbirary topological charge *Opt. Express* **19** 4085–90

[56] Senthilkumaran P 2003 Optical phase singularities in detection of laser beam collimation *Appl. Opt.* **42** 6314–20

[57] Malacara D 1978 *Optical Shop Testing* (New York: Wiley)

[58] Ghai D P, Senthilkumaran P and Sirohi R S 2008 Shearograms of optical phase singularity *Opt. Commun.* **281** 1315–22

[59] Ghai D P, Vyas S, Senthilkumaran P and Sirohi R S 2008 Detection of phase singularity using a lateral shear interferometer *Opt. Laser Eng.* **46** 419–23

[60] Ghai D P, Vyas S, Senthilkumaran P and Sirohi R S 2008 Shearograms of a singular beam using wedge plate lateral shear interferometer Opt *Lasers Eng.* **46** 791–801

[61] Hecht E and Zajac A 1974 *Optics* (Reading, MA: Addison-Wesley)

[62] Ghai D P, Senthilkumaran P and Sirohi R S 2009 Single-slit diffraction of an optical beam with phase singularity *Opt. Laser Eng.* **47** 123–6

[63] Sztul H I and Alfano R R 2006 Double-slit interference with Laguerre–Gaussian beams *Opt. Lett.* **31** 999–1001

[64] Berry M V 2005 Phase vortex spirals *J. Phys. Math. Gen.* **38** L745–51

[65] Berry M V 2008 Waves near zeros *Conf. on Coherence and Quantum Optics* (Washington, USA: OSA), pp 37–41

[66] Senthilkumaran P and Bahl M 2015 Young's experiment with waves near zeros *Opt. Express* **23** 10968–73

[67] Moreno I, Davis J A, Melvin B, Pascoguin L, Mitry M J and Cottrell D M 2009 Vortex sensing diffraction gratings *Opt. Lett.* **34** 2927–9

[68] Dai K, Gao C, Zhong L, Na Q and Wang Q 2015 Measuring OAM states of light beams with gradually-changing-period gratings *Opt. Lett.* **40** 562–5

[69] Goodman J W 2007 *Introduction to Fourier Optics* (Englewood, CO: Roberts)

[70] Hariharan P 1996 *Optical Holography: Principles, Techniques, and Applications* (Cambridge: Cambridge University Press)

[71] Zhang N, Yuan X C and Burge R E 2010 Extending the detection range of optical vortices by Dammann vortex gratings *Opt. Lett.* **35** 3495–7

[72] Zheng S and Wang J 2017 Measuring orbital angular momentum (OAM) states of vortex beams with annular gratings *Sci. Rep.* **7** 40781

[73] Hickmann J, Fonseca E and Chavez Cerda S 2010 Unveiling a truncated optical lattice associated with a triangular aperture using light's orbital angular momentum *Phys. Rev. Lett.* **105** 053904

[74] Yongxin L, Hua T, Jixiong P and Baida L 2011 Detecting the topological charge of vortex beams using an annular triangle aperture *Opt. Laser Technol.* **43** 1233–6

[75] Bahl M and Senthilkumaran P 2015 Energy circulations in singular beams diffracted through an isosceles right triangular aperture *Phys. Rev.* A **92** 1–6

[76] Singh S, Ambuj A and Vyas R 2014 Diffraction of orbital angular momentum carrying optical beams by a circular aperture *Opt. Lett.* **39** 5475–8
[77] Liu Y, Sun S, Pu J and Lu B 2013 Propagation of an optical vortex beam through a diamond-shaped aperture *Opt. Laser Technol.* **45** 473–9
[78] Condell W J 1985 Fraunhofer diffraction from a circular annular aperture with helical phase factor *J. Opt. Soc. Am.* **2** 206–8
[79] Mei Z, Zhao D and Gu J 2004 Propagation of elegant Laguerre–Gaussian beams through an annular apertured paraxial *ABCD* optical system *Opt. Commun.* **240** 337–43
[80] Liu Y and Pu J 2011 Measuring the orbital angular momentum of elliptical vortex beams by using a slit hexagon aperture *Opt. Commun.* **284** 2424–9
[81] Ambuj A, Vyas R and Singh S 2014 Diffraction of Laguerre–Gauss vortex beams by regular polygons *Frontiers in Optics 2014* (Washington DC: Optical Society of America) p JTu3A.10
[82] Taira Y and Zhang S 2017 Split in phase singularities of an optical vortex by off-axis diffraction through a simple circular aperture *Opt. Lett.* **42** 1373–6
[83] Chen R, Zhang X, Zhou Y, Ming H, Wang A and Zhan Q 2017 Detecting the topological charge of optical vortex beams using a sectorial screen *Appl. Opt.* **56** 4868–72
[84] Singh R K, Senthilkumaran P and Singh K 2007 Effect of astigmatism on the diffraction of a vortex carrying beam with Gaussian background *J. Opt. A: Pure Appl. Opt* **9** 543–54
[85] Singh R K, Senthilkumaran P and Singh K 2007 Influence of astigmatism and defocusing on the focusing of a singular beam *Opt. Commun.* **270** 128–38
[86] Wada A, Ohtani T, Miyamoto Y and Takeda M 2005 Propagation analysis of the Laguerre Gaussian beam with astigmatism *J. Opt. Soc. Am.* A **22** 2746–55
[87] Ya B A and Karamoch A I 2008 Astigmatic telescopic transformation of a high-order optical vortex *Opt. Commun.* **281** 5687–96
[88] Kotlyar V V, Kovalev A A and Porfirev A P 2017 Astigmatic transforms of an optical vortex for measurement of its topological charge *Appl. Opt.* **56** 4095–104
[89] Reddy S G, Prabhakar S, Aadhi A, Banerji J and Singh R P 2014 Propagation of an arbitrary vortex pair through an astigmatic optical system and determination of its topological charge *J. Opt. Soc. Am.* A **31** 1295–302
[90] Wada A, Ohminato H, Yonemura T, Miyamoto Y and Takeda M 2005 Effect of comatic aberration on the propagation characteristics of Laguerre–Gaussian beam *Opt. Rev.* **12** 451–5
[91] Singh R K, Senthilkumaran P and Singh K 2007 Effect of coma on the focusing of an apertured singular beam *Opt. Laser Eng.* **45** 488–94
[92] Singh R K, Senthilkumaran P and Singh K 2008 Focusing of a vortex carrying beam with Gaussian background by an apertured system in presence of coma *Opt. Commun.* **281** 923–34
[93] Bahl M, Singh B K, Singh R K and Senthilkumaran P 2015 Internal energy flows of coma-affected singular beams in low-numerical-aperture systems *J. Opt. Soc. Am.* A **32** 514–21
[94] Vaity P, Banerji J and Singh R P 2013 Measuring the topological charge of an optical vortex by using a tilted convex lens *Phys. Lett.* A **377** 1154–6
[95] Wyrowski F and Bryngdahl O 1988 Iterative Fourier-transform algorithm applied to computer holography *J. Opt. Soc. Am.* A **5** 1058–65
[96] Senthilkumaran P and Wyrowski F 2002 Phase synthesis in wave optical engineering: mapping and diffuser type approaches *J. Mod. Opt.* **49** 1831–50
[97] Bahl M and Senthilkumaran P 2012 Helmholtz-Hodge decomposition of scalar optical fields *J. Opt. Soc. Am.* A **29** 2421–7

**IOP** Publishing

Singularities in Physics and Engineering
Properties, methods, and applications
**Paramasivam Senthilkumaran**

# Chapter 4

# Propagation characteristics

## 4.1 Introduction

An optical vortex is a phase screw dislocation and is a stationary curve in three dimensions. In two dimensions it is seen as a stationary point, where the phase of the light wave is undetermined and its amplitude is zero. Hence the intensity zero points in successive planes perpendicular to the general propagation direction can be tracked [1] to study the geometry of the vortex lines in three dimensions. The intensity and phase gradients appear to exert attractive or repulsive forces on the vortices. But before venturing into such studies, some of the general ideas about the wave equations and the types of wave functions of singular beams are introduced. The role of Gouy phase, divergence, near and far field structures of singular beams are presented.

## 4.2 Wave equations and solutions

### Wave equation

The wave equation for electromagnetic waves in a medium of refractive index $n$ with velocity $\frac{c}{n}$ is given by

$$\nabla^2\vec{u}(x, y, z, t) - \frac{n^2}{c^2}\frac{\partial^2}{\partial t^2}\vec{u}(x, y, z, t) = 0 \tag{4.1}$$

The function $\vec{u}(x, y, z, t)$ is complex and is a vector function. Since $\vec{u}$ represents the electric field of the electromagnetic waves, it is a vector quantity. By writing the wave equation for the components of the electric fields in the transverse plane we can see that the wave equation can be split into two—one for each of the transverse components, and if the medium is isotropic they look alike except for the components of the electric fields. Assuming there is no change in the state of

doi:10.1088/978-0-7503-1698-9ch4

polarization during propagation, the electric field can be considered as a scalar variable and we will have the scalar form of the wave equation.

$$\nabla^2 \tilde{u}(x, y, z, t) - \frac{n^2}{c^2}\frac{\partial^2}{\partial t^2}\tilde{u}(x, y, z, t) = 0 \tag{4.2}$$

Here $\tilde{u}(x, y, z, t)$ is complex. Further for monochromatic waves, we can write the wave field as

$$\tilde{u}(x, y, z, t) = \tilde{U}(x, y, z)\exp\{i\omega t\} \tag{4.3}$$

where the space dependent and time dependent parts are given separately. The spatial dependence of the electric fields of the EM waves can be of any functional form. The function $\tilde{U}(x, y, z)$ is complex and can be written as

$$\tilde{U}(x, y, z) = U(x, y, z)\exp\{-i\phi(x, y, z)\} \tag{4.4}$$

$\tilde{U}$ is called the phasor amplitude. This is the solution of the time independent wave equation, given by

$$(\nabla^2 + k^2)\tilde{U} = 0 \tag{4.5}$$

This is referred as the Helmholtz equation. The time independent wave equation (4.5) can be obtained from the wave equation (4.2) by substituting equation (4.3) in it.

**Paraxial waves**

If the propagation vectors of the wave subtend to a small angle with the nominal propagation direction, then the waves are called paraxial as the rays are at smaller angles to the optical axis of the system. This means that in equation (4.4), the phase is written as $\phi(x, y, z) = kz$, which means that the wavefront is plane. By absorbing dependence of phase on $x$, $y$ into $U(x, y, z)$ and making it complex valued, equation (4.4) can be written as

$$\tilde{U}(x, y, z) = \tilde{A}(r)\exp\{-ikz\} \tag{4.6}$$

where $\tilde{A}(r)$ is complex valued which has spatially varying amplitude and phase in such a way that the variations are smooth and slow. The spatially varying amplitude and phase are such that, they do not introduce generation of rays traveling at large angles with respect to the optical axis, thereby violating the definition of paraxial rays. Rapid spatial variation in phase indicates large phase gradient which is related to the propagation vector at larger angle. Introduction of this waveform requires that the Helmholtz equation also needs to be modified accordingly. We will come back to that later. Nevertheless starting from the Helmholtz equation, we can see some of the forms of $\tilde{U}(x, y, z)$ that are used in singular optics. Numerous solutions of equation (4.5) are possible [2] and one of them is $\tilde{U}(x, y, z) = \vec{F}(x, y)\exp(-ikz)$ where $\vec{F}(x, y)$ satisfies Laplace's equation

$$\frac{\partial^2 F}{\partial x^2} + \frac{\partial^2 F}{\partial y^2} = 0 \tag{4.7}$$

Here, by dropping the polarization, $F$ is treated as a scalar field. The solutions of Laplace's equation are smooth and do not have extremum points, which fits well with the slowly varying envelope function. One such solution is $F(x, y) = (x + iy)$. Therefore, the function $F(x, y)\exp\{-ikz\}\exp\{i\omega t\}$ satisfies the wave equation (4.2). This solution describes a single optical phase dislocation at the origin and traveling along $z$. Writing in cylindrical coordinates $x + iy = r\exp\{i\theta\}$ so that the phase $\phi$ equals the polar angle $\theta$. Applying Cauchy–Riemann conditions [3] one can see that the function $x + iy$ is analytic. The non-analytic function $x - iy$ also satisfies the Laplace's equation, and this represents the singularity with its phase varying opposite to that of $x + iy$ and hence can be used to represent a negatively charged vortex. Multiple degenerate solutions of Laplace's equation, such as $(x \pm iy)^m$ representing vortex of charge $\pm m$ are also used. The phase contours for a vortex is a star like pattern. The phases of the optical field at different $z$ planes are given by $\phi = m\theta - kz + \omega t$. Hence as the vortex propagates, the star pattern rotates and advances in space and time. This means that the helical wavefront propagates like a spiral shaped rotating helical blade (that makes a hole) during propagation.

The complex amplitude for a vortex [1] situated away from origin can be given by

$$\psi(r, \phi) = c(r)\exp\{im\phi_a\} \tag{4.8}$$

where the phase of the vortex at position $r_{00} = (a, 0)$ is

$$\phi_a = \frac{1}{m}\,\mathrm{Arctan}\left\{\frac{y}{x - a}\right\} \tag{4.9}$$

Depending on the core function $c(r)$ different types of vortices are there as introduced in chapter 2. For a point vortex $c(r) = 1$ and such a vortex can be produced by sending a plane wave through a spiral phase plate and field just behind the element which has uniform amplitude, however, this uniform amplitude cannot be sustained and a dark core is generated due to diffraction. For an $r$-vortex $c(r) = |r - r_{00}|$, which has high amplitude away from the core and this can be limited by envelope functions as described earlier. Another type of vortex is the tanh vortex [1] in which $c(r) = \tanh\{\frac{|r - r_{00}|}{r_c}\}$, where $r_c$ is the vortex core radius. The point vortex and the $r$-vortex are the limiting cases of the tanh vortex [4]. For a Laguerre–Gaussian (LG) background beam with beam radius $w$, for small values of $r_c$ such that $r_c \ll w$, the tanh vortex behaves like a point vortex and behaves like an $r$-vortex for large values of $r_c$.

## 4.3 Slowly varying envelope approximation—paraxial Helmholtz equation

Assuming a slowly varying envelope for the phasor amplitude $\tilde{U} = \tilde{A}(r)\exp\{-ikz\}$, the Helmholtz equation is

$$(\nabla^2 + k^2)\tilde{U} = (\nabla_T^2 + \partial_z^2 + k^2)\tilde{U} = 0 \tag{4.10}$$

where the Laplacian operator $\nabla^2$ has been divided into transverse and longitudinal parts. Let us see the longitudinal derivative

$$\begin{aligned}\partial_z^2\tilde{U} &= \partial_z^2[\tilde{A}(r)\exp\{-ikz\}] = \partial_z[\tilde{A}(r)(-ik)\exp\{-ikz\} + \exp\{-ikz\}\partial_z\tilde{A}(r)] \\ &= [\partial_z^2\tilde{A}(r) - 2ik\partial_z\tilde{A}(r) - k^2\tilde{A}(r)]\exp[-ikz]\end{aligned} \tag{4.11}$$

Using this in equation (4.10) and noting $\partial_z^2\tilde{U} = 0$ as $\partial_z\tilde{A}(r)$ is small, we have

$$\nabla_T^2\tilde{U} - 2ik\frac{\partial}{\partial z}\tilde{U} = 0 \tag{4.12}$$

This is called the paraxial Helmholtz equation. It is also called the slowly varying envelope approximation of the Helmholtz equation.

#### Wave equation in cylindrical coordinates

The paraxial wave equation in cylindrical coordinates is given by

$$\frac{1}{r}\frac{\partial}{\partial r}\left(r\frac{\partial U}{\partial r}\right) + \frac{1}{r^2}\frac{\partial^2 U}{\partial \theta^2} - 2ik\frac{\partial U}{\partial z} = 0 \tag{4.13}$$

The LG beams given below is the solution of the above paraxial Helmholtz equation in cylindrical coordinates.

$$\begin{aligned}U(r, \theta, z) = {}& E_0\left(\sqrt{2}\frac{r}{w(z)}\right)^{|l|} L_p^l\left(2\frac{r^2}{w^2(z)}\right)\exp\left[-\frac{r^2}{w^2(z)}\right] \\ &\times \exp\left[-i\left(k\frac{r^2}{2R(z)} + \Phi(z) + l\theta + kz\right)\right]\end{aligned} \tag{4.14}$$

Here $L_p^l$ is the Laguerre polynomial of radial index $p$ and azimuthal index $l$. The last term describing the phase distribution has a quadratic phase (spherical phase under approximation), Gouy phase, helical phase term and a plane wave term. The helical phase term is important as this phase distribution has the singularity. The amplitude of the beam has $r$ dependence and hence the beam has circularly symmetric amplitude distribution. $w(z)$ describes the beam radius at a given $z$ plane.

## 4.4 Gouy phase

A converging spherical wave passing through its focus undergoes a phase change of $\pi$. Gouy [5], who first observed this phase anomaly, used Huygen's principle to explain this phase change. The continuous change in the curvature of the beam due to diffraction [6] results in the acquisition of Gouy phase. There are many explanations available in the literature to understand the Gouy phase [7, 8]—and two of them are presented here.

The Gouy phase shift explains the phase advance for the secondary Huygens wavelets emanating from a primary wavefront. It also determines the resonant frequencies of transverse modes in laser cavities.

By defining an effective axial propagation constant for a finite beam through second moment [8] as

$$\bar{k}_z = k - \frac{\left\langle k_x^2 \right\rangle}{k} - \frac{\left\langle k_y^2 \right\rangle}{k} \tag{4.15}$$

The first term yields the phase $kz$ of an infinite plane wave propagating along $z$. The last two terms give rise to the Gouy phase shift

$$\Phi(z) = -\frac{1}{k}\int^{z} \left\{\left\langle k_x^2 \right\rangle + \left\langle k_y^2 \right\rangle\right\} dz \tag{4.16}$$

Hence the Gouy shift is the expectation value of the axial phase shift owing to the transverse momentum spread.

**Gouy phase in Gaussian beam**

Consider a Gaussian beam whose complex amplitude is given by

$$\psi(r, z) = \frac{w_0}{w(z)} \exp\left[i[kz - \Phi(z)] - r^2\left(\frac{1}{w^2(z)} - \frac{ik}{2R(z)}\right)\right] \tag{4.17}$$

where $w(z)$ and $R(z)$ are the radius of the beam waist and radius of curvature of the beam at $z$. The Gaussian beam has plane wavefronts at $z = 0$, the minimum waist plane and at $z = \pm\infty$. In between, during propagation, the beam is spherical and the radius of curvature becomes minimum when $z = Z_R = \frac{\pi w_0^2}{\lambda}$, where $Z_R$ is the Rayleigh distance. The minimum waist radius is $w_0$. Intuitively one can see that beam has high divergence when the light is from a point source, whereas having a finite size like $w_0$ reduces the divergence of the Gaussian beam. It is interesting to note that the beam starts as a plane wave then, as it propagates, it become spherical and then becomes plane wave at very large distances. From Huygen's theory, a point source produces spherical waves, and a plane wave produces plane waves. But because of the finite extent of the plane wave, diffraction plays a crucial role and the Gaussian beam on its way becomes spherical and then plane. In equation (4.17), $\Phi(z)$ is the Gouy phase. The phase shift due to on-axis plane wave component is $kz$ and of the two terms with $r^2$, the real part is the Gaussian amplitude term and the imaginary part represents a spherical wave with radius of curvature $R(z)$. The $z$ dependence of $R$ shows that initially $R$ decreases with $z$ and later increases.

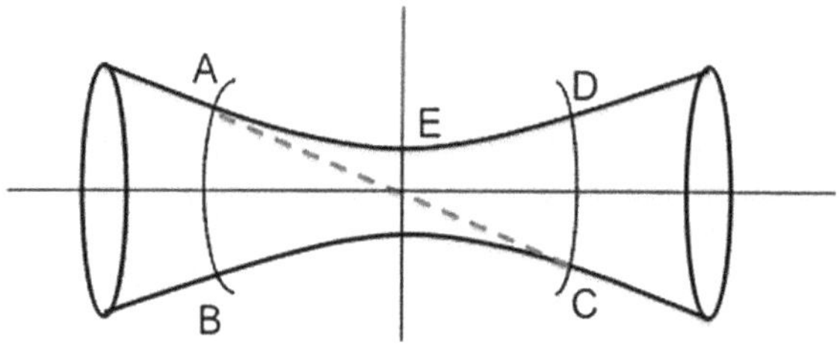


**Figure 4.1.** Shown here is the amplitude contour surface at $(1/e)$ of the maximum amplitude in a Gaussian beam. This expanding tubular surface has minimum cross-sectional radius $w_0$. AB and CD are wavefronts. The optical path traversed by a point A on the wavefront from A to D is AED, while the geometric path (ray path) by extending local wavefront normal at A to C is AC (dotted line). The path difference (AED–AC) is the phase anomaly—Gouy phase.

Referring to figure 4.1, the optical path traversed by a point A on the wavefront from A to D is AED, as on a wavefront all points have same phase and it is decided by the optical path taken by the point A located on the wavefront. Note that the wavefront undergoes a curvature change while going through the minimum waist plane. The point A on successive wavefronts, can be seen to draw the path AED. But the geometric path (ray path) by extending local wavefront normal at A to C is AC (dotted line). Point C lies on the wavefront CD. Since both C and D are on the same spherical wavefront they should have same phase, but they turned out to be different, while finding the optical path from A. The path difference (AED–AC) is the phase anomaly—Guoy phase. It arises due to the optical path difference between the ray and wave pictures. This interesting intuitive explanation is due to Boyd [7].

### Gouy phase in a Hermite–Gaussian beam

The complex amplitude due to a Hermite–Gaussian beam is given by [9]

$$\psi(x, y, z) = A_{l,m}\left[\frac{w_0}{w(z)}\right]G_l\left[\frac{\sqrt{2}x}{w(z)}\right]G_m\left[\frac{\sqrt{2}y}{w(z)}\right] \times \exp\left[-ikz - ik\frac{r^2}{2R(z)} + i(l+m+1)\Phi(z)\right] \tag{4.18}$$

where

$$G_l(u) = H_l(u)\exp\left(\frac{-u^2}{2}\right) \tag{4.19}$$

is the Hermite–Gaussian function of order $l$ and $A_{l,m}$ is a constant and $\Phi(z) = \arctan(z/z_R)$. The Hermite function of zero order is given as $H_0(u) = 1$, the first order is $H_1(u) = 2u$, the second order is $H_2(u) = 4u^2 - 2$, the third order is $H_3(u) = 8u^3 - 12u$ and so on. As you can see in these functions the number of zeros is equal to the order number $l$. For example, for $l = 2$ the function is quadratic and has two solutions and for $l = 3$ the function is cubic and has three zeros and so on.

These functions at large values of $u$ assume large values and the presence of a Gaussian envelope limits these functions. In equation (4.18), the third term in the phase distribution $(l + m + 1)\Phi(z)$ is the Gouy phase term which depends on the mode numbers. The wave function $\psi$ has a phase term that linearly varies with $z$, which is the on-axis plane wave component, and a spherical wave component given by the second term. The third term is the Gouy phase term.

**Gouy phase in a LH beam**

The complex amplitude for a LG beam is given by

$$\psi(\rho, \phi, z) = A_{lm}\left[\frac{w_0}{w(z)}\right]\left(\frac{\rho}{w(z)}\right)^l L_m^l\left(\frac{2\rho^2}{w^2(z)}\right)\exp\left(\frac{-\rho^2}{w^2(z)}\right) \cdot \exp\left[-ikz - ik\frac{\rho^2}{2R(z)} - il\phi + i(l + 2m + 1)\Phi(z)\right] \tag{4.20}$$

The first term in the phase part is the linear phase variation due to propagation of the plane wave component of on-axis plane wave. The second term is the spherical wave (quadratic approximation) term and the last term is the Gouy phase term given by $(l + 2m + 1)\Phi(z)$ where $\Phi(z) = \mathrm{Arctan}\left(\frac{z}{z_R}\right)$. Like the Hermite polynomials in HG beams the Laguerre polynomials decide the amplitude variation in the LG beam. Some of the Laguerre polynomials are $L_0(r) = 1$, $L_1(r) = 1 - r$, $L_2(r) = \frac{1}{2}(r^2 - 4r + 2)$ and all the polynomials have value 1 at $r = 0$. The Gaussian term provides the envelope and ensures that the beam has finite energy and is spatially restricted.

**Gouy phase in a Bessel beam**

A Bessel beam as introduced by Durnin [10, 11] and the beams studied later [12] do not suggest the presence of any Gouy phase in Bessel beams. Later, there is a report [13] on the presence of a Gouy phase in Bessel beams. When the Bessel beam was introduced [10, 11], the field corresponding to a Bessel beam was given by

$$\psi(\rho, t) = \exp[i(\beta z - \omega t)]J_0(\alpha\rho) \tag{4.21}$$

where $\alpha^2 + \beta^2 = \left(\frac{\omega}{c}\right)^2$ and $\rho^2 = x^2 + y^2$. Here $\alpha$ is the transverse component and $\beta$ is the axial component of the propagation vector.

In the initial reports, the field expression used also did not have terms involving vortices, but in later reports, use of a vortex along with Bessel amplitude can be seen. A Bessel beam in cylindrical coordinates can be given by [12]

$$E(r, \phi, z) = A_0 \exp(ik_z z)J_n(k_r r)\exp(\pm in\phi) \tag{4.22}$$

where $J_n(k_r\, r)$ is the $n$th order Bessel function and the terms inside the bracket are the arguments for $J_n$.

In both equations (4.21) and (4.22), there is no reference to a Gouy phase for a Bessel beam. However, later the presence of a Gouy phase shift has been reported [13].

When compared with a Gaussian beam of the same central core size, the central core of a Bessel beam is propagation invariant for the same propagation distance. Except for the zero order Bessel function all the other orders of Bessel functions have a zero at $r = 0$ or $k_r\, r = 0$ and hence a Bessel beam can have a dark or bright central spot. The Fraunhofer diffraction of a circular aperture is given by Bessel functions and hence it is easy to intuitively feel that a Bessel beam can be formed by the diffraction of beams by apertures having circular symmetry. The general one that is discussed often is the diffraction of a ring aperture, which is nothing but circ($R_2$)–circ($R_1$) where $R_2 - \mathrm{R}_1 > 0$, the function circ($R$) is a circular aperture of radius $R$ that is commonly used in Fourier optics. Also note, that by using a Fourier transforming element like a lens, one can see that a spherical wave focuses to a point, a cylindrical wave focuses to a line and a Bessel beam focuses to a ring. Since a Bessel beam is the superposition of plane waves whose propagation vector forms a cone, they have finite propagation distance [10, 11, 12] beyond which the overlapping plane waves walk away.

Bessel beams can also be generated by illuminating an axicon by a plane wave. If we use a Gaussian beam (which can propagate to infinity) for illumination of the axicon, one can get Bessel–Gaussian beam and the Gouy phase terms associated with Gaussian beam can come into play. By bringing in a Gaussian, a super-Gaussian and a circular apodization into the Bessel beams, modified Bessel beams were introduced [14]. Here we describe the presence of a Gouy phase in a Bessel beam as given by [13]

$$\psi(\rho, t) = AJ_l(\alpha\rho)\exp\{i(l\theta - \beta z + \omega t)\} \tag{4.23}$$

where $k^2 = \alpha^2 + \beta^2$ and $\alpha$ are positive. It is the transverse component of the propagation vector that measures the spatial confinement of the beams in the transverse plane, while $\beta$ is axial component of $k$. The Gouy phase shift $\Phi_{\mathrm{Gouy}}$ for the Bessel beams, is given by

$$\Phi_{\mathrm{Gouy}} = (k - \beta)z = \frac{\alpha^2}{k^2 + \sqrt{k^2 - \alpha^2}} z \tag{4.24}$$

The Gouy phase shift accumulates proportionally to the propagation distance $z$. There is no explicit dependence on the order of the Bessel beam. The phase shift being linear repeats after accumulating $2\pi$ shift and the distance at which this happens is given by a period $\Lambda$. After a distance $\frac{\Lambda}{2}$ the Gouy phase shift of the Bessel beam can be seen by noting that the bright and dark rings of the interference pattern mutually exchange, and, after a period $\Lambda$ the resulting interference pattern repeats and is practically identical to the initial pattern.

## 4.5 Divergence of singular beams

The phase singular beam is known to carry orbital angular momentum of light which will be the subject matter in chapter 7. The multiplicity of orbital angular momentum states offered by singular beams makes them attractive for use in communication and hence there has been interest recently. Vortices of higher charges have larger core diameters, and increased divergence. Since the transmitter and receiver are of a particular size, there is a need to study the divergence of singular beams.

The contribution to the divergence comes from the skew angle of the Poynting vector [15, 16] of the vortex beam $\frac{|l|}{kr_0}$, and normal diffractive spreading arising from the finite beam diameter [17]. Since the beam has a donut intensity distribution, rather than the beam waist $w_0$, or the donut ring radius, i.e. $r\ (I_{\max})$, the standard deviation of its spatial intensity distribution, $r_{\rm rms}$ [8, 9] can be used as parameters for finding the divergence [17, 18].

The lowest-order radial mode ($p = 0$), of LG beam intensity distribution (normalized) is given by

$$I_l(r, \phi, z) = \frac{2}{w(z)^2 \pi |l|!} \left( \frac{\sqrt{2}\, r}{w(z)} \right)^{2|l|} \exp\left( \frac{-2r^2}{w(z)^2} \right) \tag{4.25}$$

For this beam the maximum intensity occurs at a radial position where $\frac{dI_l}{dr} = 0$ and is given by

$$r(I_{\max}) = \sqrt{\frac{|l|}{2}}\, w(z) \tag{4.26}$$

and the $r_{\rm rms}$ occurs at

$$r_{\rm rms}(z) = \sqrt{2\pi \int_0^{\infty} r^2 I_l r dr} = \sqrt{\frac{|l| + 1}{2}}\, w(z) \tag{4.27}$$

At $z = 0$ we have

$$r_{\rm rms}(0) = \sqrt{\frac{|l| + 1}{2}}\, w_0 \tag{4.28}$$

The divergence angle of $p = 0$ LG mode is given by

$$\alpha_l(z) = \arctan\left( \frac{\partial r_{\rm rms}(z)}{\partial z} \right) = \arctan\left( \sqrt{\frac{|l| + 1}{2}} \frac{w_0^2}{z_R^2} \frac{z}{w(z)} \right) \tag{4.29}$$

where $z_R$ is the Rayleigh range. In the paraxial regime we can write

$$\lim_{z \to \infty} \frac{z}{w(z)} = \frac{z_R}{w_0} \tag{4.30}$$

By writing tan of the angle is equal to the angle, (for small angle) the divergence angle is given by [17]

$$\alpha_l = \sqrt{\frac{|l| + 1}{2}} \frac{w_0}{z_R} = \frac{r_{\text{rms}}(0)}{z_R} \tag{4.31}$$

The Rayleigh distance $z_R = \frac{1}{2}k_0 w_0^2$, and by using equation (4.28), we have $\frac{1}{z_R} = \frac{|l|+1}{k_0 r_{\text{rms}}^2(0)}$ Hence it has been shown by [17] that divergence scales linearly with, or with the square root of, the beam's OAM. Both these scaling laws are valid, depending upon whether it is the radius of the waist of the beam's Gaussian term or the radius of rms intensity of the beam that is kept constant while varying the OAM as shown in figure 4.2.

The divergence of singular beams was studied theoretically by using radial variance of intensity [19]. Vortices generated by mode converters diverge as the square root of the order [20] and varies linearly [21] with the order, if the beams are generated by using diffractive optical elements [22] as shown in figure 4.3. By considering the width of the host Gaussian beam as $w(0)$ and defining the inner and outer radii of the donut beam ($r_1$ and $r_2$) at the source plane, their variation as a function of propagation distance has been studied by [23]. The inner and outer radii $r_1$ and $r_2$ are the radial distances at which the intensity falls to $\frac{1}{e^2}$ of the maximum intensity at $r = r_0$.

## 4.6 Near core vortex structure and propagation

Vortex beams are commonly assumed to have a helical wavefront. In the near core region, arbitrarily quick phase variation leads to a significant phase dip and it

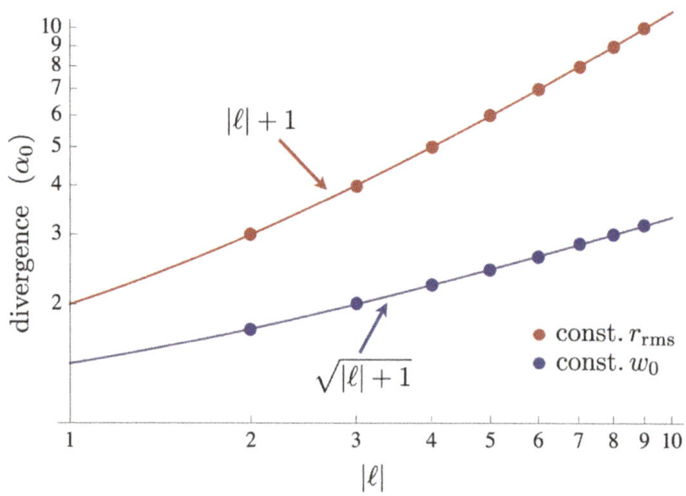


**Figure 4.2.** LG beam divergence scales linearly with OAM if the radius of the waist of the beam is kept constant, and divergence varies with the square root of the beam's OAM if the radius of rms intensity of the beam is kept constant while varying the OAM. Reproduced from [21]. © IOP Publishing Ltd. All rights reserved.

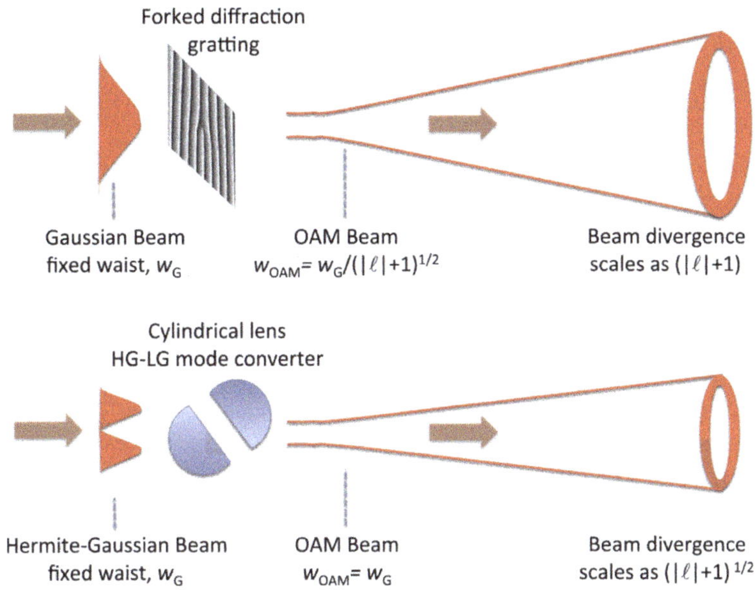


**Figure 4.3.** Divergence of OAM carrying beams created by two different ways: top, using fork grating, bottom, cylindrical mode converter. Note the starting beam waists are the same and the change in the beam waist occurs with the point singularity generation method. Reproduced from [21]. 

remains in the wavefront. The existence of this dip has been verified by using the angular spectrum method [24]. The angular spectrum method can inherently take into account the propagating as well as evanescent spatial frequencies and is able to provide the detailed wavefront structure as the vortex wavefront evolves. The phase distribution function for an optical vortex of charge $m$ is

$$\theta(r, \phi, z) = m\phi + \varphi(r, z) \tag{4.32}$$

where $\varphi$ is a function of both $r$ and $z$, whereas in many studies its dependence on $r$ is not recognized or known. The phase gradient is

$$\nabla\theta = k_r\hat{r} + k_\phi\hat{\phi} + k_z\hat{z} \tag{4.33}$$

The azimuthal component of the propagation vector in a vortex beam was studied by a few groups [15, 25, 26]. The radial component ($k_r$), arises due to the high phase gradient region [27–29] surrounding the vortex core.

In the angular spectrum of the plane waves method, the angular spectrums of waves at two different planes $z_1$ and $z_2$ represented by $A(f_x, f_y; z_1)$ and $A(f_x, f_y; z_2)$ are related by

$$A(f_x, f_y; z_2) = A(f_x, f_y; z_1)\exp\{i\alpha(z_2 - z_1)\} \tag{4.34}$$

where $\alpha$ is the $z$ component of the propagation vector and is given by

$$\alpha = \sqrt{k^2 - 4\pi^2(f_x^2 + f_y^2)} \tag{4.35}$$

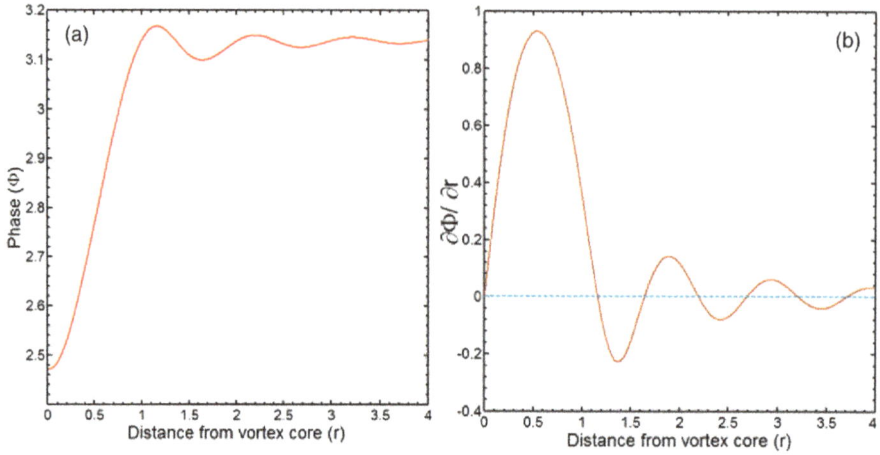


**Figure 4.4.** Plot of (a) phase $\varphi$ of the integral $I(r, z)$ (in radians) and (b) phase gradient $\frac{\partial \Phi}{\partial r}$ versus radial distance $r$ (in units of $\lambda$) from the vortex core at $z = 0.5\lambda$. Reproduced with permission from [24]. Copyright (2016) by OSA.

where $f_x$ and $f_y$ are the spatial frequency components which are related to the $x$ and $y$ components of the propagation vector. When all the spatial frequency components are included in the angular spectrum, the contributions from frequencies higher than $\frac{1}{\lambda}$ are evanescent in nature. This happens when $\alpha$ is imaginary. Equation (4.34) can be written as

$$A(\rho, \xi; z) = A(\rho, \xi: 0)\exp\left\{iz\sqrt{k^2 - 4\pi^2\rho^2}\right\} \tag{4.36}$$

where $\rho$, $\xi$ are Fourier plane polar coordinates whereas $r$, $\phi$ are polar coordinates in the field plane. The field at the $z$ plane $U(r, \phi; z)$ can be obtained by inverse Fourier transforming $A(\rho, \xi; z)$ and is given by

$$U(r, \phi; z) = \exp(i\phi)\int_0^\infty \exp\left\{\frac{iz\sqrt{k^2 - 4\pi^2\rho^2}}{\rho}\right\}J_1(2\pi r\rho)d\rho \tag{4.37}$$

where $J_1(\ldots)$ is the Bessel function of the first kind. By choosing different values of the upper limit for integration over $\rho$ it can be seen that for small angles (for example, for $\rho = \frac{0.25}{\lambda}$), the curve for $\varphi$ is close to that for the plane wave, but as the contribution from higher propagating spatial frequencies is included, phase $\varphi$ starts to lag as compared to the plane wave. When the evanescent components are also included the phase curve significantly dips further. The phase and its radial gradient component as a function of radial distance from the center of the core is plotted and shown in figure 4.4 and the shape of the dip in the wavefront after removing the helical phase part is shown in figure 2.10.

## 4.7 Propagation dynamics of optical phase singularities

Optical vortices can be introduced into a beam by using a spiral phase plate that can be inserted into a beam that may not contain a vortex. There may be many other ways of introducing a vortex in a beam as discussed in chapter 3. The position of vortex introduction in the beam cross section can also be selective. Their propagation dynamics and some of the studies reported in the literature are presented here.

### Network of vortices

Paraxial propagation of an array of optical vortices in a Gaussian host beam was studied [30] using numerical simulation. If the array of vortices are of the same charge, their relative positions, as well as their relative positions within the beam are invariant and the array simply expands or contracts and rotates rigidly. But when the array of vortices are of opposite charges propagation results in attraction, collision and annihilation of charges as shown in figures 4.5 and 4.6.

### Vortex at off-axis positions in LG beams

Flossmann *et al* [1] used two LG $LG_l^p$ beams with the azimuthal index $l = 0$ and the radial index $p \neq 0$ which have a maximum in the center of the beam. The radial index $p$ determines the number of bright or dark rings in the radial intensity distribution. They have introduced different types of vortices ($r$-type, tanh type) into the background beam by multiplying the vortex function with the background field. The computation of intensity distributions at different planes is done by two methods. One method is similar to the method of angular spectrum of plane waves and in the other method, the vortex embedded LG beam is decomposed in the LG basis, and the series expansion coefficients are found analytically. In the series expansion by substituting the various beam and vortex parameters, the intensity distributions are obtained as a function of propagation distance $z$. For the analysis

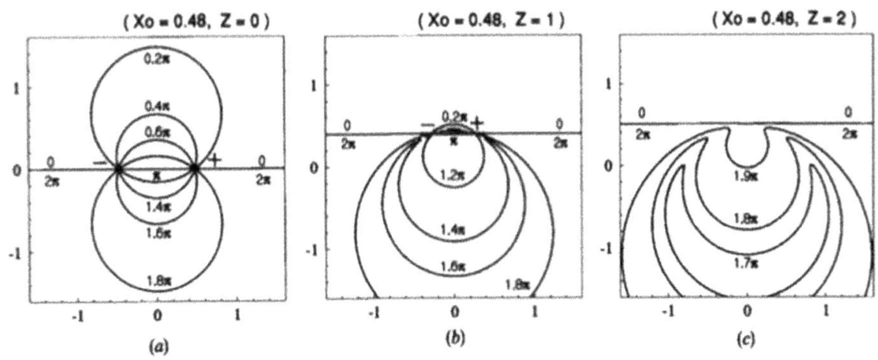


**Figure 4.5.** Phase contours of a Gaussian beam hosting two opposite charged vortices initially separated by a distance of $0.96w_0$. (a) At the waist ($z = 0$), (b) at one Rayleigh distance from the minimum waist, the two vortices are moving on a circle and are attracted to each other and (c) at two Rayleigh distances away, the two vortices have collided and vanished. Reprinted with permission from [30]. Copyright (1993) Taylor & Francis Ltd.

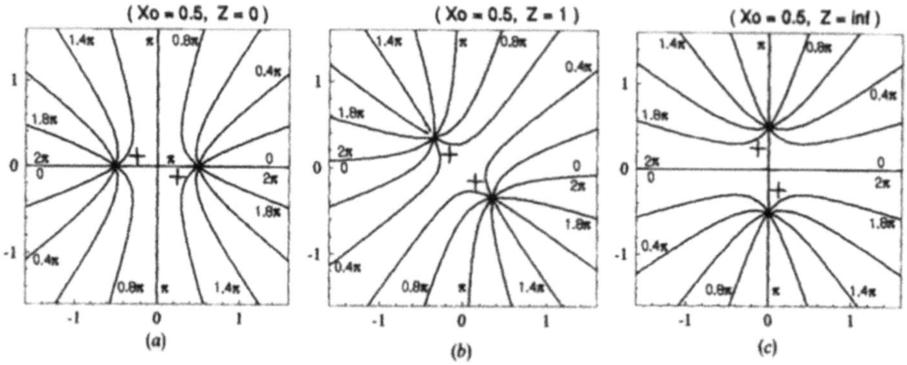


**Figure 4.6.** Phase contour plots for two vortices of equal charge (+1) initially separated by one beam radius $w_0$. (a) At the waist, (b) at one Rayleigh distance away, the two vortices have rotated 45°. (c) In the far field, the two vortices have rotated 90°. Reprinted with permission from [30]. Copyright (1993) Taylor & Francis Ltd.

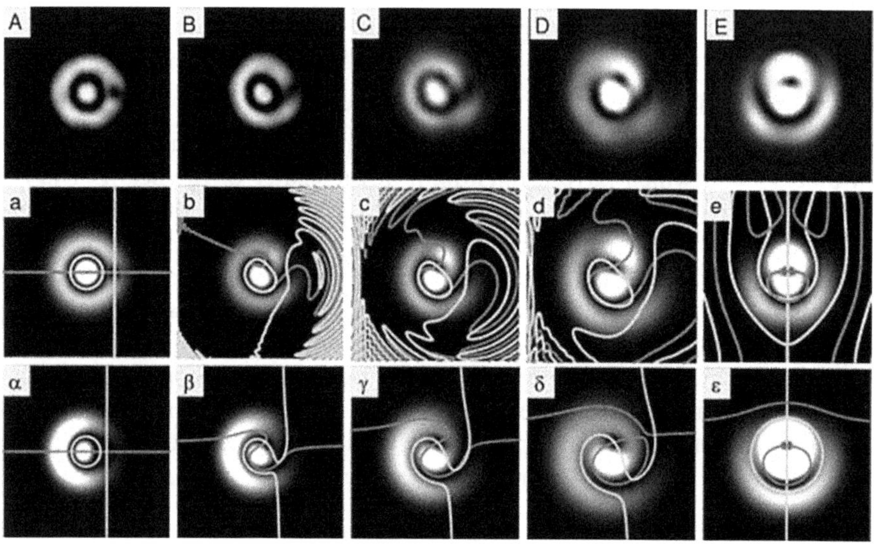


**Figure 4.7.** Propagation dynamics of an off-axis vortex ($m$ = 1, $a$ = 1.0 $w_0$) in a $LG_0^1$ background beam ($w_0$ = 0.388 mm, $z_R$ = 75 cm). First row: measured intensity distributions; second and third rows: calculated intensity distributions for a tanh vortex ($r_c$ = 0.15 mm) and an $r$-vortex, respectively. Propagation distances: (A) $z$ = 0 cm, (B) 30 cm, (C) 60 cm, (D) 90 cm, (E) far field. The light and dark gray curves correspond to the zeros of the real and imaginary part of the light field. The closely spaced light and dark gray curves in the margin of (b), (c), and (d) are artifacts of the numerical calculations. The boxes have a width of approximately 8.5 $w_0$. Reprinted from [1]. Copyright (2005), with permission from Elsevier.

$LG_0^1$ is taken and a tanh vortex and an $r$-vortex are introduced one at a time at off-center position and their propagation results are depicted in figure 4.7. Similarly another $LG_0^7$ beam with a tanh vortex at off-center position is considered for study and the propagation dynamics is shown in figure 4.8.

The propagation dynamics is determined by the $z$ dependent Gouy phases of the LG functions of the expansion. Note that the Gouy phase shift has dependence on

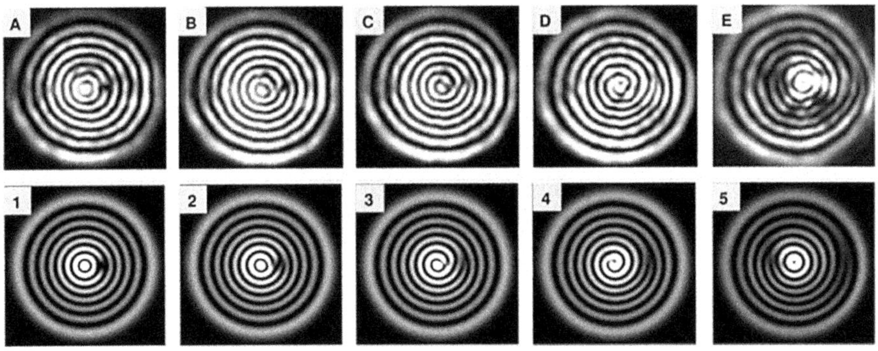


**Figure 4.8.** Propagation dynamics of an initial off-axis vortex ($m = 1$, $a = 0.9w_0$) in a $LG_0^7$ background beam ($w_0 = 0.48$ mm, $z_R = 114$ cm). First row: measured intensity distribution. Second row: calculated intensity distribution for a tanh vortex ($r_c = 0.2$ mm). Propagation distances: (A) $z = 0$ cm, (B) 7 cm, (C) 15 cm, (D) 30 cm, (E) 70 cm. The boxes have a width of approximately $8w_0$. Reprinted from [1]. Copyright (2005), with permission from Elsevier.

the indices $l$ and $p$ of the LG beam. The coefficients that are components of the vortex beam on the LG basis functions are determined by the inner product and they are different. The superposition of the Gouy phases of the LG beams causes a rotation of the phase structure, which determines the azimuthal position of the vortices.

**Vortices introduced in a plane wave using CGH**

Optical vortices can be introduced in a plane wave with uniform amplitude by many ways. For example, a computer generated hologram or a spiral phase plate inserted in a plane wave can embed a vortex in it. Using rigorous diffraction theory, Roux [31] has made some observations of the dynamical behavior of optical vortices in propagating electromagnetic waves. From these observations it can be concluded that a single optical vortex in a wavefront moves directly along the direction of propagation of the wave. The presence of another optical vortex in the beam influences the motions of each other. Two optical vortices with the same charges (an isopolar vortex pair) will gyrate around each other, and their centroid will propagate in the direction of propagation of the wave. Two optical vortices with opposite charges (a bipolar vortex pair) will not gyrate around each other, and their centroid will drift away from the direction of propagation of the wave.

**Off-axis vortices in a Gaussian beam**

The output of the laser may contain a vortex in itself as in some LG modes. A vortex can also be introduced in a Gaussian beam at off-axis positions. In a linear medium, when a vortex is placed off-center, there is no rotation of it as sometime described. Instead, a single vortex of any type introduced in an off-axis position moves across a Gaussian beam in a straight line as shown in figure 4.9. Rozas *et al* [32] have also shown that in a Gaussian background field, there is orbital motion of identical

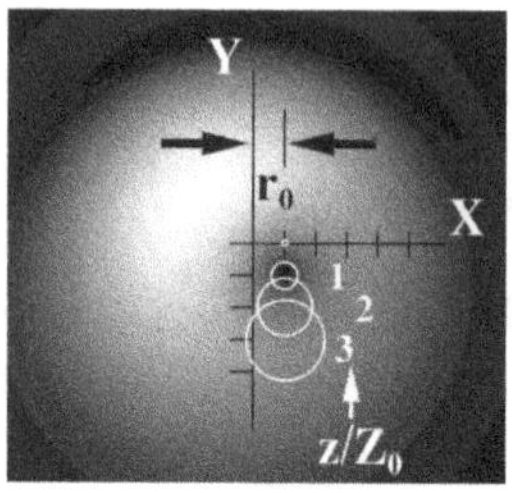


**Figure 4.9.** A tanh vortex of charge +1 in a Gaussian beam placed off-center at point (244 $\mu$m, 0). The projection of the vortex trajectory in the transverse plane, shown by the white open circles in a straight line. The vortex advance at a uniform rate given by $\frac{r0}{Z_0}$. This behavior is same for both an *r*-vortex and a tanh vortex. Reproduced with permission from [32]. Copyright (1997) by OSA.

quasi-point vortices in linear and non-linear media and demonstrated how the rotation can slow and reverse its direction.

It has been observed that *r*-vortices (unlike point vortices) exhibit no fluid-like motion. In fact, Indebetouw [30] showed that any number of identical *r*-vortices within a Gaussian beam will propagate independently of the others.

In one study [33], it was shown through numerical study that optical vortices with anisotropic phase profiles during propagation in linear and non-linear media exhibit rotation in which the rate of rotation is proportional to the anisotropy. In composite beams formed by collinear superpositions of LG beams, off-axis vortices appear [6]. These vortices upon propagation, move about the beam axis consistent with a rotation of the beam profile. The motion of these off-axis vortices are attributed to the Gouy phase.

## 4.8 Propagation of vortices in non-linear media

When a vortex is introduced into a self-defocusing medium, the characteristic core size decreases, owing to the law of refraction. The opposing effects of refraction and diffraction result in a stationary core size in a Kerr non-linear refractive medium. Thus, the vortex forms a filament of constant radial size, which has been shown to serve as a self-induced graded-index optical fiber, capable of guiding a probe beam [34]. Vortices in Kerr medium are also the subject of study [35, 36]. A non-linear optical vortex is not only stationary, but also stable to severe perturbations; therefore it is a soliton.

Optical vortex solitons in a defocusing saturable medium are analyzed [37] in the framework of the (2+1)-dimensional generalized non-linear Schrodinger equation. The numerical study shows stationary, radially symmetric localized solutions with non-vanishing asymptotics and a phase singularity (vortex solitons) are possible for varying saturation parameter. Similarly the existence of a stable black self-guided beam of circular symmetry in a bulk self-defocusing Kerr medium is reported [38]. They are highly stable which is in sharp contrast to the analogous bright beams in a self-confusing medium which are unstable. The experimental demonstration of the steering of an optical vortex soliton by the superposition of a weak coherent background field is reported by Christou *et al* [39].

There are also reports on vortex solitons in cubic-quintic medium [40], perturbed media [41], multi-vortex solitons [42] and a detailed report on vortex soliton can be found in [43].

## References


[1] Flossmann F, Schwarz U T and Maier M 2005 Propagation dynamics of opatical vortices in Laguerre–Gaussian beams *Opt. Commun.* **250** 218–30

[2] Freund I, Shvartsman N and Freilkher V 1993 Optical dislocation network in highly random media *Opt. Commun.* **101** 247–64

[3] Brown J W and Churchill R V 1996 *Complex Variables and Applications* (New York: McGraw Hill)

[4] Swartzlander G A Jr 1999 *Optical vortices, Horizons in World Physics* vol 228 (New York: NOVA Science Publishers)

[5] Gouy L G 1890 Sur une propriete nouvelle des ondes lumineuses *C R Acad Sci Paris* **110** 1251–3

[6] Baumann S M, Kaib D M, MacMillan L H and Galvez E J 2009 Propagation dynamics of optical vortices due to Gouy phase *Opt. Express* **17** 9818–27

[7] Boyd R W 1980 Intuitive explanation of the phase anomaly of focused light beams *J. Opt. Soc. Am.* **70** 877–80

[8] Feng S and Winful H G 2001 Physical origin of the Gouy phase shift *Opt. Lett.* **26** 486487

[9] Saleh B E A and Teich M C 2007 *Fundamentals of Photonics* (New York: Wiley)

[10] Durnin J, Miceli J J Jr and Eberly J H 1987 Diffraction-free beams *Phys. Rev. Lett.* **58** 1499–501

[11] Durnin J 1987 Exact solutions for nondiffracting beams. I the scalar theory *J. Opt. Soc. Am. A* **4** 651–4

[12] McGloin D and Dholakia K 2005 Bessel beams: diffraction in a new light *Contemp. Phys.* **46** 15–28

[13] Martelli P, Tacca M, Gatto A, Moneta G and Martinelli M 2010 Gouy phase shift in nondiffracting bessel beams *Opt. Express* **18** 7108–20

[14] Ruschin S 1994 Modified Bessel nondiffracting beams *J. Opt. Soc. Am. A* **11** 3224–8

[15] Leach J, Keen S, Padgett M J, Saunter C and Love G D 2006 Direct measurement of the skew angle of the Poynting vector in a helically phased beam *Opt. Express* **14** 11919–24

[16] Bliokh K Y, Bekshaev A Y, Kofman A G and Nori F 2013 Photon trajectories anomalous velocities and weak measurements: a classical interpretation *New J. Phys.* **15** 073022

[17] Padgett M 2008 On the focusing of light, as limited by the uncertainty principle *J. Mod. Opt.* **55** 3083–9

[18] Stelzer E H K and Grill S 2000 The uncertainty principle applied to estimate focal spot dimensions *Opt. Commun.* **173** 51–6

[19] Philips R L and Andrews L C 1983 Spot size and divergence for Laguerre–Gaussian beams of any order *Appl. Opt.* **22** 643–4

[20] Beijersbergen M W, Allen L and van der Veen H E L 1993 Astigmatic laser mode converters and transfer of orbital angular momentum *Opt. Commun.* **96** 123–32

[21] Padgett M J, Miatto F M, Lavery M P J, Zeilinger A and Boyd R W 2015 Divergence of an orbital-angular-momentum-carrying beam upon propagation *New J. Phys.* **17** 023011

[22] Carpentier A V, Michinel H and Salguerio J R 2008 Making optical vortices with computer-generated holograms *Am. J. Phys.* **76** 916–21

[23] Reddy S G, Prabhakar S, Permangatt C, Anwar A, Banerji J and Singh R P 2015 Divergence of optical vortex beams *Appl. Opt.* **54** 6690–3
[24] Lochab P, Senthilkumaran P and Khare K 2016 Near-core structure of a propagating optical vortex *J. Opt. Soc. Am. A* **33** 2485–90
[25] Arlt J 2003 Handedness and azimuthal energy flow of the optical vortex beams *J. Mod. Opt.* **50** 1573–80
[26] Singh B K, Bahl M, Mehta D S and Senthilkumaran P 2013 Study of internal energy flows in dipole vortex beams by knife edge test *Opt. Commun.* **293** 15–21
[27] Berry M V 2008 Waves near zeros *Conf. on Coherence and Quantum Optics* (Washington, USA: OSA), 37–41
[28] Berry M V 1994 *Faster than Fourier in Quantum coherence and Reality* (Singapore: World Scientific) pp 55–65
[29] Senthilkumaran P and Bahl M 2015 Young's experiment with waves near zeros *Opt. Express* **23** 10968–73
[30] Indebetouw G 1993 Optical vortices and their propagation *J. Mod. Opt.* **40** 73–87
[31] Roux F S 1995 Dynamical behaviour of optical vortices *J. Opt. Soc. Am. B* **12** 1215–21
[32] Rozas D, Law C T and Swartzlander G A 1997 Propagation dynamics of optical vortices *J. Opt. Soc. Am. B* **14** 3054–65
[33] Kim G-H, Lee H J, Kim J-U and Suk H 2003 Propagation dynamics of optical vortices with anisotropic phase profiles *J. Opt. Soc. Am. B* **20** 351–9
[34] Swartzlander G A Jr and Law C T 1993 The optical vortex soliton *Opt. Photon. News* **4** 10
[35] Carlsson A H, Malmberg J N, Anderson D, Lisak M, Ostrovskaya E A, Alexander T J and Kivshar Y S 2000 Linear and nonlinear waveguides induced by optical vortex solitons *Opt. Lett.* **25** 660–2
[36] De La Fuente R, Barthelemy A and Froehly C 1991 Spatial-soliton-induced guided waves in a homogeneous nonlinear Kerr medium *Opt. Lett.* **16** 793–5
[37] Tikhonenko V, Kivshar Y S, Steblina V V and Zozulya A A 1998 Vortex solitons in a saturable optical medium *J. Opt. Soc. Am. B* **15** 79–86
[38] Snyder A W, Poladian L and Mitchell D J 1992 Stable black self-guided beams of circular symmetry in a bulk Kerr medium *Opt. Lett.* **17** 789–91
[39] Christou J, Tikhonenko V, Kivshar Y S and Luther-Davies B 1996 Vortex soliton motion and steering *Opt. Lett.* **21** 1649–51
[40] Reyna A S and de Araújo C B 2016 Guiding and confinement of light induced by optical vortex solitons in a cubic-quintic medium *Opt. Lett.* **41** 191–4
[41] Chen Y and Atai J 1994 Dynamics of optical-vortex solitons in perturbed nonlinear media *J. Opt. Soc. Am. B* **11** 2000–3
[42] Buccoliero D, Desyatnikov A S, Krolikowski W and Kivshar Y S 2008 Spriraling multi-vortex solitons in nonlocal nonlinear media *Opt. Lett.* **33** 198–200
[43] Desyatnikov A S, Kivshar Y S and Torner L 2005 Optical vortices and vortex solitons *Prog. Optics* **47**

**IOP** Publishing



# Chapter 5

# Internal energy flows

## 5.1 Energy flow

The energy flow direction can be given by the Poynting vector. Normally for the fields traveling in isotropic medium the Poynting vector is in the direction of the propagation vector, which means that the flow of energy is in the forward direction. This energy flow can occur in the transverse direction to the general propagation direction of the beam and in some cases a backward flow of energy is also possible. Energy flow in singular beams and non-singular beams, spin and orbital contributions and the types of flow are discussed in this chapter.

**Poynting vector**: It is well known that magnetic and electric fields store energy. This energy can be transported by electromagnetic waves as they propagate through space. The energy the wave carries per unit time per unit area is given by the Poynting vector $\vec{S}$ and is in the direction of the velocity of the wave [1].

$$\vec{S} = \frac{1}{\mu_0}\vec{E} \times \vec{B} = \vec{E} \times \vec{H} \tag{5.1}$$

where $\vec{E}$ and $\vec{B}$ are electric and magnetic field components of the electromagnetic waves, $\mu_0$ is the free space permeability. It thus, represents the energy flux density of an electromagnetic field.

Another expression that can be used for the Poynting vector is

$$\vec{S} = cg\,\mathrm{Re}[\vec{E}^* \times \vec{H}] \tag{5.2}$$

where $c$ is the speed of light in vacuum and $g = (8\pi)^{-1}$.

**Optical current**: The optical current, depicted as $\vec{j}$ [2] equals the time averaged energy flow. Considering a complex scalar optical field $\psi = A\exp(i\Phi)$, the optical current $\vec{j}$ is given by

$$\vec{j} = I\nabla\Phi \tag{5.3}$$

where $I = \psi\psi^*$ is the intensity and $\psi^*$ is the conjugate of $\psi$. The phase of the field can be written as $\Phi = \text{Im}[\log\psi]$. Thus the phase gradient is

$$\nabla\Phi = \text{Im}[\nabla \log\psi] = \frac{\text{Im}[\psi^*\nabla\psi]}{I} \tag{5.4}$$

Hence the optical current can be given by

$$\vec{j} = I\nabla\Phi = \text{Im}[\psi^*\nabla\psi] \tag{5.5}$$

By writing the complex scalar optical field as $\psi = \xi + i\eta$, one can write $\psi^*\nabla\psi = (\xi\nabla\xi + \eta\nabla\eta) + i(\xi\nabla\eta - \eta\nabla\xi)$. Hence the optical current is given by

$$\vec{j} = \text{Im}[\psi^*\nabla\psi] = \xi\nabla\eta - \eta\nabla\xi \tag{5.6}$$

This optical current is known as the probability current in quantum mechanics and it is synonymously used as the Poynting vector in electromagnetic fields.

**Phase and phase gradients**: The phase $\Phi(x, y, z)$ of an optical field is a scalar quantity. The complex field with this phase distribution satisfies the paraxial Helmholtz equation. Since $\nabla\Phi$ is the phase gradient, it points in the direction where there is a maximum change in phase. Gradients are normal to the contour surfaces and hence the phase gradient is always normal to the wavefront. The magnitude of the phase gradient is equal to the propagation constant and the phase gradient $\nabla\Phi(x, y, z)$ is a vector field that can provide information about the propagation vector distribution as $\nabla\Phi(x, y, z) = \vec{k}$ and the energy flow as given by equation (5.1). Further, the momentum density $\vec{p} = \epsilon_0\mu_0\vec{S}$, can also be obtained from the phase gradient.

**Flow in the isotropic and anisotropic medium**: The Poynting vector $\vec{S}$ points in the direction of $\vec{E} \times \vec{H}$ whereas the propagation vector $\vec{k}$ points in the direction of $\vec{D} \times \vec{H}$. Here $\vec{D}$ is the electric displacement vector and $\vec{H}$ is the magnetic field produced by the free currents. In a linear isotropic medium $\vec{D} = \epsilon\vec{E}$ where $\epsilon = \epsilon_0\epsilon_r$ is a constant. Hence the direction of $\vec{D}$ and $\vec{E}$ are the same. As a result $\vec{S}$ and $\vec{k}$ point in the same direction. But for an anisotropic medium, $\epsilon$ can be written as a tensor and hence the direction of $\vec{D}$ and $\vec{E}$ are not the same in all directions. Hence, in certain directions in anisotropic medium where $\vec{D}$ and $\vec{E}$ are not collinear, the Poynting vector and the propagation vector will be oriented in different directions. In uniaxial crystals these are the directions where double refraction occurs. In our discussion, since we deal with linear isotropic medium, $\vec{S}$ is in the direction of $\vec{k}$.

**Energy flow in plane waves**: Consider an example of a plane electromagnetic wave whose electric field $\vec{E}$ is given by $\vec{E} = E_0\cos(kz - \omega t)\hat{x}$ and magnetic field $\vec{B}$ given by $\vec{B} = B_0\cos(kz - \omega t)\hat{y}$, where the direction of propagation is along the $z$ axis. The Poynting vector is thus, given as

$$\vec{S} = \frac{1}{\mu_0} E_0 \cos(kz - \omega t)\hat{x} \times B_0 \cos(kz - \omega t)\hat{y} = \frac{E_0 B_0}{\mu_0} \cos^2(kz - \omega t)\hat{z}$$
$$= \frac{EB}{\mu_0}\hat{z} \tag{5.7}$$

The energy flow can be seen in the direction of the propagation vector. It is easier to work out the optical current by using the phase distribution of a given optical field. The flow direction is the wavefront's local normal direction. For example, a plane wave traveling at an angle with respect to the $z$ direction with field distribution $E_0\hat{y}\exp\{i(k_x x + k_z z - \omega t)\}$ has the energy flow given by

$$\vec{j} = I\nabla\{k_x x + k_z z\} = I\{k_x\hat{x} + k_z\hat{z}\} \tag{5.8}$$

Note in this case the general propagation direction is $z$ and the energy flow also has a component in the transverse plane to the $z$ direction.

**Energy flow in spherical waves**: Consider a spherical wave whose complex amplitude is given by $\frac{\exp(ikr)}{r}$, expressed in spherical polar coordinates. The phase of the spherical wave is $\Phi = kr$ and the phase gradient is $\nabla\Phi = k\hat{r}$. Hence the optical current is given by

$$\vec{j} = \frac{1}{r^2}k\hat{r} \tag{5.9}$$

which is diverging. For a converging wave the flow of energy is converging towards a point.

**Energy flow in a singular beam**: The complex amplitude for a singular beam can be written as $r\exp[i(m\theta + kz)]$ and the phase distribution is $\Phi = m\theta + kz$ in cylindrical coordinate system. Hence the energy flow in a singular beam is given by

$$\vec{j} = I\left\{\frac{m}{r}\hat{\theta} + k\hat{z}\right\} \tag{5.10}$$

The energy flow in a singular beam follows a helical path winding about the core of the vortex.

## 5.2 Internal energy flows

Poynting vector field distribution describes the total energy flow of a light beam. This flow can be classified into longitudinal and transverse flow components with respect to the general flow direction. These internal flow structures can have angular momentum (AM) components that can be intrinsic or extrinsic. The spin angular momentum component [3] is intrinsic which is independent of the origin, whereas the orbital angular momentum component arising due to phase vortices in the beam can be intrinsic or extrinsic [4]. Further, there can be an orbital angular momentum component arising from spatially varying polarization distribution. In a spatially varying polarization distribution, called an inhomogeneously polarized beam the presence of polarization vortices gives rise to the orbital angular momentum (OAM) components. In the study of polarization singularities, there are C-point and V-point singularities that have OAM components. These aspects will be the subject matter for chapter 9. Hence there are three structural summands namely, one extrinsic AM and two intrinsic AMs. The two intrinsic AM are intrinsic orbital AM and intrinsic spin AM of the beam. The extrinsic AM owes to transverse displacement of the light energy of the beam as a whole. The intrinsic energy redistribution arising due to the vector nature (polarization) of electromagnetic waves, consists of (1) solenoidal spin current arising due to spatially varying polarization distribution and (2) spin associated with circular polarization. Both intrinsic structural components of the energy flow can be attributed to polarization in which the fields oscillate with optical frequency and possess no direct mechanical meaning. For the spin flow, this motion is obviously the instantaneous rotation of the field vectors (circular polarization) and for the orbital flow, it is the 'running' component in the pattern of instantaneous oscillation of the spatially inhomogeneous electromagnetic field.

In this chapter we concentrate on energy flows in the transverse plane arising due to OAM due to phase and polarization distributions. Since OAM refers to the running component of energy flow in the beam as a whole, the discussion on the flow due to phase vortices is also applicable to intrinsic part of OAM.

## 5.3 Visualizing internal energy flow

### 5.3.1 Bekshaev–Bliokh–Soskin method

The energy flow in spatially coherent monochromatic optical fields can be obtained from the electric and magnetic fields using this method [5]. The electric and magnetic fields of spatially coherent monochromatic waves are given by

$$\begin{aligned}\vec{E}(\vec{R}, t) = \mathrm{Re}[\vec{E}(\vec{R})\exp(-i\omega t)]\\ \vec{H}(\vec{R}, t) = \mathrm{Re}[\vec{H}(\vec{R})\exp(-i\omega t)]\end{aligned} \tag{5.11}$$

Here $\vec{R}$ is the 3D radius vector $\vec{R} = \vec{r} + z\hat{e}_z$, where $z$ is the predominant longitudinal component of $\vec{R}$ in the general propagation direction and $\vec{r}$ is the transverse component. The time averaged Poynting vector, which is the measure of the energy flow (more precisely, the flow density), is given by

$$\vec{S} = cg\,\mathrm{Re}[\vec{E}^* \times \vec{H}] \tag{5.2}$$

Here $g = (8\pi)^{-1}$ is a constant factor, in the Gaussian system, and $c$ is the speed of light. The Poynting vector also expresses the momentum density of the field

$$\vec{p} = \frac{1}{c^2}\vec{S} = \frac{g}{c}\mathrm{Re}[\vec{E}^* \times \vec{H}] \tag{5.13}$$

The density of angular momentum of light is given by

$$\vec{R} \times \vec{p} = \frac{g}{c}\mathrm{Re}[\vec{R} \times (\vec{E}^* \times \vec{H})] \tag{5.14}$$

The momentum density can again be rewritten as

$$\vec{p} = \frac{g}{\omega}\,\mathrm{Im}[\vec{E}^* \times (\nabla \times \vec{E})] \tag{5.15}$$

Note the momentum density is written only in terms of electric fields. Similarly in terms of $\vec{H}$ only, the momentum density is

$$\vec{p} = \frac{g}{\omega}\,\mathrm{Im}[\vec{H}^* \times (\nabla \times \vec{H})] \tag{5.16}$$

This momentum density $\vec{p}$ can be decomposed into two parts namely 'spin flow density ($\vec{p_S}$)' and 'orbital flow density ($\vec{p_O}$)'. They are given by

$$\vec{p_S} = \frac{g}{4\omega}\,\mathrm{Im}[\nabla \times (\vec{E}^* \times \vec{E} + \vec{H}^* \times \vec{H})] \tag{5.17}$$

and

$$\vec{p_O} = \frac{g}{2\omega}\,\mathrm{Im}[\vec{E}^* \cdot (\nabla)\vec{E} + \vec{H}^* \cdot (\nabla)\vec{H}] \tag{5.18}$$

In these equations

$$\vec{E}^* \cdot (\nabla)\vec{E} = E_x^* \nabla E_x + E_y^* \nabla E_y + E_z^* \nabla E_z \tag{5.19}$$

In the absence of electric charges and currents, the total energy flow

$$\nabla \cdot \vec{p} = \nabla \cdot (\vec{p_S} + \vec{p_O}) = 0 \tag{5.20}$$

In fact both its spin and orbital parts, are solenoidal $\nabla \cdot \vec{p_S} = 0$ and $\nabla \cdot \vec{p_O} = 0$. The energy flow lines are always continuous. Being solenoidal, the angular momentum can be associated with vorticities defined by

$$\Omega_S = \nabla \times \vec{p_S} = -\frac{g}{4\omega}\,\mathrm{Im}[\nabla^2(\vec{E}^* \times \vec{E} + \vec{H}^* \times \vec{H})] \tag{5.21}$$

and

$$\Omega_O = \nabla \times \vec{p_O} = \frac{g}{2\omega}\,\mathrm{Im}[\nabla(\vec{E}^*) \cdot (\times\nabla)\vec{E} + \nabla(\vec{H}^*) \cdot (\times\nabla)\vec{H}] \tag{5.22}$$

where

$$\nabla(\vec{E}^*) \cdot (\times\nabla)\vec{E} = \nabla E_x^* \times \nabla E_x + \nabla E_y^* \times \nabla E_y + \nabla E_z^* \times \nabla E_z \tag{5.23}$$

The angular momentum (per unit $z$-length) is

$$\vec{J} = \int \vec{R} \times \vec{p}\, d^2\vec{r} = \int \vec{R} \times (\vec{p_O} + \vec{p_S}) d^2\vec{r} = J_O + J_S \tag{5.24}$$

### 5.3.2 Helmholtz–Hodge decomposition method

Helmholtz–Hodge decomposition method (HHD) has been demonstrated for scalar optical fields [6]. This method ignores the vectorial nature of light and only the linear and orbital angular momentum part of the energy flow are visualized. In this method, the phase distribution of the optical field at a plane is converted into a vector field by finding the phase gradient from the phase distribution. Since the phase gradient connects the propagation vector and the flow of energy direction (Poynting vector), various energy flow components can be decomposed using the Helmholtz–Hodge decomposition.

The Helmholtz–Hodge decomposition method is based on the Helmholtz theorem which says that a vector field $\vec{F}(x, y)$ can be decomposed into three parts: a curl-free part of the field, a divergence-free part and both curl- and divergence-free parts of the field [6–8]. The component of the field which has both zero curl and zero divergence is called the harmonic part of the field. Non-zero curl represents the presence of vortices in the field, non-zero divergence represents the presence of sources or sinks in a field and the harmonic component represents a source free region. This harmonic part of the field $H$ is basically continuous and differentiable and hence analytic

$$\vec{F}(x, y) = f_1 + f_2 + f_3 = \nabla\varphi + \nabla \times \vec{A} + H \tag{5.25}$$

where $\varphi(r)$ and $\vec{A}$ are the scalar and vector potentials that can be obtained by solving the Poisson equations as given in section 3.3.4.

$f_1$ is the irrotational component, whose curl is zero, $f_2$ is the solenoidal component, whose divergence is zero, and for $f_3$ both curl and divergence are zero. The component $f_2$, represents the singularities that relates to the OAM of the field. The component $f_1$ is curl-free and carries only the sources and sinks of energy. The actual energy flow happens in 3D, while the vector fields $\vec{F}$, $\nabla\varphi$, $\nabla \times \vec{A}$ and $H$ are presented in transverse or 2D planes. Such transverse flow patterns are known as internal energy flows. Since the transverse planes in a volume form a continuum, the observation plane divides the neighborhood planes into two sets that are on either side of it. With respect to the general flow direction, we refer to the immediate neighborhood planes as preceding and succeeding. The strength of these energy flows is sought from the intensity distribution and the magnitude of the phase gradient $\nabla\Phi(x, y, z)$. In the 2D flow patterns, a flow line can terminate when the transverse component of the flow pattern vanishes, i.e. when the flow becomes purely longitudinal. It can also terminate at intensity nulls. Since no real sources or sinks of electromagnetic energy are present during propagation, they represent only interplanar flows in the immediate neighborhood of the edge dislocations. The boundary conditions imposed in HHD ensure a normal curl-free component and a tangential divergence-free component. The HHD method works on discrete fields where the surface considered is a simply connected simplicial surface with a piecewise constant vector field $\vec{F}$. The gradient is thus for a piecewise linear function. The divergence and rotation can be found by applying Green's integration by path, and this solves the issues for fields with singular points.

## 5.4 Focusing of singular beams—effect of aberrations

Specific internal energy flows in the focusing of singular beams are discussed in this section. Under loose focusing, the effect of aberrations on the transverse and longitudinal energy flows are affected. Internal energy flow patterns in the presence of two specific aberrations namely astigmatism and coma are considered for the study.

**Astigmatism**: Singular beams have circulating energy components. Focusing by a low numerical aperture system suffering from astigmatic aberration modifies these energy flow components [9]. The phase gradient arising due to these aberration splits the energy circulation due to higher charge vortex into multiple circulations. Using the Helmholtz–Hodge decomposition method, the transverse components of the Poynting vector fields are separated into solenoidal and irrotational components. The solenoidal components relate to the orbital angular momentum of the beams [5], and the irrotational components are useful in the transport of intensity equations [10–12] for phase retrieval.

The phase variation of astigmatism is expressed as $kA_a\rho^2\cos^2\theta$, where $\rho$ and $\theta$ are polar coordinates at the exit pupil plane of the lens system. The phase gradient field for astigmatism is $kA_a\rho(2\cos^2\theta\hat{\rho} - \sin 2\theta\hat{\theta})$. The divergence of this field is $2\mathrm{kA}_a$, and it has zero curl. But the curling field lines of the vortex field are modified by astigmatism. The transverse phase map shown in figure 5.1(a) is that of a vortex of charge $m = -3$ and its phase gradient map is shown in figure 5.1(b). These are the energy flow lines and for a vortex they are in the form of closed concentric circles. These closed circular flow lines are shown modified in figures 5.1(c)–(e) when different amounts of astigmatisms $A_a = 0.5$, 1.0 and 1.5 are added. Starting from predominantly circular flow for low aberration, the singular beam is subjected to shearing flow as the astigmatism is increased. The oppositely directed flow in the upper and lower part tend to tear apart the higher order vortex.

To find the focal plane intensity distribution and the flow pattern, the astigmatism imbued vortex field is propagated using the Fresnel–Kirchhoff diffraction integral. The coordinate system employed in the diffraction integral to evaluate the complex amplitude at the focal plane is shown in figure 5.2. At the focal plane $r$, $\phi$, $z$ are the coordinates. The field distribution $U(r, \phi, z)$ at the focal plane evaluated using the Fresnel–Kirchhoff integral is expressed as

$$U(r, \phi, z) = \int_0^{\rho_0}\int_0^{2\pi} A\exp(im\theta)\exp(i(2\pi/\lambda)\, \times A_a\rho^2\cos^2(\theta)\exp(i2\pi r\rho\cos(\theta - \phi))\rho d\rho d\theta \tag{5.26}$$

This field at the focal plane is complex, and the corresponding phase gradient field is computed. This complex field is subjected to Helmholtz–Hodge decomposition and the solenoidal part of the field is analyzed.

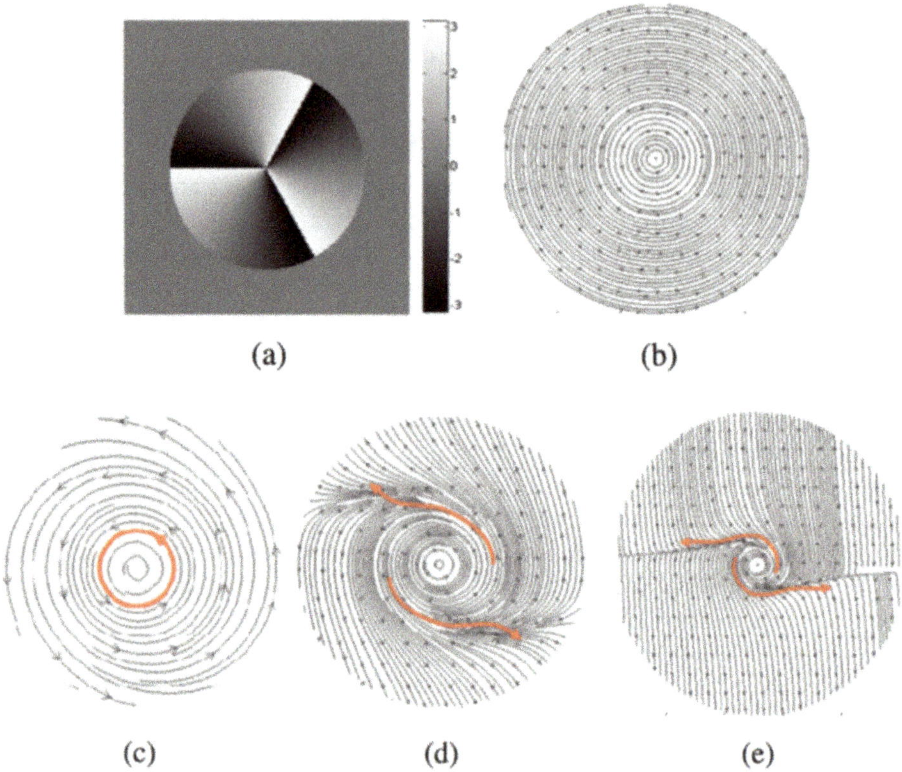


**Figure 5.1.** Phase profile and (b)–(e) phase gradient field streamlines for a circular aperture-limited vortex beam when astigmatism is added: (b) $A_a = 0$; (c) $A_a = 0.5$; (d) $A_a = 1.0$; and (e) $A_a = 1.5$. Reproduced with permission from [9]. Copyright (2014) by OSA.

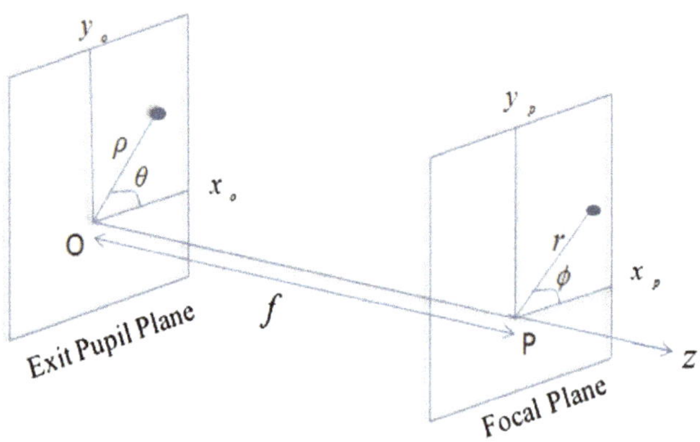


**Figure 5.2.** Coordinate system used in evaluating Fresnel–Kirchhoff diffraction integral. Reproduced with permission from [9]. Copyright (2014) by OSA.

A vortex beam that is focused by a system suffering from astigmatism reveals the presence of energy circulation and divergence. The solenoidal and irrotational components of optical currents are precisely visualized in figure 5.3 for the

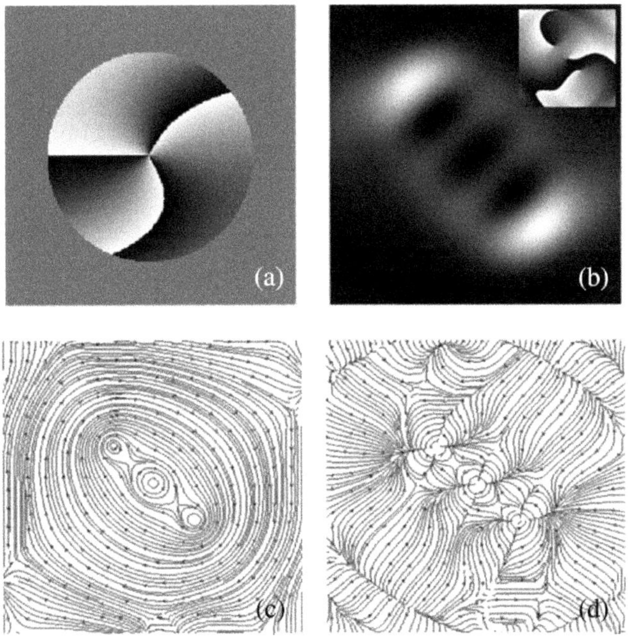


**Figure 5.3.** Focal plane distributions for a $m = +3$ vortex beam is focused using an astigmatism affected lens ($A_a = 0.5$). (a) Phase profile of the beam at the exit pupil of a lens. (b) Intensity profile at the focal plane evaluated using the Fresnel–Kirchhoff diffraction integral. (Inset) Phase profile at the focal plane. (c) Hodge solenoidal component that shows optical currents swirling counter-clockwise about three singular points. (d) Irrotational component. Reproduced with permission from [9]. Copyright (2014) by OSA.

astigmatic ($A_a = 0.5$) focusing of the $m = 3$ vortex beam. Focusing of a vortex beam preserves the vortex in which the focal field has a donut shape and the optical currents circulate about the phase defect. But astigmatism perturbs this circulating flow of Poynting vector field. The internal energy in a focused field of a higher order vortex gets completely redistributed by vortex breakup. A higher order vortex breaks up into single charged vortices due to astigmatic focusing. This remodeling of energy and the nature of splitting depends upon the aperture size, the charge of the focused vortex beam, and the aberration coefficient. It can be clearly seen that the strong circulation of the $m = 3$ vortex beam has its phase gradient field split into three separate circulations about the intensity nulls. Further, the polarity of the separate circulations and that of the focused vortex are the same. In such a case, $\oint \nabla\Phi \cdot dl = 6\pi$ if the line integral encloses all the vortices.

The longitudinal intensity derivative of a wavefront is related to its transverse phase gradient and is used in transport of intensity equations. The divergence component of the Poynting vector field along the propagation direction relates to the flow of longitudinal optical energy.

**Coma**: The circulating phase gradient component of a singular beam under coma afflicted low numerical aperture focusing system also splits the higher charged vortices into fragments. The phase variation of coma is expressed as $kA_c\rho^3 \cos\theta$, and

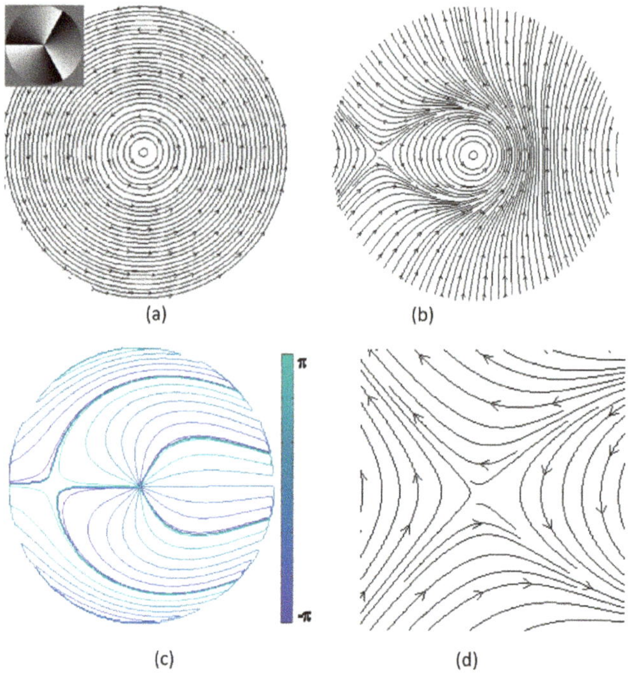


**Figure 5.4.** (a) Phase profile (inset) and the phase gradient field streamlines for a circular aperture-limited vortex beam for $A_c = 0$ and (b) for $A_c = 1.5$, (c) phase contour plot for $A_c = 1.5$, and (d) magnified view of the streamlines, near the saddle point. Reproduced with permission from [13]. Copyright (2015) by OSA.

the phase gradient field is $kA_c\rho^2(3\cos\theta\hat{\rho} - \sin\theta\hat{\theta})$. The divergence of this phase gradient field is $8k\rho A_c \cos\theta$, and it is irrotational. However, these curling field lines of the vortex field are affected by the addition of a minor amount of coma. This splits the higher charge vortex into elementary vortices and distributes them spatially in the focal field. This splitting depends on the charge, polarity and aberration coefficient. Using HHD the transverse component of the Poynting vector field distribution at the focal plane is decomposed into the curl or solenoidal component and divergence or irrotational component [13].

Considering a beam limited by circular aperture, figure 5.4 shows the internal flows for the gradient field of the phase map for an $m = -3$ vortex field shown in the figure 5.4(a) inset. Introduction of coma of $A_c = 1.5$ modifies the closed circular energy flow lines as depicted in figure 5.4(b). There is a saddle point in the phase distribution of a coma-aberrated vortex field that tends to split the vortex. Figure 5.4(c) depicts the phase contour plot in which the saddle is on the left side. The magnified view of the streamlines just near the saddle shows that the energy flow lines move out in opposite directions in both the horizontal and vertical directions. Due to this the higher order vortex is split into unit charge vortices and pushed to one side during focusing.

The focal plane field distribution of a coma-imbued vortex field is evaluated using the Fresnel–Kirchhoff diffraction integral.

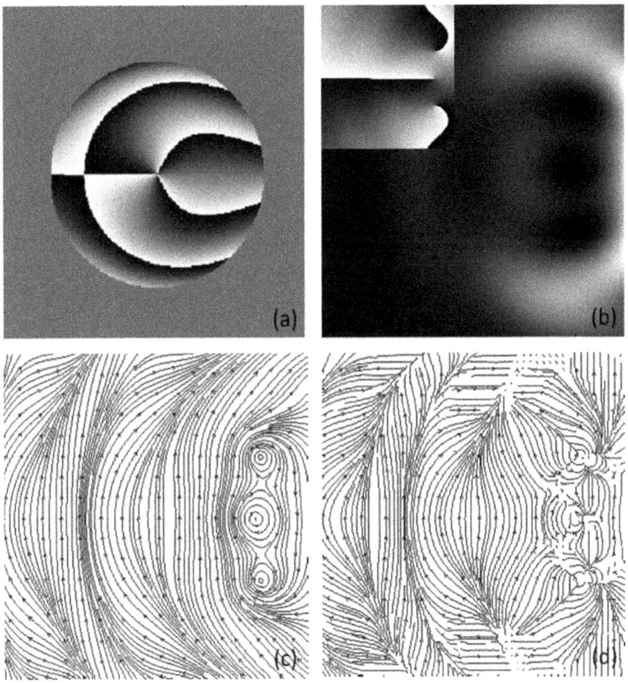


**Figure 5.5.** Splitting of vortex beam ($m = -3$) under focusing by a coma affected lens. (a) Phase profile at the exit pupil of a lens, (b) intensity profile and (inset) phase profile at the focal plane, (c) the Hodge solenoidal component that shows optical currents swirling clockwise about three singular points, (d) the irrotational component. Reproduced with permission from [13]. Copyright (2015) by OSA.

$$
\begin{aligned}
U(r, \phi, z) = \int_0^{\rho_0} \int_0^{2\pi} & A \exp(im\theta)\exp(i(2\pi/\lambda)) \\
& \times A_c \rho^3 \cos(\theta)\exp(i2\pi r\rho \cos(\theta - \phi))\rho d\rho d\theta
\end{aligned}
\tag{5.27}
$$

A typical focal field in a coma afflicted system is shown (figure 5.5) where a vortex beam of charge $m = -3$ is focused. The phase of the optical field at the exit pupil plane of the lens is given in figure 5.5(a). The intensity distribution at the focal field is given in figure 5.5(b) with the inset showing the phase distribution at the focal plane. The HHD decomposed focal field flows are shown in figures 5.5(c) and (d).

## 5.5 Experimental detection

There are many diffraction experiments that prove the transverse circulating flow of energy in singular beam. Some of them are presented here. It is possible to determine the handedness and visualize the flow by monitoring the beam propagation after a knife edge or a similar opaque object. In the geometrical shadow region just behind the obstacles, the intensity pattern rotates or intrudes into the darker region only in one direction, depending on the polarity of the vortex beam. This rotation is faster in the regions closer to the vortex core and slower at the outer regions.

### Half-plane diffraction

Masajada has shown [2] that a single vortex in a Gaussian beam synthesized by a computer generated hologram was diffracted using a half-plane and observed the intensity patterns. The beam was converged by an objective and diffracted by a half-plane and the observations are made for positive and negative vortices. Unfortunately, the synthetic holograms do not generate pure Gaussian beams, but one can still observe that the sense of intensity intrusion into the shadow region is dependent on the polarity of the charges.

### Knife-edge method

There is also another experiment by Arlt [14] in which a knife-edge or a similar opaque object is used to diffract the singular beam. Due to the transverse circulating energy flow in the beam, in the geometrical shadow region just behind the obstacle the intensity ring is found to rotate depending on the handedness of the beam. By this method it is possible to determine unambiguously the sign of the topological charge of the vortex. Evolution of the intensity profile directly visualizes the rotation of the beam's Poynting vector.

The propagation of an obstructed LG beam, by using a knife-edge that obscures half of the beam profile at its beam waist, was studied. After propagation, the distributions at the Rayleigh range $Z_R$ were scrutinized and the intensity patterns are shown in figure 5.6. Single and multiple ringed LG beams (different radial index) with various azimuthal index were taken up for the study. It was found that the sense of rotation in the intensity pattern depends on the sign of the azimuthal index and the amount of rotation of the pattern is independent of the magnitude of azimuthal index. For multiple ringed LG beam, it was found that the inner rings rotate faster than the outer rings.

### Knife-edge test on pair of vortices

So far diffraction experiments with single vortex embedded in beams were presented. But when the beam contains two vortices (dipoles considered here) different energy flow patterns are possible. The schematic shown in figure 5.7 explains the possible flow patterns. Moreover, in a conventional knife-edge test, the knife-edge is moved around (the focus) where the test beam is focused. The rotation of Poynting vector in dipole vortex beam (DVB) during propagation has been experimentally detected in a knife-edge test [15].

A knife-edge test or Foucault test [16] consists of a point source that illuminates the curved test surface. This test surface can be a spherical mirror, or a converging lens that can form an image. A collimated beam passes through a lens (say) that is being tested, and forms an image at the focal plane of the lens. A knife-edge is introduced before the focus of the lens where it blocks half of the field distribution or the rays (say below optical axis). This knife is moved along the beam propagation direction, i.e. moved first to the focal plane and then beyond that. When an obstacle is moved beyond the image plane (here the focal plane), the rays above the optical

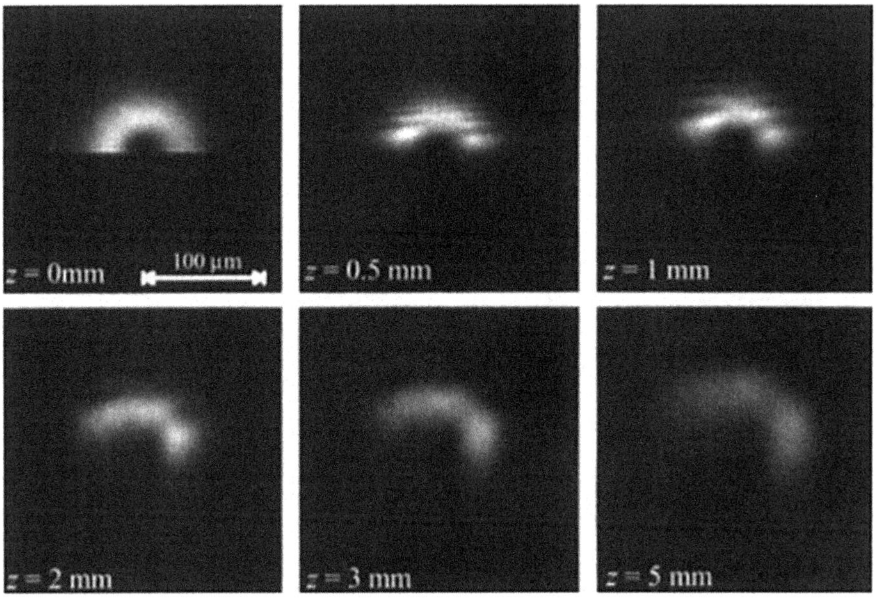


**Figure 5.6.** Experimental results for a LG beam with radial and azimuthal index $p = 0$, $l = 2$, $\lambda = 632$ nm and $w_0 = 30$ $\mu$m. Presented are intensity cross-sections for different propagation distances covering about one Rayleigh range. Reprinted with permission from [14]. Copyright (2003) Taylor & Francis Ltd.

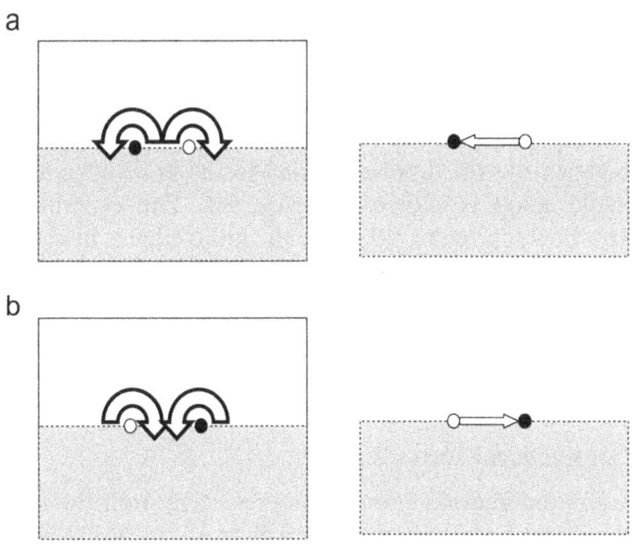


**Figure 5.7.** Schematics showing the orientation of the dipole with the knife-edge. Sense of rotation of the Poynting vector around each vortex is shown. Positive and negative vortices are shown in black and white respectively. The dipole pointing from negative to positive vortex is shown by white arrow. Reprinted from [15], with permission from Elsevier.

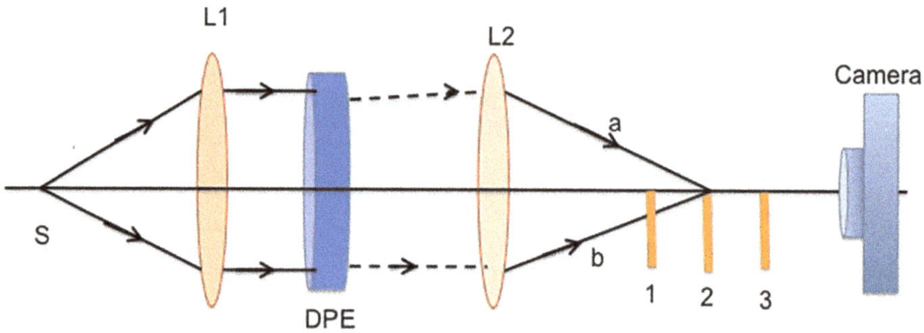


**Figure 5.8.** Schematic of the knife-edge experimental setup. L1, L2 lenses, SPP—spiral phase plate. When the knife-edge is at position 1, ray b is blocked and when it is at position 3, ray a is blocked.

axis are blocked. At these three positions, observation of the intensity distribution is made some distance after the focal plane. This gives an idea about the wavefront aberrations. The knife-edge partially blocks the optical rays from the test surface to form a shadow pattern that reveals the surface profile of the test surface. The observed shadow patterns are different depending upon the position of the inserted knife-edge in the beam and the test surface. The knife-edge can be inserted before, at, and after the image.

As shown in figure 5.11, by keeping the two lenses L1 and L2 aberration free, the test element, namely the spiral phase plate is inserted into the collimated beam. The dipole vortex beam coming out of the dipole producing element (DPE) is focused using the second lens and the knife-edge is introduced in one of the three positions marked as 1, 2 and 3 in the setup. The camera can record the shadow patterns in these three cases. Another modified setup is shown in reference [15].

The knife-edge test reveals that the orientation of the vortex dipole significantly modifies the intensity intrusions in the geometrical shadow region. There are two possible ways of orienting the dipoles parallel to the knife-edge as shown in figure 5.7. The schematic setup is shown in figure 5.8. The experimentally obtained shadowgrams are shown in figure 5.9. Here the knife-edge is inserted after the focal plane and the shadowgrams are presented. In the space between the dipoles in one orientation there is more flow of energy whereas in the other orientation the flow is outward from the space between the dipoles. This flow pattern is as explained in the schematic shown in figure 5.7.

### Knife-edge test on fractional vortex dipoles

During propagation the internal energy flows resulting from the azimuthal component of optical currents in beams carrying a pair of fractional vortices show four possible cases [17]. Fractional charge vortices in a beam pair-up with each other by means of a connection dark intensity line between the cores of the vortices. This pairing process occurs independent of the polarity of the vortices [18]. This is in contrast to the isolated intensity null points of vortices of integer charges and thus calls for a study on the internal energy flows in such beams.

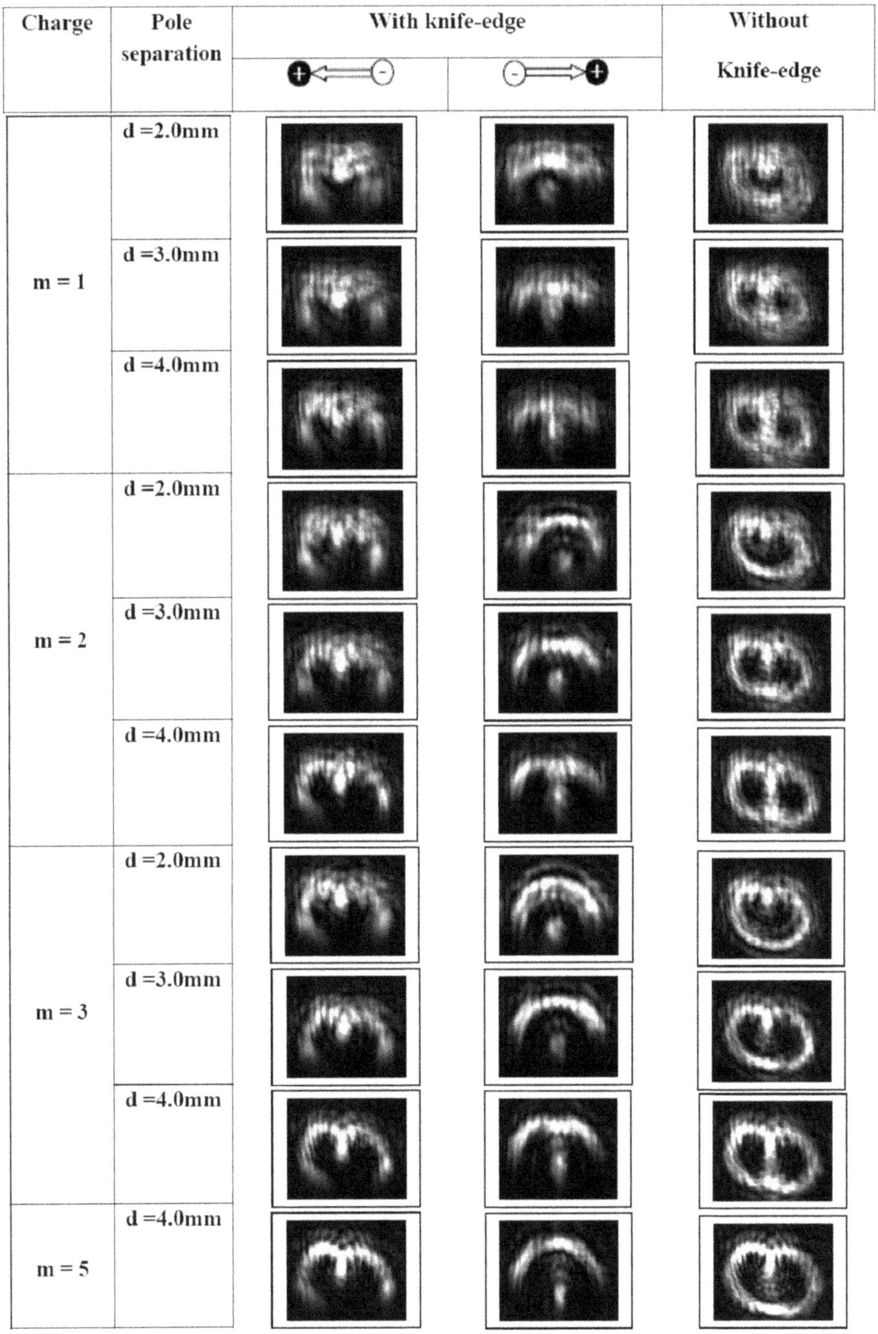


**Figure 5.9.** Experimental results: intensity profile when knife-edge is inserted into DVB. Results for when the dipole moment is increased are shown from top to bottom. The first two columns of patterns are for two different orientations of the dipole. The last column shows patterns without knife-edge. Reprinted from [15], with permission from Elsevier.

**Figure 5.10.** Schematics of the flow patterns for four configurations of fractional charge pairs. Unlike integer charge vortices, fractional charged vortices are connected by a dark intensity curve starting from one fraction vortex and ending at another vortex. Charges and flow directions are colored to distinguish the two types. The knife-edge is shown by the light blue horizontal strip.

In the schematic shown in figure 5.10, four possible current flow patterns arising due to pairs of vortices are shown. In the knife-edge experiment, after the introduction of the knife-edge, the transverse plane energy flow in such beams reveals itself by intensity intrusions in the geometric shadow region in a similar fashion as that of vortices of integer charges.

**Near-field diffraction through a slit**

The near-field diffraction patterns generated by a single slit illuminated by dark core beams such as Laguerre-Gauss and Mathieu beams are studied by Roland A Terborg *et al* [19]. The dark core is made to go through the center of the single slit. Because of donut structure of the beam, bright portions of the donut illuminate the slit only at two parts that are on either side of the vortex core. These two intensity lobes in the near field undergo diffraction as the beam propagates and rotation of these two lobes can be clearly seen. This allows quantification of the energy circulation and skew angle of the wavefronts in an optical vortex.

## 5.6 Energy circulations in diffraction patterns

The diffraction of a vortex beam through different types of apertures that were presented in an earlier chapter can also be subjected to HHD and the curl component that relates to orbital angular momentum can be studied. But there is not much work reported in the literature on the use of the HHD method on diffracted fields except the one [20] that is presented here.

Finally, before winding up this chapter, the diffraction patterns of singular beams going through a right angle isosceles triangular aperture were captured and the segregated curl part of the energy flows are shown in figure 5.11. On the intensity pattern, the circulating energy flow directions are plotted. It can be seen that when the vortex is of charge $m$ there are $m + 1$ slanted bright fringes are observed in the diffraction pattern. The curl part of the flow however shows that in each of the dark fringes there is a vortex. Further, there are also many circulations seen outside the straight fringe region.

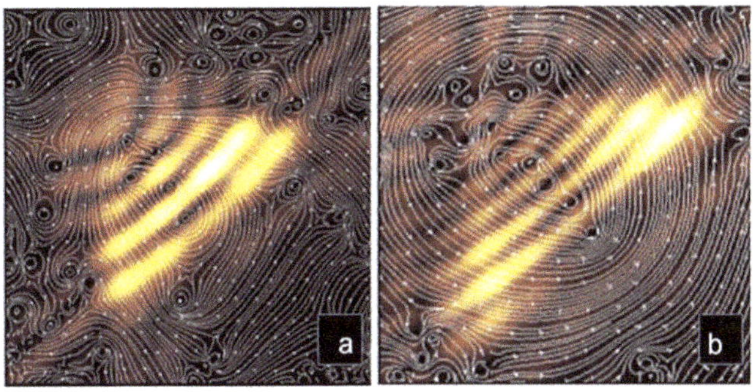


**Figure 5.11.** Hodge decomposed phase gradient fields depicting energy flows for singular beams with $m = 4$ and 6, when diffracted through a right triangular aperture. Reprinted with permission from [20]. Copyright (2015) by the American Physical Society.

# References

[1] Poynting J H 1909 The wave motion of a revolving shaft, and a suggestion as to the angular momentum in a beam of circularly polarised light *Proc. R. Soc. Lond. Ser.* A **82** 560–7

[2] Masajada J 2000 Half-plane diffraction in the case of gaussian beams containing an optical vortex *Opt. Commun.* **175** 289–94

[3] Beth R A 1936 Mechanical detection and measurement of the angular momentum of light *Phys. Rev.* **50** 115–27

[4] O'Neil A T, MacVicar I, Allen L and Padgett M J 2002 Intrinsic and extrinsic nature of the orbital angular momentum of a light beam *Phys. Rev. Lett.* **88** 053601

[5] Bekshaev A Y, Bliokh R Y and Soskin M S 2011 Internal flows and energy circulation in light beams *J. Opt.* **13** 053001

[6] Bahl M and Senthilkumaran P 2012 Helmholtz-Hodge decomposition of scalar optical fields *J. Opt. Soc. Am.* A **29** 2421–7

[7] Petronetto F, Paiva A, Lage M, Tavares G, Lopes H and Lewiner T 2010 Meshless Helmholtz-Hodge decomposition *IEEE Trans. Vis. Comput. Graph.* **16** 338–42

[8] Globus A, Levit C and Lasinki T 1991 A tool for visualizing the topology of three-dimensional vector fields *Proc. IEEE Visualization* pp 33–40

[9] Bahl M and Senthilkumaran P 2014 Focal plane internal energy flows of singular beams in astigmatically aberrated low numerical aperture systemss *J. Opt. Soc. Am.* A **31** 2046–54

[10] Petruccelli J C, Tian L and Barbastathis G 2013 The transport of intensity equation for optical path length recovery using partially coherent illumination *Opt. Express* **21** 14430–41

[11] Streibl N 1984 Phase imaging by the transport equation of intensity *Opt. Commun.* **49** 6–10

[12] Gureyev T and Nugent K 1997 Rapid quantitative phase imaging using the transport of intensity equation *Opt. Commun.* **133** 339–46

[13] Bahl M, Singh B K, Singh R K and Senthilkumaran P 2015 Internal energy flows of coma-affected singular beams in low-numerical-aperture systems *J. Opt. Soc. Am.* A **32** 514–21

[14] Arlt J 2003 Handedness and azimuthal energy flow of the optical vortex beams *J. Mod. Opt.* **50** 1573–80

[15] Singh B K, Bahl M, Mehta D S and Senthilkumaran P 2013 Study of internal energy flows in dipole vortex beams by knife edge test *Opt. Commun.* **293** 15–21

[16] Malacara D 1978 *Optical Shop Testing* (New York: Wiley)

[17] Singh B K, Mehta D S and Senthilkumaran P 2013 Visualization of internal energy flows in optical fields carrying a pair of fractional vortices *J. Mod. Opt.* **60** 1027–36
[18] Vyas S, Singh R K, Ghai D P and Senthilkumaran P 2012 Fresnel lens with embedded vortices *Int. J. Opt.* **2012** 249695
[19] Terborg R A and Volke-Sepulveda K 2013 Quantitative characterization of the energy circulation in helical beams by means of near-field diffraction *Opt. Express* **21** 3379–87
[20] Bahl M and Senthilkumaran P 2015 Energy circulations in singular beams diffracted through an isosceles right triangular aperture *Phys. Rev.* A **92** 1–6

# Chapter 6

# Vortices in computational optics

## 6.1 Introduction

In many of the iterative algorithms in wave optical engineering, the initial guess is a random field and this random field is subjected to conditions/constraints so that, during each iteration, the random field is modified and is driven to settle into a final solution that satisfies the required constraints to the best possible extent. This best possible extent is defined by a target level or acceptance level and once it is reached the iteration is stopped. If the solution to a given problem is not unique, the initial guess field undergoes modifications during each iteration and settles down to one of the many possible solutions to the problem. If the function does not converge to a solution and reaches a level beyond which any further iteration does not improve the situation then stagnation occurs. Random complex field distributions are used in encryption, holography, speckle metrology and in scattering experiments.

Speckle fields have random complex field distributions. Random speckle fields are known to contain optical vortices [1, 2] in the complex field distribution. Each of the dark regions of the speckle pattern may contain a phase vortex. Since in a vortex, high spatial frequency components are from the regions near the core, the presence of vortices poses problems in iterative procedure, because in many iterative procedures high frequency components may not survive. In this chapter the role of vortices in computational optics is discussed.

The topics that are presented in different sections are chosen in the following way. First the milestone development in holography namely, the use of diffused illumination is presented in section 6.2. This is called phase randomization in holography that improves certain attributes of the recorded holograms. But this comes with a speckle problem. Hence, refinement in the diffused illumination is needed and this is presented in section 6.3. In the next section, a hologram with diffused illumination in computer generated holography (CGH) is described. The iterative procedure used for making a hologram using CGH and the stagnation problem in the process due to optical vortices are presented in section 6.5. Solution

doi:10.1088/978-0-7503-1698-9ch6

to the speckle and stagnation problems in CGH phase synthesis is described in the next section. The role played by vortices in phase unwrapping, geometric transforms and in phase retrieval are presented in the subsequent sections.

## 6.2 Diffused illumination in holography

Use of diffused illumination in holography, which is one of the important milestones in holography was proposed in 1964 by Leith and Upatnieks [3]. Diffused light can be produced by passing a coherent laser beam through a diffuser made of ground glass. The intensity distribution of diffused light has a grainy structure of scattered light called speckles. Since diffused illumination has a wide range of randomly oriented propagation vectors, it increases the spreading of the object field to be recorded onto the holographic plate. Especially when Fourier transform holograms are recorded, most of the objects have strong zero spatial frequency components and this leads to the exposure problems such as saturation. But by mixing random phase with the object distribution this central peak is spread across the recording plane thereby homogenizing the field distribution that leads to better recording. This spreading of object information across the recording plane leads to improved redundancy in the hologram recording. This is called phase randomization in holograms.

Because of phase randomization, each point on the holographic plate receives information from all points of the object. Hence, such holograms, even when broken into pieces, retain the capability of reconstruction of the original object that was recorded. Each piece of the hologram would have received a stronger contribution from every part of the object and due to the redundant recording of object information, the object can be reconstructed even with a broken piece of the hologram. A hologram where the object is not subjected to phase randomization during recording shows feeble redundancy. To explain this phenomenon, two recorded holograms—one using diffused illumination and another using plane wave illumination of the object are considered. Both the holograms are recorded using off-axis plane wave as reference beam. During reconstruction, when both the holograms are illuminated by the same reference plane waves, the objects are reconstructed perfectly. But when a portion of the holograms are used in reconstruction instead of the whole recorded holographic plate, the quality of reconstruction in the two cases are conspicuously of different quality.

Phase randomization is suitable only for intensity objects, such as the page of a book, or a photograph. These are objects used for display. The phase of such an object is a free parameter. This means that for intensity objects, the phase distributions $\psi(x, y)$ are immaterial as all the phase distributions will produce the same intensity distribution.

$$I(x, y) = |a(x, y) \exp\{i\psi(x, y)\}|^2 = a^2(x, y) \tag{6.1}$$

This means that for display of an object it is important to produce only the intensity distribution $I(x, y)$ at the observation plane and the phase distribution corresponding to the object is unimportant. Because of this freedom in the phase distribution, diffused illumination is used to phase randomize the object field distribution. One of the problems of phase randomization is that during reconstruction, the object is infested with speckles that degrades the quality of the reconstructed image. This is

**Figure 6.1.** Object with different phase distributions were used to construct band-limited spectra by iterative method. From the spectra, normalized reconstructed intensities obtained when (a) a random phase, (b) a constant phase, and (c) an iterated phase were used to phase randomize the object. Reproduced with permission from [11]. Copyright (1998) by OSA.

depicted in figure 6.1. Random phase improves the redundancy in recording at the cost of poor image quality whereas constant phase has poor redundancy with good image quality. Hence, to achieve good phase randomization without the speckle problem in the reconstructed image, efforts were made to design diffusers.

## 6.3 Synthesized diffusers

In this section, the detrimental role played by optical phase singularities, and the improved performance of new diffusers are highlighted. The methods of diffuser synthesis is not elaborated here. The existence of many methods are mentioned to highlight the research activities on diffusers. During this period of time (prior to 1974), since the subject of singular optics had not occurred, a lot of effort went into the design of diffusers. Recall that the phase defects by Berry and Nye were introduced in 1974 whereas the subject of diffused illumination was introduced in the 1960s and some of the studies on diffusers were in later years due to advances in computational capabilities. Moreover the communities working on holography and singularity were not aware of each other and were working in isolation.

Efforts have been made to synthesize diffusers [4–7] that can be used for phase randomization in holography. A good diffuser is expected to spread the incident field distribution uniformly over a finite area. Normally, random phase distribution scatters the light in all directions so that the scattered field has all the spatial

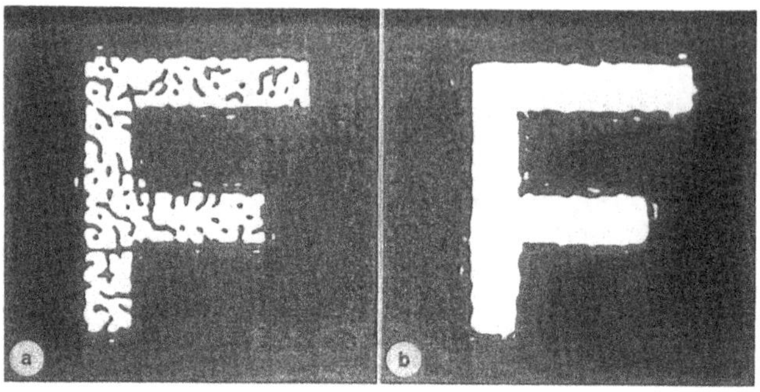


**Figure 6.2.** Optically generated diffraction patterns of digital holograms: These holograms were made with two types of random (diffused) illumination of the object. The reconstructed images when (a) a diffuser with a random phase was used; (b) an object dependent diffuser synthesized by iterative procedure was used. Reproduced with permission from [5]. Copyright (1991) by OSA.

frequency components. But what is desired is that there should be scattering, but the scattered light should be band-limited, or in other words, it should spread only over a finite region in the Fourier plane, or in the holographic plate which has finite size. Another aspect of scattered light is that it contains speckles. Speckle, which is a granular structure of the optical wavefront, affects both the hologram and the reconstruction and is a nuisance. Speckle free reconstruction [8] was also desired.

Efforts had been made to generate a new class of general purpose diffusers [7] that are suitable to use with any objects, unlike a particular diffuser for each object [9]. Figure 6.2 shows the difference between the reconstructed image of binary amplitude object when holograms are generated by random phase diffuser and synthesized object dependent diffusers. Using different constraints, diffusers with a series of pseudorandom phase sequences [6] were designed and a theoretical analysis of their power spectra was carried out. Analysis of pseudorandom phases with typical forms of the power spectrum of almost rectangular shape were achieved.

## 6.4 Phase synthesis in computer generated holograms

Mathematical computation is used either in the recording process or in the reconstruction process in computer generated holography. When the recording is done optically in the laboratory the reconstruction is done computationally and when the recording is done using data synthesized by computation, the reconstruction is done optically. Computer generated holograms are very useful to generate wavefronts that are difficult to realize by using conventional optical elements.

In Lohmann's detour holograms [10] the complex amplitude at the hologram plane is recorded by placing a large number of micro slits of various sizes at specific locations in the hologram plane such that the diffracted field from this detour hologram produces the desired object field in its first diffraction order. Each slit can occupy a certain position inside a cell. The length and width of each cell are decided by the sampling interval at the recording plane as decided by the sampling theorem. Each slit whose size encodes the amplitude and whose position within a cell encodes the phase of the complex field ensures the whole recording (amplitude and phase) of the information. Each cell is of

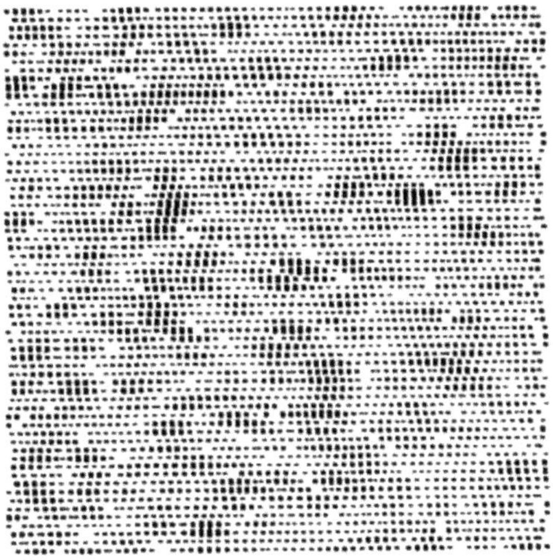

**Figure 6.3.** Lohmann–Paris binary amplitude detour phase hologram transmittance function. The object field is mixed with ground glass phase distribution and the transmittance function at the hologram plane is computed and used for hologram generation. Reproduced with permission from [10]. Copyright (1967) by OSA.

identical size and occupies a finite area, and their center distances are chosen according to the sampling interval required for recording the complex field, which is normally band-limited. In detour holograms there is no explicit use of a reference beam, but it is hidden there in the process of placing slits in each of the cells. This sampling interval decides the carrier frequency of the (hidden) reference beam and this decides the angle at which the reconstructed image is diffracted from the hologram during reconstruction.

In Fourier transform holograms, since the zero frequency component of the object is strongly peaked, in comparison to other frequency components, the number of slits placed in making a detour hologram, is very small and these slits are clustered only around the zero and low spatial frequency regions of the hologram plane, which results in very poor quality reconstruction. This is because the higher frequency components are weak, and hence to record the amplitude the size of the slit also becomes zero because of quantization. Therefore, there is a need to spread or homogenize the Fourier transform of the object field using random phase. In other words phase randomization is required. Hence iterative procedures such as GS algorithms are recommended to synthesize the complex field at the hologram plane [11, 12]. The size of the individual slits can be made identical, like in a phase hologram for which the full information of the object has to be contained in the phase distribution. Iterative procedures can help in the phase synthesis for making computer generated holograms. Phase randomization also leads to speckled reconstructed images. The Lohmann–Paris binary amplitude hologram and its reconstruction are shown in figures 6.3 and 6.4 respectively.

## 6.5 Stagnation problem in IFTA

Iterative Fourier transform algorithms (IFTA) due to Gerchberg and Saxton [13] are used to recover phase from intensity measurements, and are used extensively in

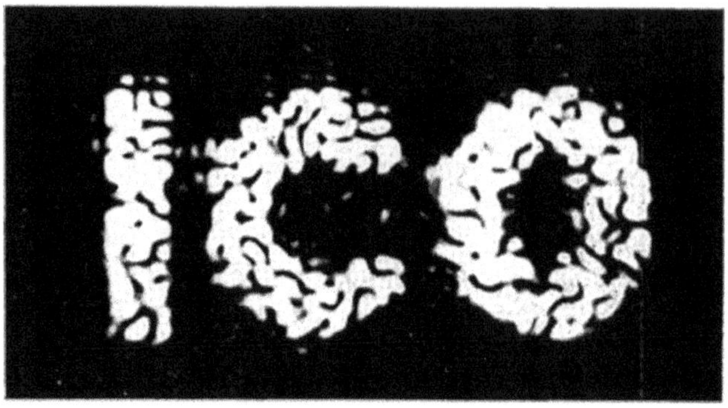

**Figure 6.4.** Reconstructed image from the hologram shown in figure 6.3. Reproduced with permission from [10]. Copyright (1967) by OSA.

computational optics. They are also used in electron microscopy, wavefront sensing, astronomy, crystallography, digital holography and in diffractive optics. The intensity distribution at the Fourier plane is available and the phase distribution in the object plane is to be found. These distributions are connected by Fourier transformation in relation with appropriate constraints on them.

For certain types of objects, constraints and initial phase distributions, the iterative algorithm fails to converge to a good solution and stagnates. A good solution satisfies the constraints in both the domains. One of the reasons attributed to this stagnation is optical phase singularities. These singularities have rotating phase gradients, and near the singularity, the gradient has very high magnitude. These are robust features and are not removable by further iterations.

Let us consider the IFTA procedure that computes a complex field distribution of an intensity object whose phase variation is such that it produces a Fourier transform (FT) that is band-limited and the intensity distribution in the FT plane is homogenized without the strong zero spatial frequency component [12, 14]. Such a Fourier transform also must have the property of redundancy as that of the one obtained by phase randomization. To start the iteration, from the intensity distribution of the object $I(x, y)$ the amplitude $A(x, y)$ is computed by $A(x, y) = \sqrt{I(x, y)}$. But $\mathcal{F}\{A(x, y)\}$ will have strong zero frequency and lacks redundancy. Hence, to improve redundancy and to homogenize the Fourier transform, random phase distribution $\psi_{ra}(x, y)$ is imposed on the object amplitude $A(x, y)$. This is safe as the object is an intensity object. But it also produces undesirable effects on the Fourier transform. $\mathcal{F}[A(x, y)\exp\{i\psi_{ra}(x, y)\}]$ spreads all across the frequency plane. To contain the FT distribution in a restricted region, for e.g., over an rectangular aperture region (onto a holographic plate of finite size), it is required that the spectra component outside the aperture should be made zero. Such a field will not correspond to the original object with which we started as we have now dropped a considerable portion of the Fourier transform. To put this simply, we want to have an intensity object whose field distribution has a spectrum which is restricted (band-limited) but should have homogenized field distribution in the Fourier plane. To achieve this, adding a random phase in the object domain causes the spectrum to spread out and any effort to restrict the spectrum by an aperture will not produce the same object intensity distribution if we do an inverse Fourier transform. We must ensure the amplitude $A(x, y)$ at the object plane and finite size of the spectrum at the Fourier plane. These two restrictions are called object and frequency domain constraints. In an iterative procedure the complex fields are sent back and forth between the object and FT planes and in each domain, the complex field is modified during each step of the iteration, i.e., in the FT domain the field is subjected to multiplication with a rectangular aperture and in the object domain the developed phase distribution is retained and only the amplitude distribution is replenished with $A(x, y)$. Basically one simply transforms the complex fields back and forth between the two domains, imposing the constraints in one before returning to the other. This way after a finite number of iterations, field distributions that satisfy both the object and Fourier domain constraints can be achieved.

The use of random phase as the initial choice in the IFTA, (after large number of iterations) leave behind isolated dark spots on the object field distribution

$$\tilde{A}_j(x, y) = A_j(x, y) \exp\{i\psi_j(x, y)\} \tag{6.2}$$

which is not a good solution to the problem stated, i.e., $|\tilde{A}_j(x, y)|^2 \neq I(x, y)$. A stagnated image with many dark spots on it is shown in figure 6.5. Increasing the number of iterations improves the signal to noise ratio very little and the program stagnates. This stagnation is related to the dark spots that are phase singularities. Considerable time and effort has been put into the investigation of the appearance of dark spots. The accumulated phase around a closed path about the dark spot is $\pm 2\pi$. To overcome the occurrence of dark spots and to avoid stagnation in IFTA, these phase jumps must be eliminated. Since these phase jumps are due to optical vortices any local modification to the phase distribution around the dark amplitude point will not eliminate the problem. A global restructuring of the phase distribution is required to eliminate this stagnation problem. These vortices are robust and are impossible to remove during iteration by any local phase manipulation.

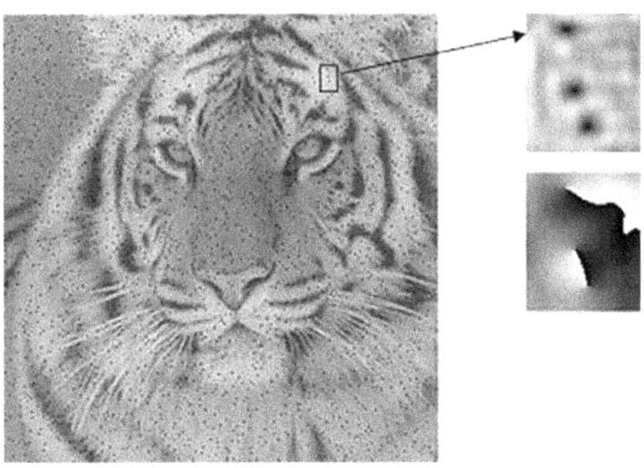

**Figure 6.5.** The stagnated image in which many dark spots can be seen. The insets show the magnified view of amplitude and phase corresponding to the dark spots. Amplitude zero and helical phase structures of these dark spots indicate that these are optical vortices. Reprinted from [16]. Copyright (2005), with permission from Elsevier.

## 6.6 Solution to the speckle problem

The speckle problem has been there for decades in the field of computer generated holography. These speckles are due to phase singularities, and they cannot be removed by any local phase modification in and around the singular point. Removal of a singular point by any local modification to phase does not help and during further iterations these dark spots reappear. To remove them globally changes to the phase distribution are needed. To remove a singularity from a complex field, it is multiplied with a field that contains a vortex whose charge should be opposite to the one that is being removed. Both the positive and negative vortex fields are conjugate to each other. In IFTA, the phase function obtained at stagnation is subjected to this vortex removal procedure. In this procedure, the charge and location of each vortex is identified and using a vortex of opposite charge this vortex is removed from the phase function, one by one. This will leave a vortex free wavefront whose resultant phase distribution is devoid of any circulating phase gradients.

During the initial stages of iteration, the number of vortices present in the random phase function is very large. When such a field is subjected to low pass filtering, the core size of each of the vortex becomes bigger and interaction between this vortex and the vortices in its neighborhood increases. Oppositely charged vortices tend to attract each other and annihilate. This self-annihilation process dominates in the initial stage [15–17] where the vortex density is high. During this self-annihilation process the signal to noise ratio increases and that dominates the initial stages of iterations. The program stagnates when this self-annihilation process is over. Once stagnation occurs, the complex field is subjected to a forced vortex annihilation process. In this process the complex field is spatially scanned for vortices and at each of the vortex location an oppositely charged vortex is introduced to destroy the vortex. This can eliminate the speckles completely from the phase distribution that is developed during iteration. Holograms made using this synthesized phase, produce speckle free images during reconstruction as shown in figure 6.6.

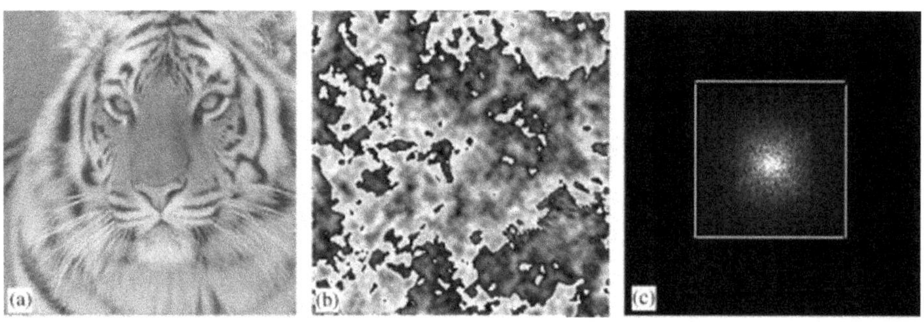


**Figure 6.6.** (a) Amplitude distribution of the vortex removed object field after iterations, (b) vortex removed phase distribution after iterations, (c) band-limited spectrum of the object satisfying the Fourier domain constraint. Reprinted from [16]. Copyright (2005), with permission from Elsevier.

## 6.7 Phase unwrapping in the presence of vortices

In computational optics the phase distributions are presented always in the phase interval $-\pi \leqslant \psi \leqslant \pi$ whereas phase values are in general not restricted to this range. In the actual phase distribution, a modulo $2\pi$ operation is carried out and the remainder is depicted in the phase distributions. Such a reduced phase distribution is known as a wrapped phase distribution. By unwrapping this phase distribution, a single wavefront can be constructed. In this phase distribution there will be no $2\pi$ phase discontinuity. This operation is called phase unwrapping.

According to a fundamental theorem of calculus, the value of a function $f(t)$ between any two points can be obtained by evaluating the integral

$$f(b) - f(a) = \int_a^b \frac{df}{dt} dt \tag{6.3}$$

The first term in the integrand $\frac{df}{dt}$ is the slope and $dt$ is small interval in $t$. The term $\frac{df}{dt}dt$ actually gives the incremental change in $f$ when the variable $t$ is increased by $dt$. Including the integration also in the interpretation, it is the total increment in $f$ when $t$ is varied from $a$ to $b$ which is obtained by adding small increments in $f$. For functions of two variables, like in the phase distributions $\psi(x, y)$, the phase values are given by

$$\psi(b) - \psi(a) = \int_a^b \nabla\psi \cdot dl \tag{6.4}$$

Hence the phase at $b$ is given by $\psi(b) = \int_a^b \nabla\psi \cdot dl + \psi(a)$. The path integral from $a$ to $b$ is path independent only when the phase distribution is devoid of any curl component of phase gradient. In other words when $\nabla \times \nabla\psi = 0$ the integral is path independent. When $\nabla \times \nabla\psi \neq 0$ the integral becomes path dependent. Hence phase unwrapping is tantamount to integrating the gradients.

In phase unwrapping jargon, the presence of rotating curl components of phase gradient and path dependence are due to residues. These residues are actually phase residues that are different from the residues that occur in complex variable theory. Actually these phase residues are phase singularities in the complex field distribution. Ghiglia *et al* [18] noticed these phase inconsistencies in two-dimensional phase unwrapping in 1987. These phase singularities were named residues by Goldstein, Zebker and Werner [19] in 1988. Huntley *et al* called them sources of discontinuity [20].

### 6.7.1 Residue theorem

There is a close analogy between the residues in complex variables and the phase residues that are encountered in the phase unwrapping problem. The residue theorem [21] helps in evaluating the residues and is given by

$$\oint f(z)dz = 2\pi i \times S_c \tag{6.5}$$

where $z$ is the complex variable $z = x + iy$, symbol $S_c$ is the sum of residues enclosed by the contour (integration path) and $i$ is the fourth root of unity. In the Laurent series expansion [22] of $f(z)$ about the various singular points $z_j$, the residues appear as coefficients of $(z - z_j)^{-1}$.

**Phase residue theorem**: The closed path integral around a phase residue equals the integer multiple of $2\pi$

$$\oint \nabla\psi(r) \cdot dr = 2\pi \times S_p \tag{6.6}$$

where $S_p$ is the sum of the enclosed phase residues inside the contour. This definition is exactly the same as the one used in defining the charge of the vortex equation (2.5). The topological charge of a vortex can be a positive or negative integer and in a similar way the phase residue can be positive or negative. One can clearly see the analogy between the phase unwrapping and singular optics theories where the underlying concepts are the same but the two communities work in isolation.

When the phase residue charges are balanced in a region, which means that inside the closed path of the integral, where there are an equal number of positive and negative residues, the line integral is zero and by Stokes theorem converting the line integral to surface integral

$$\oint \nabla\psi(r) \cdot dr = \iint (\nabla \times \nabla\psi(r)) \cdot \overrightarrow{da} = 0 \tag{6.7}$$

which means that the phase gradient field becomes curl-less and hence the vector field is irrotational or conservative. The path dependency of the integral also vanishes. In equation (6.7), $\overrightarrow{da}$ is the area vector.

This balancing of the phase residue charges can be carried out by first identifying the presence of residues, and an opposite charged residue can be introduced inside the closed loop by numerical methods. Identification of residues can be done by going point by point on the field distribution and evaluating the closed loop integral each time for non-zero accumulated phase as is done in singular optics [15–17] or by using the Helmholtz–Hodge decomposition method [23] explained earlier in this book.

## 6.8 Non-Bryngdahl transforms using branch points

**Geometric phase transforms**: Geometric phase transforms are general types of map transforms [24] that can be used to modify the image. These modifications can be of map transforms such as local translation, inversion, reflection, stretching, and rotation. This requires highly space-variant systems.

A conventional space-invariant imaging system can be converted to a space-variant system by adding one refractive or reflective element such as fun-house mirrors or aberrating media into the system to distort the field. Computer generated holograms can also be designed to perform such geometric phase transforms [25].

**Branch points**: Branch-point phase singularities in optical waves was predicted [26] and observed [27] in laser modes. Branch-point phase singularities are nothing but phase singularities. A branch cut in the lines of constant phase arises from the presence of a wavefront dislocation in the region [28]. They have helical wavefront, and the singularity is present only at a point. They have been used in diffractive optics [29] to perform non-Bryngdahl geometric transformations. Vortices can be detected using numerical means [30] and the other methods of detection were presented in chapter 3.

**Bryngdahl and non-Bryngdahl transforms**: The phase function of the computer generation hologram that performs the geometric phase transformations can be obtained from a set of differential equations given below

$$\begin{aligned} \frac{\partial \psi(x, y)}{\partial x} &= \frac{k}{f} u(x, y) \\ \frac{\partial \psi(x, y)}{\partial y} &= \frac{k}{f} v(x, y) \end{aligned} \tag{6.8}$$

where $x$, $y$ are input plane coordinates and $u$, $v$ are output plane coordinates, $\psi(x, y)$ is the phase function of the computer generation hologram, $k$ is the wavenumber and $f$ is the focal length of the lens. $u(x, y)$ and $v(x, y)$ are the transformation equations of the geometrical transform. Bryngdahl used the stationary phase method [31] to derive these equations. In the Fourier configuration, the input and output planes are, respectively, the front and back focal planes of a Fourier-transforming lens. This is a typical $4f$ Fourier filtering setup in which the filters are introduced in the Fourier plane and the filtered images are obtained at the image plane. These equations (6.8) are called Bryngdahl differential equations [29] and can be written in compact form as $\nabla\psi = 2\pi F = \frac{k}{f}\vec{U}$, where $\vec{U} = u\hat{x} + v\hat{y}$. This equation shows that in a Fourier configuration the output position vector is proportional to the spatial frequency of the input plane.

Bryngdahl transformation equations obey the continuity condition

$$\frac{\partial u}{\partial y} = \frac{\partial v}{\partial x} \tag{6.9}$$

which means that the phase function is continuous and differentiable. One such transformation is shown in figure 6.7 where the object is distorted using a filter that does not contain any phase singularity. Transformations that do not obey this continuity condition are called non-Bryngdahl transforms. One of the simplest non-Bryngdahl transforms is the rotation transform [32] and another example is the ring-to point transform [25]. These transforms use phase singularities that are called branch points.

## 6.9 Diffraction of singular beams

Diffraction of singular beams through different types of apertures was presented in chapter 3, in the context of vortex detection mechanisms. Vortices of higher charge disintegrate in unit charge vortices and their placement in the diffraction pattern depends on the polarity of the singularities. A comparison between the diffraction of

**Figure 6.7.** Frequency plane mask for performing Bryngdahl transforms, and the image that underwent (Bryngdahl type) transform using the filter.

singular and non-singular beams through an aperture shows that the diffraction patterns are significantly different.

**Non-redundant information from vortex diffraction patterns**

In phase retrieval the phase distribution of a complex field is extracted from intensity distribution(s). Depending on the method, one or many intensity measurements are made and from the measured data, algorithms are designed to extract the unknown phase of the complex field. Hence, it is natural to think that each of the intensity measurement should provide different information that helps in the recovery process. If the information offered from many of the measurements provide little or no new information, it is obvious that some more intensity measurements may be required. This may be the case when defocusing of the system is employed between measurements. From this point of view, capturing two intensity measurements—one with a phase singularity introduced in the field and the other without a phase singularity, offer completely different diffraction patterns and hence they are sure to contain a good amount of non-redundant information. This is one of the benefits of using optical phase singularities in the phase retrieval problem.

## 6.10 Phase retrieval

Phase measurement is an important problem in many areas of optics, e.g., microscopy, imaging, wavefront sensing, imaging through turbulence, metrology, etc. Phase measurement is carried out using optical interferometry where the unknown phase is detected using a reference wave whose phase distribution is known. There are also many non-interferometric approaches based on single or multiple intensity measurements. All non-interferometric phase measurement systems require iterative numerical solution for phase estimation.

The iterative algorithm for phase retrieval from single intensity measurement and object constraints is based on the Gerchberg and Saxton method [13]. There are also non-interferometric methods based on transport of intensity equation (TIE) [33–35], where two or more longitudinally shifted intensity measurements are used for phase estimation. Two or more intensity measurements in the Fresnel zone are also used in iterative schemes [36, 37] for phase reconstruction. Intensity measurements up to 20 are also suggested for robust recovery. Defocus diversity near the focal plane has also been used where two defocus measurements are typically used [38] for complex field construction. When a limited number of intensity measurements are made, it would be better that each of the intensity measurements provide different information that would be helpful in phase reconstruction. This means that the acquisition of non-redundant information from multiple intensity measurement is useful. From the point of view of providing non-redundant information, multiple intensity distributions obtained from longitudinally shifted planes will not differ from each other drastically. Around the focus the curvature of the wavefront flips its sign in going through the exact focus, and hence can offer better non-redundant information.

The diffraction patterns obtained with and without a phase vortex plate placed at the diffracting apertures are known to produce completely different intensity distributions. This non-redundant information between the two intensity patterns is very useful in phase recovery and this has been successfully demonstrated [39]. Typically, it is desirable that each intensity measurement should offer diverse information content that is useful in the phase recovery. But there is a close spatial similarity between two consecutive defocused images and hence defocus diversity may require more images to be captured and used in iteration.

For an unknown field, $\exp(i\psi(x, y))$ is to be estimated, non-redundant information can be obtained through intensity measurements from the quadrature components $\cos(\psi(x, y))$ and $\sin(\psi(x, y))$. The definition of a quadrature relation in two dimensions is however not a trivial problem. The quadrature transform in one dimension is the Hilbert transform corresponding to the Fourier domain filter denoted by a step phase function $\mathrm{sgn}(x, y)$. This step phase filter gives $\psi$ phase shift to negative spatial frequency components (say $f_x$) and leaves the positive frequency components unchanged. Such an operation in the frequency domain results in quadrature transform in the space domain. The spiral or vortex phase filter is an appropriate generalization of the Hilbert transform to two dimensions [40, 41]. The charge-one spiral-phase filter is a generalization of the one-dimensional Hilbert transform in the sense of Mandel's theorem [42, 43] and provides the most efficient envelope-carrier representation. The intensity pattern corresponding to the spiral-phase transform is the Fraunhofer diffraction pattern of an object that is illuminated by a vortex beam.

Consider a complex object wavefront denoted by $g(r, \theta)$ is incident on the aperture $C(r, \theta)$ of a thin convex lens. At the back focal plane two intensity measurements are made. Two intensity measurements on the diffraction pattern obtained—one with a plane wave and the other with vortex-beam illumination for the phase retrieval problem.

$$\begin{aligned} I_1(\rho, \phi) &= |\mathcal{F}\{g(r, \theta)C(r, \theta)\}|^2 \\ I_2(\rho, \phi) &= |\mathcal{F}\{g(r, \theta)C(r, \theta) \exp(i\theta)\}|^2 \end{aligned} \tag{6.10}$$

where $\mathcal{F}$ denotes Fourier transform. The first and second intensity measurements are the Fourier transform of the object field through the aperture without and with the spiral-phase mask at the aperture plane respectively. Let $A_1(\rho, \phi)$ and $A_2(\rho, \phi)$

denote the complex amplitudes corresponding to these two intensity distributions $I_1(\rho, \phi)$ and $I_2(\rho, \phi)$ respectively and the complex amplitudes may be represented as

$$\begin{aligned} A_1(\rho, \phi) &= \sqrt{I_1(\rho, \phi)}\exp\,(i\psi_1(\rho, \phi)) \\ A_2(\rho, \phi) &= \sqrt{I_2(\rho, \phi)}\exp\,(i\psi_2(\rho, \phi)) \end{aligned} \tag{6.11}$$

where $\psi_1(\rho, \phi)$ and $\psi_2(\rho, \phi)$ are the unknown phase functions at the detector plane. The knowledge of the two intensity measurements in equation (6.10) along with the aperture function $C(r, \theta)$ constraint may be used to estimate the unknown complex function $g(r, \theta)$ by an iterative algorithm similar to GS algorithm [13].

The basic iteration flow chart is shown schematically in figure 6.8. The iteration may be explained briefly in the following steps.

1. We initiate the Fourier plane field for the first measurement with a uniform random phase function with phase distributed in $[0, 2\pi]$. The amplitude of the field is taken to be the square root of the intensity that was recorded without the spiral-phase mask at the aperture.
2. The corresponding field is inverse Fourier transformed and the lens aperture constraint is applied. This means that the inverse Fourier transformed field due to the addition of random phase at the recording plane tends to spread. At the lens aperture this field is multiplied with $C(r, \theta)$ which is the aperture constraint.
3. The resultant field is multiplied by a spiral-phase function $\exp(i\theta)$ and Fourier transformed to get an estimate of the field in the plane of the second

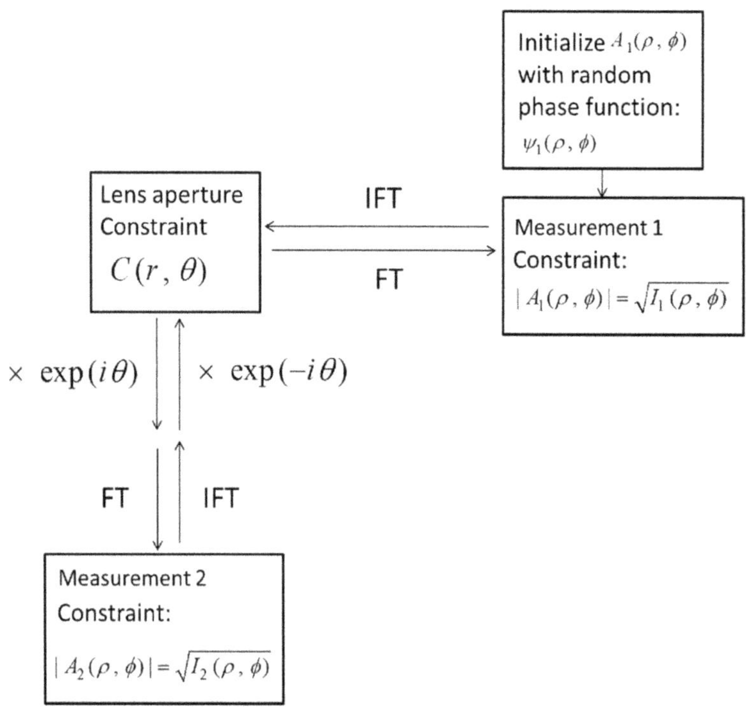


**Figure 6.8.** Block diagram of iteration used for phase retrieval with spiral-phase diversity. Reproduced with permission from [39], copyright (2015) by OSA.

measurement. This means that after spatially limiting the function by the aperture, the diffraction pattern of this function with a spiral phase introduced at the aperture is obtained.

4. The amplitude constraint for the second measurement is applied leaving the phase of the field unchanged. This means that from the diffraction intensity the amplitude is computed by taking the square root of intensity, while the phase that has developed so far is retained as such.
5. The estimated field in the second plane of measurement is then inverse Fourier transformed, multiplied by conjugate spiral-phase function $\exp(-i\theta)$, and once again the lens aperture constraint is applied. Use of conjugate spiral function is tantamount to making the clear aperture without spiral-phase function for the next intensity measurement at the detector plane.
6. The above procedure is continued back and forth between the two measurement planes until a stable phase solution is obtained.

In the simulations the spiral-phase function is defined such that the pixel representing the singular point has unit amplitude and zero phase. The zero amplitude and unknown phase at the singular point is however developed automatically during the back and forth propagation between the aperture and measurement planes.

A collimated laser beam illuminates the object. Two Fraunhofer diffraction patterns of the object were captured at the focal plane of a lens—one pattern without and the other pattern with a spiral-phase mask introduced near the object plane as shown in figure 6.9. From these two diffraction patterns, using the algorithm described above, the phase is recovered.

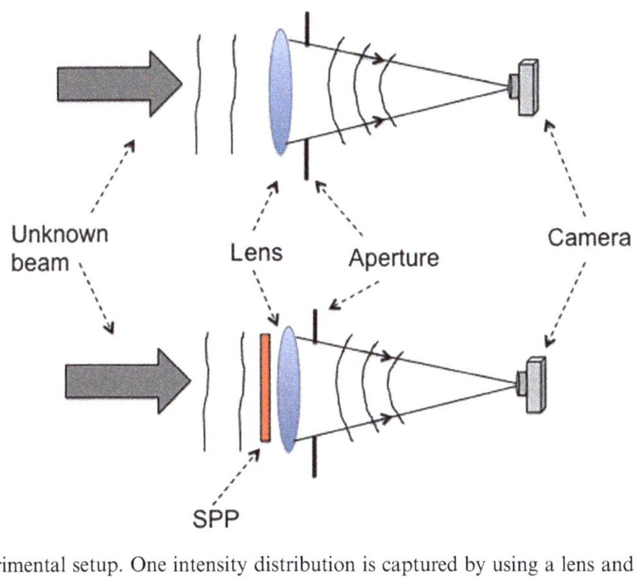


**Figure 6.9.** Experimental setup. One intensity distribution is captured by using a lens and aperture (top) and the other intensity distribution is captured with a spiral-phase plate introduced in the setup (bottom). These two intensity pattern have the diversity for good recovery of unknown field through the algorithm.

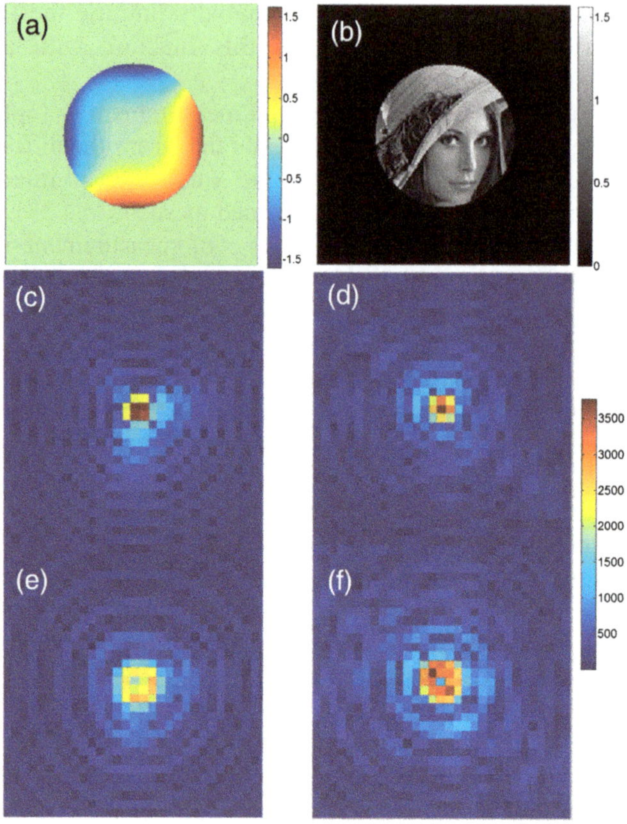


**Figure 6.10.** Simulated data for phase imaging using spiral-phase diversity. (a) Cubic-phase object, (b) Lena phase object. Both images have 256 × 256 pixels and the circular mask of radius 70 pixels mimics a lens aperture. (c), (d) Central portion of Fourier transform intensity patterns $I_1(\rho, \phi)$ corresponding to the two phase objects. (e), (f) Central portion of Fourier transform intensity patterns $I_2(\rho, \phi)$ with spiral-phase mask in lens aperture. Reproduced with permission from [39], copyright (2015) by OSA.

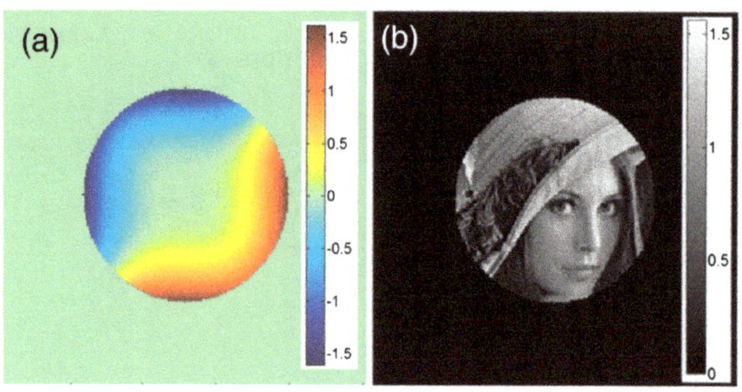


**Figure 6.11.** (a), (b) Recoveries of phase objects using 200 iterations of the spiral-phase diversity method as shown in figure 6.8 applied to the intensity data in figure 6.10. Reproduced with permission from [39], copyright (2015) by OSA.

Two phase objects one consists of cubic-phase variation and the other, a Lena phase object were used to test the phase retrieval scheme described here (figure 6.10). The recovered phase and the input phase were found to be in good agreement with each other indicating the usefulness of phase singularity in phase recovery problems as shown in figure 6.11.

## References


[1] Baranova N B, Zel'dovich B Y, Mamaev A V, Pilipetski N F and Shkunov V V 1981 Dislocations of the wavefront of a speckle-inhomogeneous field *JETP Lett.* **33** 195–9

[2] Freund I, Shvartsman N and Freilkher V 1993 Optical dislocation network in highly random media *Opt. Commun.* **101** 247–64

[3] Leith E N and Upatnieks J 1964 Wavefront reconstruction with diffused illumination and three-dimensional objects *J. Soc. Am.* **54** 1295–301

[4] Bräuer R, Wojak U, Wyrowski F and Bryngdahl O 1991 Digital diffusers for optical holography *Opt. Lett.* **16** 1427–9

[5] Bräuer R, Wyrowski F and Bryngdahl O 1991 Diffusers in digital holography *J. Opt. Soc. Am.* A **8** 572–8

[6] Nakayama Y and Kato M 1979 Diffuser with pseudorandom phase sequence *J. Opt. Soc. Am.* **69** 1367–72

[7] Dallas W J 1973 Deterministic diffusers for holography *Appl. Opt.* **12** 1179–87

[8] Wyrowski F and Bryngdahl O 1989 Speckle-free reconstruction in digital holography *J. Opt. Soc. Am.* A **6** 1171–4

[9] Hirsch P M, Jordan J A and Lesem L B 1971 Method of making an object-dependent diffuser *U.S.Patent* 3,619,022

[10] Lohmann A W and Paris D P 1967 Binary Fraunhofer holograms, generated by computer *Appl. Opt.* **6** 1739–48

[11] Wyrowski F and Bryngdahl O 1988 Iterative Fourier-transform algorithm applied to computer holography *J. Opt. Soc. Am.* A **5** 1058–65

[12] Aagedal H, Schmid M, Teiwes S, Beth T and Wyrowski F 1996 Theory of speckles in diffractive optics and its application to beam shaping *J. Mod. Opt.* **43** 1409–21

[13] Gerchberg R and Saxton W 1972 A practical algorithm for the determination of phase from image and diffraction plane pictures *Optik* **35** 237–46

[14] Fienup J R and Wackerman G C 1986 Phase retrieval stagnation problems and solutions *J. Opt. Soc. Am.* A **3** 1897–907

[15] Aoki T, Sotomaru T, Ozawa T, Komiyama T, Miyamoto Y and Takeda M 1998 Two dimensional phase unwrapping by direct elimination of rotational vector fields from phase gradients obtained by heterodyne techniques *Opt. Rev.* **5** 374–9

[16] Senthilkumaran P, Wyrowski F and Schimmel H 2005 Vortex stagnation problem in iterative Fourier transform algorithms *Opt. Laser Eng.* **43** 43–56

[17] Senthilkumaran P and Wyrowski F 2002 Phase synthesis in wave optical engineering: mapping and diffuser type approaches *J. Mod. Opt.* **49** 1831–50

[18] Ghiglia D C, Mastin G A and Romero L A 1987 Cellular automata method for phase unwrapping *J. Opt. Soc. Am.* **4** 267–80

[19] Goldstein R M, Zebker H A and Werner C L 1988 Satellite radar interferometry: Two dimensional phase unwrapping *Radio Sci.* **23** 713–20

[20] Huntley J M and Buckland J R 1995 Characterization of $2\pi$ phase discontinuity in speckle interferograms *J. Opt. Soc. Am.* A **12** 1990–6

[21] Ghiglia D C and Pritt M D 1998 *Two-dimensional Phase Unwrapping: Theory, algorithms and Software* (New York: Wiley)

[22] Brown J W and Churchill R V 1996 *Complex Variables and Applications* (New York: McGraw Hill)

[23] Bahl M and Senthilkumaran P 2012 Helmholtz-Hodge decomposition of scalar optical fields *J. Opt. Soc. Am.* A **29** 2421–7

[24] Bryngdahl O 1974 Geometrical transformations in optics *J. Opt. Soc. Am.* **64** 1092–9

[25] Cederquist J and Tai A M 1984 Computer-generated holograms for geometrical transformations *Appl. Opt.* **23** 3099–104

[26] Lugiato L A, Oldano C and Narducci L M 1988 Cooperative frequency locking and stationary spatial structures in lasers *J. Opt. Soc. Am.* B **5** 879–88

[27] Brambilla M, Battipede F, Lugiato L A, Penna V, Prati F, Tamm C and Weiss C O 1991 Transverse laser patterns. I. phase singualrity crystals *Phys. Rev.* A **43** 5090–113

[28] Scivier M S and Fiddy M A 1985 Phase ambiguities and the zeros of multidimensional band-limited functions *J. Opt. Soc. Am.* A **2** 693–7

[29] Roux F S 1994 Branch point diffractive optics *J. Opt. Soc. Am.* A **11** 2236–43

[30] Mobashery A, Hajimahmoodzadeh M and Fallah H R 2015 Detection and characterization of an optical vortex by the branch point potential method: analytical and simulation results *Appl. Opt.* **54** 4732–9

[31] Born M and Wolf E 1980 *Principles of Optics* (New York: Pergamon)

[32] Roux F S 1993 Diffractive optical implementation of rotation transform performed by using phase singularities *Appl. Opt.* **32** 3715–9

[33] Fienup J R 1982 Phase retrieval algorithms: a comparison *Appl. Opt.* **21** 2758–69

[34] Paganin D and Nugent K A 1988 Noninterferometric phase imaging with partially coherent light *Phys. Rev. Lett.* **80** 2586–9

[35] Dorrer C and Zuegel J D 2007 Optical testing using the transport-of-intensity equation *Opt. Express* **15** 7165–75

[36] Anand A, Pedrini G, Osten W and Almoro P 2007 Wavefront sensing with random amplitude mask and phase retrieval *Opt. Lett.* **32** 1584–6

[37] Almoro P, Pedrini G and Osten W 2006 Complete wavefront reconstruction using sequential intensity measurements of a volume speckle field *Appl. Opt.* **45** 8596–605

[38] Diaz-Douton F, Pujol J, Arjona M and Luque S O 2006 Curvature sensor for ocular wavefront measurement *Opt. Lett.* **31** 2245–7

[39] Sharma M K, Gaur C, Senthilkumaran P and Khare K 2015 Phase imaging using spiral-phase diversity *Appl. Opt.* **54** 3979–85

[40] Larkin K G, Bone D J and Oldfield M A 2001 Natural demodulation of two-dimensional fringe patterns. I general background of the spiral phase quadrature transform *J. Opt. Soc. Am.* A **18** 1862–70

[41] Larkin K G 2001 Natural demodulation of two-dimensional fringe patterns. II stationary phase analysis of the spiral phase quadrature transform *J. Opt. Soc. Am.* A **18** 1871–81

[42] Mandel L 1967 Complex representation of optical fields in coherence theory *J. Opt. Soc. Am.* **57** 613–7

[43] Khare K 2008 Complex signal representation, Mandel's theorem and spiral phase quadrature transform *Appl. Opt.* **47** E8–12

**IOP** Publishing

Singularities in Physics and Engineering
Properties, methods, and applications
**Paramasivam Senthilkumaran**

# Chapter 7

# Angular momentum of light

## 7.1 Introduction

Electromagnetic waves carry energy and momentum. The energy flux is given by the Poynting vector $\overrightarrow{S}$ and the momentum density is given by $\overrightarrow{\mathscr{P}} = \epsilon_0\mu_0\overrightarrow{S}$. Johannes Kepler used the concept of radiation pressure way back in the seventeenth century, to explain why the tail of a comet around the Sun's orbit always points away from the Sun. This radiation pressure is due to the momentum carried by light and it can exert a pressure on any object exposed to light. Maxwell published a paper on radiation pressure in 1862 and it was experimentally demonstrated by Lebedev in 1900.

## 7.2 Linear momentum

The linear momentum carried by electromagnetic waves is given by $\overrightarrow{p} = \hbar\overrightarrow{k} = \frac{E}{c}\hat{k}$ for a linearly polarized light. This is the momentum per photon. Here $E$ is the energy per photon, $c$ speed of light and $\overrightarrow{k}$ is the propagation vector. The linear momentum density $\overrightarrow{\mathscr{P}}$ averaged over a cycle, of a light beam can be calculated from its electric and magnetic fields

$$\overrightarrow{\mathscr{P}} = \epsilon_0\langle\overrightarrow{E} \times \overrightarrow{B}\rangle \tag{7.1}$$

where $\overrightarrow{E}$ and $\overrightarrow{B}$ are electric and magnetic fields respectively and $\epsilon_0$ is the permittivity of free space.

Consider a monochromatic plane wave traveling in the $z$ direction. The electric and magnetic fields of the wave are

$$\begin{aligned}\overrightarrow{E} &= E_0\cos(kz - \omega t)\hat{x}\\ \overrightarrow{B} &= \frac{E_0}{c}\cos(kz - \omega t)\hat{y}\end{aligned} \tag{7.2}$$

 

Hence the momentum density is given by

$$\vec{\mathscr{P}} = \frac{1}{c}\epsilon_0 E_0{}^2 \cos^2(kz - \omega t)\hat{z} = \frac{1}{c}\epsilon_0 E^2 \hat{z} \tag{7.3}$$

The energy per unit volume $\mathscr{E}$ stored in EM fields is

$$\frac{1}{2}\left(\epsilon_0 E^2 + \frac{1}{\mu_0}B^2\right) = \epsilon_0 E^2 = \mathscr{E} \tag{7.4}$$

## 7.3 Angular momentum

Poynting using the analogy with the wave motion predicted that the circularly polarized photons carry $\hbar$ units of angular momentum [1]. A circularly polarized light passing through a half wave plate can transfer its angular momentum to the plate to become a circularly polarized light with opposite handedness and this momentum transfer was detected by Beth in his experiments with suspended quarter wave plate [2]. This angular momentum carried by circularly polarized light is called spin angular momentum (SAM) of light. Apart from that, the angular momentum carried by the phase of the light is called orbital angular momentum (OAM).

In 1992, Allen *et al* recognized that light beams with an azimuthal phase dependence of $\exp(il\phi)$ carry an orbital angular momentum (OAM) that can be many times greater than the spin angular momentum, where $\phi$ is the azimuthal coordinate in the beam's cross section, and $l$ can take any integer value, positive or negative [3, 4]. The magnitude of orbital angular momentum carried by the beam is related to the topological charge of the beam and is $\pm l\hbar$ per photon. The key point was that this OAM was a natural property of all helically phased beams, and hence could be readily realizable in the laboratory. There is also great interest in these types of beams with OAM due to diverse possible applications [5–9]. Orbital angular momentum arises whenever a beam's phase fronts are not perpendicular to the general propagation direction and hence transverse components of propagation vectors are present. Termed as internal flows, the angular momentum and internal energy carried by such beams were described in chapter 5. The most common form of helically phased beam is the so-called Laguerre–Gaussian (LG) laser hybrid mode. Other beams with helical wavefronts such as Bessel beams [10], Mathieu beams [11], and Ince–Gaussian beams [12] can also carry orbital angular momentum. In chapter 9, it is shown that plane wavefronts with characteristic inhomogeneous polarization distribution can also carry orbital angular momentum. In the same way that the suspended half-wave plate acquires an angular momentum of $2\hbar$ when a right circularly polarized photon passes through it (figure 7.1), the cylindrical mode converter can acquire an angular momentum of $2l\hbar$ when a photon with orbital angular momentum $l\hbar$ passes through it.

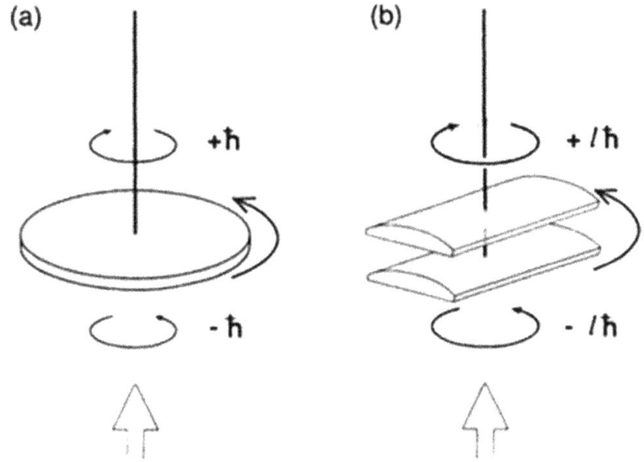


**Figure 7.1.** (a) A suspended half-wave plate undergoes torque in transforming right-handed into left-handed circularly polarized light, Beth's experiment. (b) Suspended cylindrical lenses undergo torque in transforming a Laguerre–Gaussian mode of orbital angular momentum $-l\hbar$ per photon, into one with $+l\hbar$ per photon [3]. Reprinted with permission from [3]. Copyright (1992) by the American Physical Society.

The local wavefront normal indicates the direction of energy and momentum flow in the fields. Since the vortex beam has helical shaped wavefronts the local normal to the wavefront also spirals as the beam advances. This gives rise to transverse components of propagation vector distribution which has non-zero curl. This suggests that the singular beam carries OAM. The photon trajectories for a Bessel beam with $l = 2$ are shown in figure 7.2. These trajectories are reconstructed from measurements of the transverse Poynting-vector distribution, with a constant longitudinal component added.

The angular momentum density, $\vec{\mathcal{J}}$, of a light beam can be calculated from

$$\vec{\mathcal{J}} = \epsilon_0(\vec{r} \times \langle \vec{E} \times \vec{B} \rangle) = \vec{r} \times \vec{\mathcal{P}} \tag{7.5}$$

This equation encompasses both the spin and orbital angular momentum density of a light beam. The local value of the linear momentum density [16] under paraxial approximation is given by

$$\vec{\mathcal{P}} = i\omega\frac{\epsilon_0}{2}(u^*\nabla u - u\nabla u^*) + \omega k\epsilon_0|u|^2\hat{z} + \omega\sigma\frac{\epsilon_0}{2}\frac{\partial|u|^2}{\partial r}\hat{\phi} \tag{7.6}$$

where $u = u(r, \phi, z)$ is the complex scalar function describing the distribution of the field amplitude. Here $\sigma$ describes the degree of polarization of the light; $\sigma = \pm 1$ for right- and left-hand circularly polarized light, respectively, and $\sigma = 0$ for linearly polarized light. The cross product of this linear momentum density with the radius vector $\vec{r}(r, 0, z)$ yields an angular momentum density. The angular momentum density in the $z$ direction can come from the first or the third term of the linear momentum density equation (7.6). As one can see, the first term depends on the phase gradient of the complex field and the third term depends on the polarization of

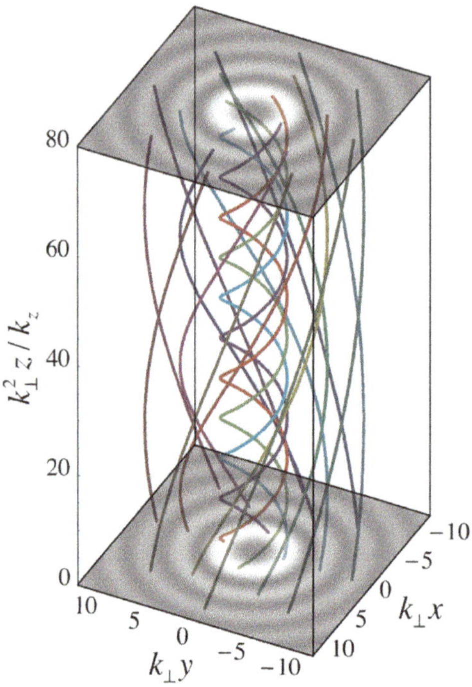


**Figure 7.2.** Photon trajectories are helical in a Bessel beam with a vortex of charge 2. This suggests that photon momentum is helical and hence singular beam carries OAM. Reproduced from [18] © IOP Publishing Ltd and Deutsche Physikalische Gesellschaft. CC BY-NC-SA 3.0.

the complex field. The angular momentum arising from the cross product of radius vector $\overrightarrow{r}$ with the phase gradient term is called the orbital angular momentum and the cross product with the third term is called the spin angular momentum of light.

## 7.4 Orbital and spin angular momentum of light

### 7.4.1 Angular momentum due to circular polarization

The spin angular momentum arising due to polarization of light is given by $\sigma_z \hbar$ per photon with $-1 \leqslant \sigma_z \leqslant 1$, where the extrema $\sigma_z = \pm 1$ corresponds to pure circular polarization.

From the azimuthal part of the linear momentum density given in equation (7.6), the $z$-component of the angular momentum density is given by

$$\mathcal{J}_z = |\overrightarrow{r} \times \overrightarrow{\mathcal{P}_\phi}| = r\mathcal{P}_\phi = r\omega\sigma \frac{\epsilon_0}{2} \frac{\partial |u|^2}{\partial r} \tag{7.7}$$

The angular momentum density in equation (7.7) integrated over the beam, gives spin angular momentum (SAM) and is given by $\sigma\hbar$.

### 7.4.2 Angular momentum due to azimuthal phase dependence in the beam

The $z$-component of the angular momentum density arising from the phase gradient terms in equation (7.6) is given by

$$\mathscr{J}_z = \vec{r} \times \left\{ i\omega\frac{\epsilon_0}{2}(u^*\nabla u - u\nabla u^*) \right\} \tag{7.8}$$

The angular momentum density in equation (7.8) integrated over the beam gives OAM and is given by $l\hbar$. Note that in chapter 5, we have seen that the optical current in terms of phase gradient is given by

$$\frac{i}{2}(u^*\nabla u - u\nabla u^*) = Im(u^*\nabla u) = I\nabla\phi \tag{7.9}$$

The orbital angular momentum is due to the azimuthal phase dependence in the beam and is equivalent to $l\hbar$ per photon. The analogy between paraxial wave equation with the Schrodinger wave equation suggests the identification of an orbital angular momentum for light beams [13]. The operator corresponding to the $z$-component of orbital angular momentum is $L_z = -i\hbar\frac{\partial}{\partial\phi}$ and applying this to beams having azimuthal phase dependence like $\exp\{il\phi\}$ gives an orbital angular momentum component to phase singular beams. This analogy between paraxial optics and quantum mechanics has been exploited to introduce an eigenfunction description of laser beams, and quantum harmonic oscillator modes [14]. Laguerre–Gaussian modes are the laser mode analog of the angular momentum eigenstates of the isotropic two-dimensional harmonic oscillator.

In the case of speckle patterns, the overall OAM is zero as there is an equal number of positive and negative charge optical vortices. The angular momentum density $\vec{\mathscr{J}}$, is related to the linear momentum density $\vec{\mathscr{P}} = \epsilon_0\langle\vec{E} \times \vec{B}\rangle$ by the relation $\vec{\mathscr{J}} = \vec{r} \times \vec{\mathscr{P}}$ where $\epsilon_0$ is the free space permittivity and $\vec{E}$ and $\vec{B}$ are the electric and magnetic fields, respectively. From this relation, for any angular momentum component directed along the $z$ direction, there must be a transverse component of linear momentum, i.e. in the $xy$-plane. This means that to have a $z$-component of angular momentum there must be a transverse component, i.e. a $z$-component of the electric and/or magnetic field. Further, such a transverse component of momentum should be circulating. The center of the vortex beam corresponds to the position of zero intensity and hence it carries neither linear nor angular momentum. Indeed, the angular momentum is associated with regions of high intensity, which is a bright annular ring for a donut vortex beam. The azimuthal component of the Poynting vector is parallel to that of the transverse linear momentum density in a helical phase front [7]. Spin angular momentum (SAM) has only two independent states corresponding to left- and right-handed circular polarization, while orbital angular momentum has an unlimited number of possible states, corresponding to all integer values of $l$.

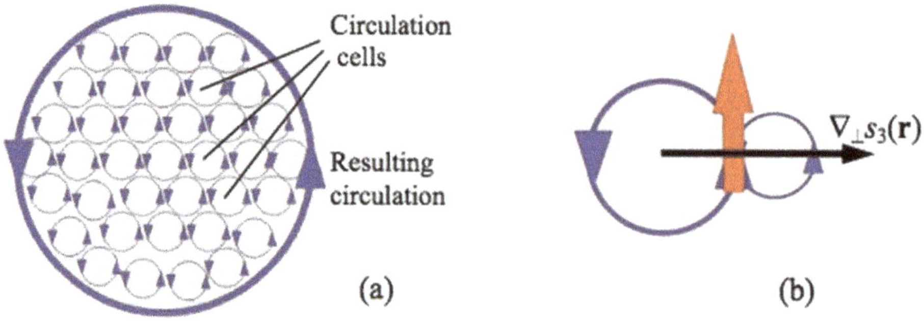


**Figure 7.3.** Orbital angular momentum due to spatially varying spin angular momentum. (a) In a uniform distribution of spin angular momentum according to Stokes theorem, the internal circulations cancel out and only circulation at the boundary remains. (b) In an non-uniform distribution of spin angular momentum, there can be a volume spin current, as there is incomplete internal cancellations of circulations. Reproduced with permission from the corresponding author [15]. © IOP Publishing. Reproduced with permission. All rights reserved.

### 7.4.3 Angular momentum due to spatially varying circular polarization

There is a possibility that a spatially varying circular polarization can result in orbital angular momentum. This can result in optical current flow across the beam cross section. Consider the fundamental theorem of curls which states that

$$\int\int(\nabla \times \vec{A}) \cdot \vec{da} = \oint \vec{A} \cdot \vec{dl} \tag{7.10}$$

All the elementary circulations, which are of the same magnitude, can result in a single large circulation around the boundary enclosing the region. This is due to the cancellation of circulating components between adjacent circulating cells. This is explained in figure 7.3(a). But when there is a spatially varying spin angular momentum distribution, there can be a volume spin current due to incomplete cancellation of circulations. As a result any larger circulation arising due to the spatially varying circular polarization is not origin-dependent as it arises due to individual spins which are origin-independent. In figure 7.3(b), the spin current arising between two unequal adjacent circular polarization components is depicted. The gradient of Stokes' $S_3$ parameter in the figure indicates spatially varying circular polarization. The Stokes' parameter $S_3 = \pm 1$ for circular states. Any state of polarization can be decomposed into right and left circularly polarized components and $\nabla S_3$ in the figure represents spatial variation of SAM component.

## 7.5 Intrinsic and extrinsic angular momenta

There are three forms of angular momenta namely intrinsic spin angular momentum, intrinsic orbital angular momentum and extrinsic orbital angular momentum [16] abbreviated as EOAM. The angular momentum that is independent of the choice of the axis is intrinsic. The spin angular momentum is therefore intrinsic and the orbital angular momentum carried by phase singularity with its core along the optical axis of the system is intrinsic. Any vortex which is situated away from the center of the beam axis gives rise to extrinsic orbital angular momentum.

These three forms of the AM of light beams correspond to different degrees of freedom of light and to different structural constituents. In free space propagation these different forms of AM are conserved. Any transverse shift of the field centroid (trajectory), leads to the conversion from intrinsic OAM to the extrinsic OAM.

The angular momentum which arises for any light beam from the product of the $z$-component of linear momentum about a radius vector, may be said to be extrinsic because its value depends upon the choice of calculation axis.

Berry showed [17] that the orbital angular momentum of a light ($J$) beam does not depend upon the lateral position of the axis and can therefore also be said to be intrinsic, provided the direction of the axis is chosen so that the transverse momentum is zero. When integrated over the whole beam the angular momentum in the $z$ direction is

$$\vec{J_z} = \epsilon_0 \int \int dxdy(\vec{r} \times \langle \vec{E} \times \vec{B} \rangle) \tag{7.11}$$

If the axis is laterally displaced by $\vec{r_0}(r_{0x}, r_{0y})$ it is easy to show that the change in the z-component of angular momentum is given by

$$\begin{aligned} \Delta J_z &= (r_{0x} \times P_y) + (r_{0y} \times P_x) \\ &= r_{0x}\epsilon_0 \int \int dxdy \langle \vec{E} \times \vec{B} \rangle_y + r_{0y}\epsilon_0 \int \int dxdy \langle \vec{E} \times \vec{B} \rangle_x \end{aligned} \tag{7.12}$$

The angular momentum is intrinsic only if $\Delta J_z$ equals zero. Any beam with a helical phase front apertured symmetrically about the beam axis has zero $\Delta J_z$ and, consequently, an orbital angular momentum of $l\hbar$ per photon, independent of the axis of calculation [16]. The orbital angular momentum of the light beam may therefore be described as intrinsic. However, when the beam is passed through an off-axis aperture, its transverse momentum is non-zero and the orbital angular momentum depends upon the choice of calculation axis and so must be described as extrinsic.

The OAM is intrinsic only if the vortex core coincides with the beam axis. In this case the OAM will turn out to be independent of the choice of the calculation axis. For apertured beams even if the vortex is positioned at the center of the aperture and the beam axis does not coincide with the center of the aperture, the computed OAM will yield different results and is dependent on the choice of the coordinate system used. Therefore, the OAM in this case is extrinsic. This means that vortices positioned away from the beam axis (also called the centroid of the beam) have extrinsic OAM. Therefore, it does not follow that the angular momentum of the apertured beam is intrinsic as the result does depend upon the choice of calculation axis. Figure 7.4 depicts the numerically calculated local spin and orbital angular momentum densities in the direction of propagation for a circularly polarized LG beam. It shows how (1) the axis of the beam, (2) the position about which angular momentum is calculated and (3) aperturing, affect the calculated OAM. Further,

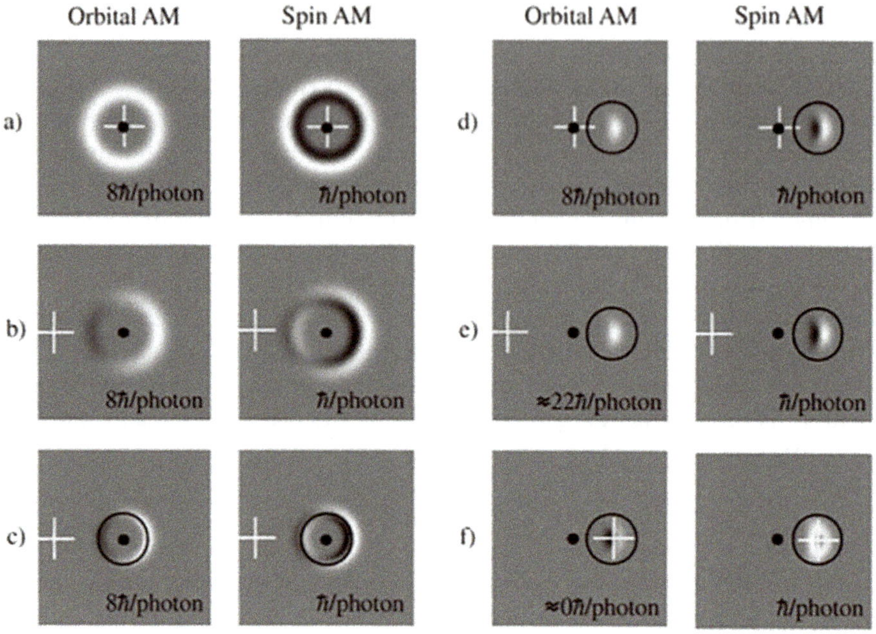


**Figure 7.4.** Numerically calculated local spin and orbital angular momentum densities in the direction of propagation for a $l = 8$ and $\sigma = 1$ Laguerre–Gaussian mode [16]. In the gray scale, white represents the positive and black the negative contributions. The axis of the beam is marked by a black spot while the axis about which the angular momenta are calculated is shown by a white cross. The black circle marks the position of a soft edged aperture. Note that the spin angular momentum is equivalent to $\sigma\hbar$ per photon irrespective of the choice of aperture or calculation axis, whereas the orbital angular momentum is only $l\hbar$ per photon if the aperture or calculation axis coincide with the axis of the original beam. Reprinted with permission from [16], Copyright (2002) by the American Physical Society.

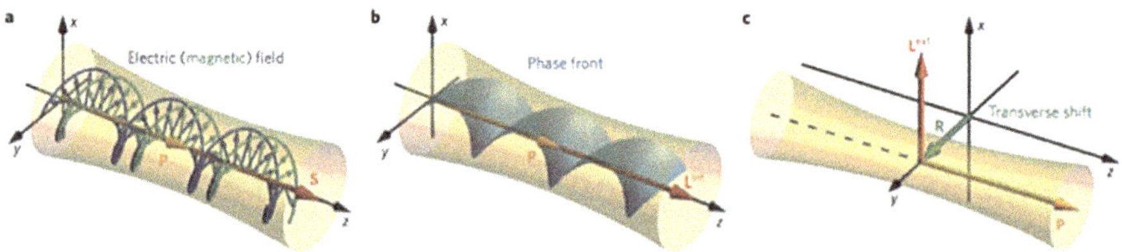


**Figure 7.5.** Angular momenta of paraxial optical beams [19]. (a) SAM for a right-hand circularly polarized beam with $\sigma = 1$. The instantaneous electric and magnetic field vectors are shown. (b) IOAM in a vortex beam with $l = 2$. The instantaneous surface of a constant phase is shown. (c) EOAM due to the propagation of the beam at a distance $R$ from the coordinate origin. Reprinted by permission from [19]. Copyright (2015) Springer Nature.

figure 7.5 shows the EOAM due to the shift in the centroid of the beam from the optical axis.

## References

[1] Poynting J H 1909 The wave motion of a revolving shaft, and a suggestion as to the angular momentum in a beam of circularly polarised light *Proc. R. Soc. Lond. Ser.* A **82** 560–7

[2] Beth R A 1936 Mechanical detection and measurement of the angular momentum of light *Phys. Rev.* **50** 115–27

[3] Allen L, Beijersbergen M W, Spreeuw R J and Woerdman J P 1992 Orbital angular momentum of light and the transformation of Laguerre–Gaussian laser modes *Phys. Rev.* A **45** 8185–9

[4] Allen L, Padgett M J and Babikar M 1999 The orbital angular momentum of light *Progress in Optics* vol 39 (Amsterdam: Elsevier)

[5] Padgett M J 2017 Orbital angular momentum 25 years on *Opt. Express* **25** 11265–74

[6] Simpson B N, Dholakia K, Allen L and Padgett M J 1997 Mechanical equivalence of spin and orbital angular momentum of light: An optical spanner *Opt. Lett.* **22** 52–4

[7] Torres J P and Torner L 2011 *Twisted Photons: Applications of Light with Orbital Angular Momentum* (New York: Wiley)

[8] Willner A E *et al* 2015 Optical communications using orbital angular momentum beams *Adv. Opt. Photon* **7** 66–106

[9] Yao A M and Padgett M J 2011 Orbital angular momentum: origins, behavior and applications *Adv. Opt. Photon* **3** 161–204

[10] McGloin D and Dholakia K 2005 Bessel beams: diffraction in a new light *Contemp. Phys.* **46** 15–28

[11] Gutiérrez Vega J C, Iturbe-Castillo M and Chav́ez Cerda S 2000 Alternative formulation for invariant optical fields: Mathieu beams *Opt. Lett.* **25** 1493–5

[12] Bandres M A and Gutiérrez Vega J 2004 Ince–Gaussian beams *Opt. Lett.* **29** 144–6

[13] van Enk S J and Nienhuis G 1992 Eigenfunction description of laser beams and orbital angular momentum of light *Opt. Commun.* **94** 147–58

[14] Nienhuis G and Allen L 1993 Paraxial wave optics and harmonic oscillators *Phys. Rev.* A **48** 656–65

[15] Bekshaev A, Bliokh K Y and Soskin M 2011 Internal flows and energy circulation in light beams *J. Opt.* **13** 053001

[16] O'Neil A T, MacVicar I, Allen L and Padgett M J 2002 Intrinsic and extrinsic nature of the orbital angular momentum of a light beam *Phys. Rev. Lett.* **88** 053601

[17] Berry M V 1998 *Int. Conf. on Singular Optics* vol 3487 (Bellingham, WA: SPIE)

[18] Bliokh K Y, Bekshaev A Y, Kofman A G and Nori F 2013 Photon trajectories anomalous velocities and weak measurements: a classical interpretation *New J. Phys.* **15** 073022

[19] Bliokh K Y, Rodrigues-fortuno F J, Nori F and Zayats A V 2015 Spin-orbit interactions of light *Nat. Photon.* **9** 796–808

# Chapter 8

# Applications

## 8.1 Metrology

### 8.1.1 Optical vortex lattice interferometer

When two beams interfere in space, constructive and destructive interference form bright and dark surfaces in the volume where the beams overlap. When viewed at an observation (two-dimensional) plane, these surfaces appear as interference fringes in a conventional interferometer. When one of the interfering beams is modified, the fringe pattern undergoes a change, and this change is tracked using conventional optical testing methods. But when three or more waves interfere, light vanishes at curves rather than on surfaces. These curves intersect an observation plane at isolated points and they appear as dark points instead of fringes. Each such dark point is a vortex point, where a phase singularity is located. The optical vortex interferometer (OVI) deals with the distribution and dynamics of these dark points [1]. The positions of these dark points are tracked during measurements. Hence the accuracy and resolution of the OVI strongly depends on the accuracy of vortex point's localization [2].

Initially a regular lattice of optical vortices is generated. This can be accomplished in many different ways—one of them is to use multiple beam interference [3–5]. In a multiple beam interference, the geometry of the arrangement of dark points in the lattice and their density can be tailored [6].

In optical testing the sample under measurement is introduced in one or more beams of the OVI. This in turn disturbs the vortex lattice geometry and these changes can be related to the sample parameters. The method of wave tilt measurement [7–8] is shown here as a basic example. An optical wedge that introduces a wave tilt is inserted in one of the interfering waves that is used for vortex lattice formation. This results in the modification in the vortex lattice. Some of the changes in the positions of the lattice points by the introduction of the wedge, are depicted in figure 8.1. One can derive formulas for the wave tilt through the $x$ and $y$ axes separately. Only one measurement step is necessary and the wedge angle

doi:10.1088/978-0-7503-1698-9ch8

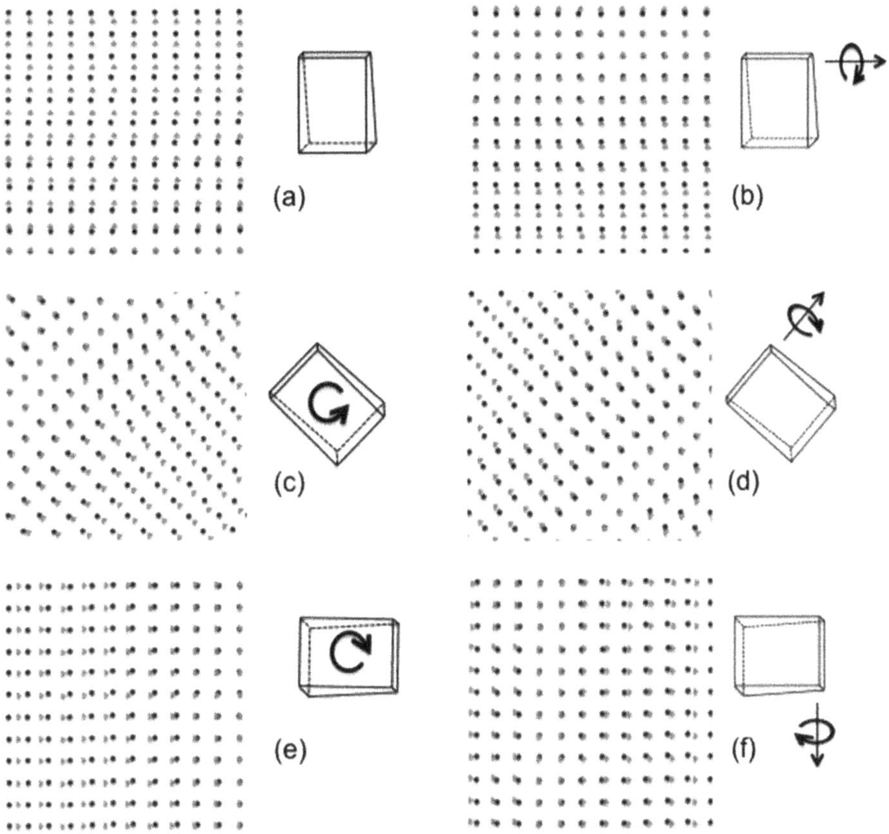


**Figure 8.1.** Optical vortex interferometer. The position of the vortex points (schematic) without the wedge (red dots) and with the wedge (black dots). (a) Wedge inserted horizontally, (b) wedge inserted at 45° and (c) vertically.

and its orientation can be obtained. OVI also enables a sound statistical analysis of the measurements, which is usually not possible in conventional interferometric measurements. Use of single pass OVI for tracing the dynamic changes of light polarization or birefringent sample properties [9] is possible. In a single measurement, both azimuth and ellipticity or ellipticity and phase difference between waves polarized along fast and slow axes of the birefringent medium can be determined.

It has been reported in the literature that OVI can be used for wavefront reconstruction [10, 11]. By knowing the charge of the vortices in the vortex lattice, we can reconstruct the wavefront geometry without any ambiguities. Further applications that are reported include phase unwrapping [12] and vortex charge determination [12–15].

## 8.2 Collimation testing

Plane waves are used everywhere in optics. Plane waves can be generated by placing a point source at the focal plane of a collimating lens. Beam collimation is always required in optical measurement systems and is related to the accuracy of measurement [16]. A number of methods have been proposed for collimation testing. Interferometric techniques such as shear interferometry [17], Talbot interferometry [18] are used to test beam collimation. The evolution of interference fringes in an interferometer [19] is monitored when the collimation of the beam is disturbed. For example, a plane wave interfering with a collinear reference plane wave yields no fringe condition at collimation and produces concentric circular fringes when the collimation is disturbed. In a shear interferometer also, no fringe condition is the criterion for collimation and its disturbance is revealed by the appearance of parallel straight fringes perpendicular to the shear direction. Hence transformation from no fringe condition to circular or straight fringe condition is the collimation disturbance criteria. A tilted plane wave upon interference with an on-axis plane wave produces straight line fringes. Hence combining shear with tilt parallel or/and perpendicular to shear, offers many other possibilities in forming new criteria for collimation testing. Fringes can be expanded or rotated depending on the setup configuration.

Optical vortices can also be used in interferometers for collimation testing [20]. The phase contour lines in a singularity are radial and for a vortex of charge $m$, the total phase variation in the observation plane is $m2\pi$. Hence, an interferogram formed between an optical vortex beam of charge $m$ and a plane beam will produce $m$ radial fringes. The fringe formation condition is given by $\Theta(x, y) = n2\pi$, where the fringe order $n$ can take maximum value $|m|$, the magnitude of the vortex charge, where the $\Theta$ is the phase difference between the test wavefront and the reference plane surface. But when the collimation is disturbed due to defocusing, i.e. the point source is moved from the front focal plane of the collimation lens by $\Delta f$ then the wavefront assumes a spherical shape instead of a plane. Then the fringe formation condition is given by

$$\frac{\Delta f}{2f^2}(x^2 + y^2) + \frac{1}{k}\theta(x, y) = m\lambda \tag{8.1}$$

This represents the presence of $m$ spiral fringes. The spiraling of fringes happens either clockwise or anti-clockwise, depending on the sign of the defocus $\Delta f$. Hence by tracking the spiraling of fringes the disturbance from collimation can be detected. In figure 8.2, the interference fringes that can be obtained when there is collimation and when it is defocused are shown.

It is also possible to use binary amplitude spiral phase gratings [21, 22] in Talbot interferometric configuration. Spiral gratings can produce vortices as their

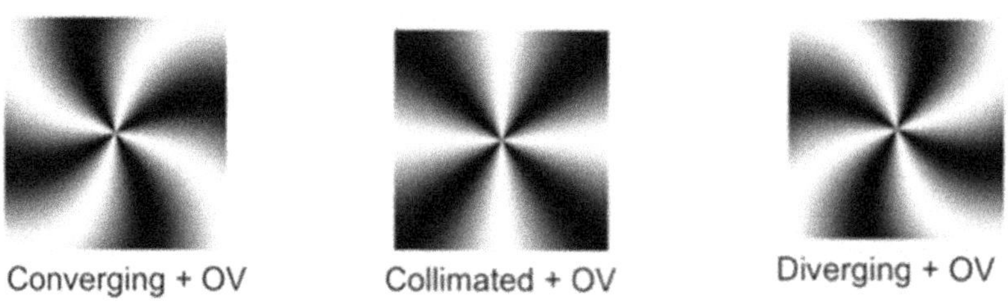


**Figure 8.2.** Collimation testing with vortices: interferograms obtained when the beam is (a) converging, (b) collimated and (c) diverging. Reproduced with permission from [20]. Copyright (2003) by OSA.

transmittance function can be considered as a superposition of vortex beams and spherical beams. Low frequency spiral gratings can be self-imaged under collimated monochromatic light illumination. By placing another spiral grating of the same periodicity but spiraling in the opposite sense at the Talbot self-image plane Moire fringes consisting of many radial fringes can be obtained. The number of fringes being equal to the sum of the magnitude of the charges in the two gratings put together. The periodicity in the spiral grating becomes larger/smaller for diverging/converging beam illumination and the spacing between the Talbot self-image planes also would accordingly change little. This change in the period makes the Moire fringes spiral in or out depending on the amount of decollimation. The spiraling of fringes in the clockwise or anti-clockwise direction depends on the sign of the vortex charge associated with the spiral grating and the curvature of the wavefront. For a given grating the sense of spiraling depends on whether the beam is converging or diverging.

## 8.3 Spiral interferometry

In most interferometers in optical testing, light reflected from or transmitted through an object interferes with a uniform plane reference wave to generate an interference pattern that carries information about surface height or phase profile of the object wave. Interferograms of sufficiently smooth objects consist of closed contour lines indicating that the closed fringes enclose an extremum point like a peak or a valley. Each contour line is indicative of height of the phase profile measured in units of optical wavelength. Phase profile measurement using these contour lines suffer from the major limitation that there is no direct way to distinguish between depressions and elevations in the optical thickness of the sample. In figure 8.3 the interference pattern formed between a convex surface and a plane reference wave is shown as the same as

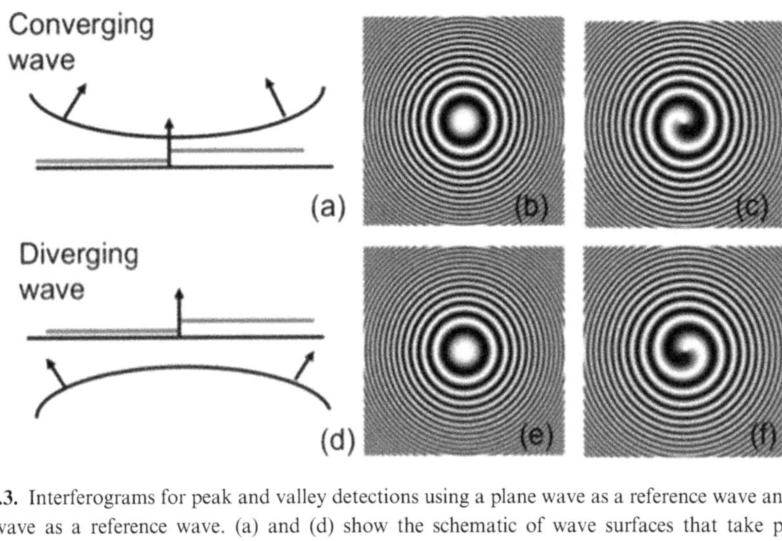


**Figure 8.3.** Interferograms for peak and valley detections using a plane wave as a reference wave and a vortex helical wave as a reference wave. (a) and (d) show the schematic of wave surfaces that take part in the interference. Circular fringes are formed by the interference of a plane wave with (b) a converging wave and (e) a diverging wave. Spiral fringes are formed by the interference of helical wave with (c) a converging wave and (f) a diverging wave.

the interference between a concave surface and a plane wave. Both surfaces give concentric ring patterns thereby making a distinction between them is difficult. Because both a peak or a valley will produce closed fringes a fringe shifting technique is required to distinguish between peak and valley. These techniques are based on generating additional information by recording multiple interferograms with different phase offsets, between object and reference wave (phase stepping) or from different directions (angular multiplexing) or with different wavelengths (wavelength multiplexing).

Spiral interferometry [20, 23–25], where a spiral phase element is used as a spatial filter, removes the ambiguity between elevation and depression in the optical thickness of a sample. As discussed in collimation testing interference of a helical wave and a spherical wave tend to produce spiral fringes and the sense of spiraling depends on the curvature of the wavefront. This helps distinguish concave and convex surfaces. In spiral interferometry spatial filtering of the object wave with a spiral phase element results in interference fringes which are spiral in shape in an interferometer. The sense of rotation of these spirals makes a distinction between elevation or depression in the surface height of the object. A spiral interference pattern obtained with single exposure of the object contains sufficient topographic information to allow the complete reconstruction of the object phase profile. This is an extremely useful feature for high speed interferometry to record interferograms with a single laser pulse or for video recording of rapidly varying surfaces so that

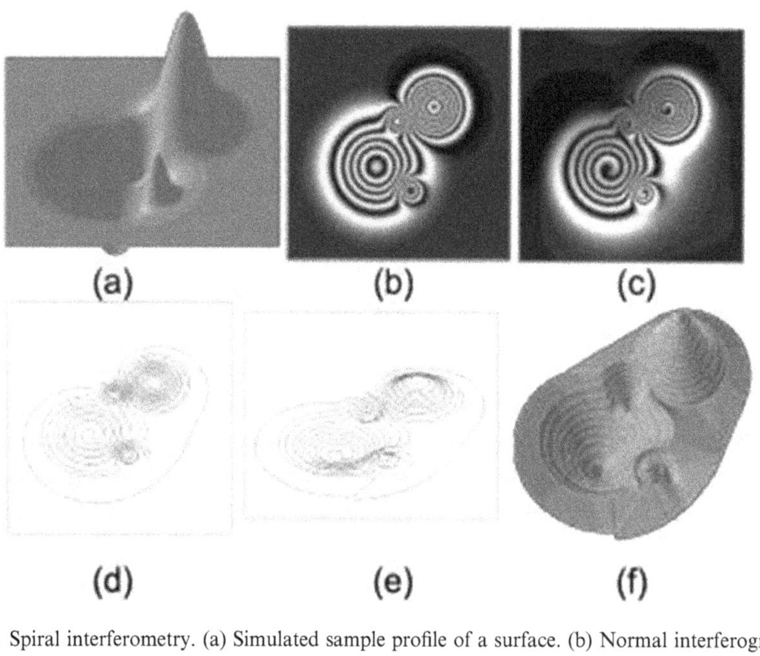


**Figure 8.4.** Spiral interferometry. (a) Simulated sample profile of a surface. (b) Normal interferogram where closed fringes do not distinguish between elevations and depressions. (c) Spiral interferogram obtained by filtering with the modified spiral phase contrast method. Depending on the topography, the spirals change their rotational direction. (d) A single contour line of the spiral interferogram. (e) Processed contour line, which allows a unique height to be assigned to each single point of the contour line. (f) Reconstructed surface profile by interpolation between the sampling points given by the contour line. Reproduced with permission from [23]. Copyright (2005) by OSA.

each video frame is processed separately. Figure 8.4 shows a normal interferogram and spiral type interferogram of a simulated phase object. In figure 8.4(a), there are two conspicuous extrema represented by closed interference fringes. In figure 8.4(c), the way the spiral fringes start (clockwise or anti-clockwise) is an indication that one of the extreme is a maximum and the other is a minimum.

## 8.4 Spatial filtering

Edge enhancement is one of the most fundamental operations in image processing, analysis, recognition, and in machine vision. Because edges play an important role in understanding images, one way to enhance the contrast is to enhance the edges. For edge enhancement, optical Fourier transform techniques to filter out low spatial frequency components are usually employed by using spatial light modulators (SLMs) or mechanical spatial filters [26, 27].

### Hilbert transform

The transmittance function of a Hilbert transform function can be given by a signum function defined as

$$\begin{aligned} f(x) &= -1 \quad \text{for} \quad x \leqslant 0 \\ &= +1 \quad \text{for} \quad x > 0 \end{aligned} \tag{8.2}$$

This function can be realized by a transparent glass plate in which half of the plate is given a coating that retards the wave by $\pi$ radians with respect to the other uncoated glass plate. This is a step profile and this plate can produce an edge dislocated wave, which is a wavefront with line phase defect. The two binary phase masks shown in figure 8.5 have a line phase defect. On either side of the line defect, the phase values are $\frac{\pi}{2}$ and $-\frac{\pi}{2}$ respectively. When such a filter is placed in the Fourier plane of the $4f$ optical processing system, it gives an extra phase of $\pi$ to spatial frequency components in the frequency half-plane depending on the orientation of the phase jump [28–30]. As a result the filtered image is edge enhanced in the direction perpendicular to the edge dislocation direction. Hence Hilbert transformation of objects can identify the edges of the object in a particular direction. To detect the other edges of the object the filter has to be rotated in its plane.

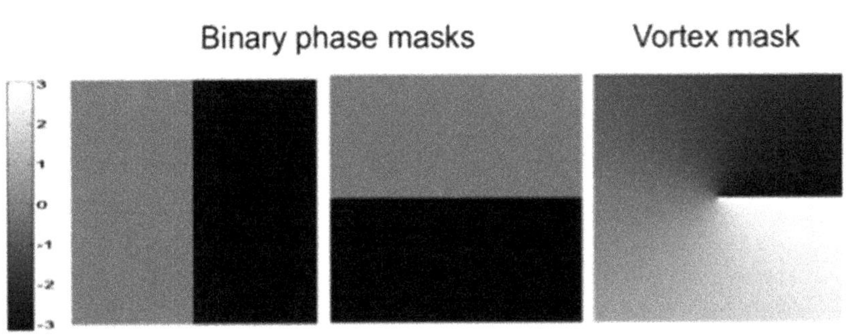


**Figure 8.5.** Phase masks for performing Hilbert transforms. The gray scale shows the phase values in radians.

**Isotropic edge enhancement**

An optical vortex with topological charge $m$ can be used to perform the $m$th order Hankel transform [26]. It has been shown initially by Khonina *et al* [27] and then by Davis *et al* [31] that a vortex phase mask can also be used as a spatial filter in $4f$ geometry to achieve symmetric edge enhancement. Vortex producing lens [32] can also be used for isotropic edge enhancement. If the phase profile of the vortex mask is analyzed, it is obvious that there is a phase difference of $m\pi$ at a symmetric position in any radial line with respect to the vortex core. Similar characteristics can be seen in the one-dimensional Hilbert transform [28–30]. Spiral phase filtering using a spiral phase plate (SPP), which is characterized by function $\exp\{im\theta\}$, is regarded as a radial Hilbert phase mask with $m$ as the order of the radial Hilbert transform. In this way, edge of the object in all directions can be highlighted. Such an operation is not possible by performing a 1D Hilbert transform sequentially in two orthogonal directions. This is depicted in figure 8.6.

**Anisotropic edge enhancement**

Sometimes edges in a particular direction require detection or to be emphasized more than other edges. This angular selection of edges is possible by an anisotropic edge enhancement operation. Generally, the edge-enhancing effect is isotropic, particularly when an SPP is used as a radial Hilbert phase mask. Using a 1D Hilbert mask for selective enhancement limits the angular selection of edges. Hence there are efforts to find methods to perform selective edge enhancement. Selective edge enhancement can be achieved by using fractional charge vortex phase plates. The radial Hilbert transform, which is effectively the vortex spatial filtering, does the edge enhancement by redistributing the intensity. The intensity is redistributed in a symmetric manner because the radial Hilbert mask is symmetric. Therefore, to enhance the selective edges in a particular desired direction, one has to break this symmetry.

The methods reported for selective edge enhancement, use the fractional vortex masks [33, 34]. The fractional Hilbert transform [29, 31, 35] method uses the modified Hilbert transform, in which the phase difference between two radial points on either side of the origin is a fractional multiple of $\pi$. A fractional spiral phase filter

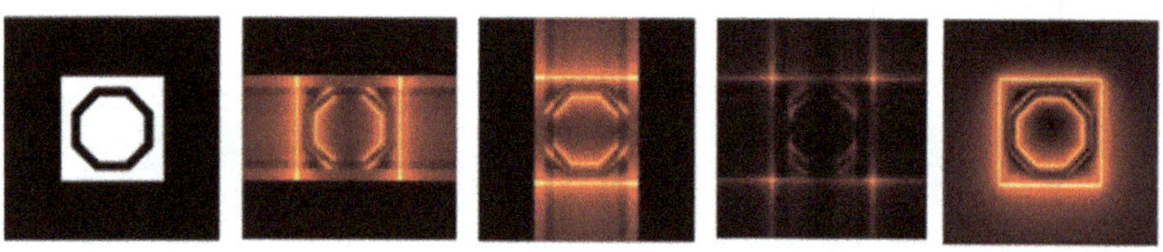

**Figure 8.6.** Two 1D Hilbert transforms one along $x$ and another along $y$ performed one after another will not result in isotropic edge detection. Isotropic edge detection is possible with a spiral phase mask used as a radial Hilbert filter. The first figure is the binary amplitude object. The second and third figures show the 1D Hilbert filtering performed along $x$ and $y$ directions respectively on the same object. The fourth figure shows the edge enhancement where $y$ filtering is done sequentially on an $x$-filtered image. The final figure shows isotropic edge enhancement by the use of a radial Hilbert transform filter.

with an additional offset angle and by shifting the singularity was also proposed for selective edge enhancement [35]. Another method is to use anisotropic vortex phase distribution in the filter. The phase distribution in a vortex is $\psi = m\theta$. The rate of change of the phase around the vortex is given by $\frac{d\psi}{d\theta} = m$, which is a constant. In an anisotropic optical vortex, this is not a constant. The anisotropic vortex [36] considered by Kim *et al* is given by

$$V_a = x + i\sigma y = r\exp\{i\psi\} \tag{8.3}$$

This filter was not used for the study of edge enhancement. The anisotropic vortex considered by Sharma *et al* [33], for selective edge enhancement of objects, is given by

$$\exp\{i\psi_s\} = \exp\left[i\theta\left\{\left|\sin^n\left(\frac{\theta}{2}\right)\right|\right\}\right] \tag{8.4}$$

Here, $\psi_s$ is the phase function corresponding to the anisotropic vortex function. The modulus of the sine functions has been taken to preserve the helical wavefront shape of the vortices, and the parameter $n$ is an integer. They have employed the modulo $2\pi$ operation to keep $\theta$ always in the interval $[-\pi, +\pi]$. In other words, the phase is wrapped between $[-\pi, +\pi]$. The azimuthal phase gradient component is

$$\frac{d\psi_s}{d\theta} = \left|\sin^{(n-1)}\left(\frac{\theta}{2}\right)\left[\sin\left(\frac{\theta}{2}\right) + \frac{n}{2}\theta\cos\left(\frac{\theta}{2}\right)\right]\right| \tag{8.5}$$

Hence in an anisotropic vortex, the azimuthal phase gradient component is not a constant. As a consequence, a Hilbert filter-like $\pi$ phase jump can be seen only at specific angles around the vortex core. Such vortex phase plates when used in a spatial filtering setup yield edge enhancement properties only in certain directions. This selective edge enhancement is possible by introducing different kinds of anisotropy in the vortex phase profile [33].

A fractional vortex dipole made up of two fractional vortices of charge ±0.5 produces a phase jump of $\pi$ along the line joining the two charges. This phase jump is equivalent to a one-dimensional Hilbert mask, but the termination of this phase jump at the two ends of the dipole gives it a mixed type dislocation topology. Hence this fractional vortex dipole filter (figure 8.7) offers selective edge enhancement capabilities with additional control parameters such as fractional charge strength and dipole charge separation distance [37]. When the distance between the two charges goes to infinity, the filter behaves exactly like a one-dimensional Hilbert mask. The variation of phase profiles in the filter plane as the dipole charge separation distance is varied is shown in figure 8.7(e).

**Spiral phase contrast imaging**

To see phase objects clearly, in phase contrast microscope a phase shift of $\pm\frac{\pi}{2}$ is given to the zero frequency component of the phase object. This is done at the Fourier plane where the spatial frequency distribution of the object is available. As a result the phase variation of a phase object is converted into an intensity variation. This way the phase

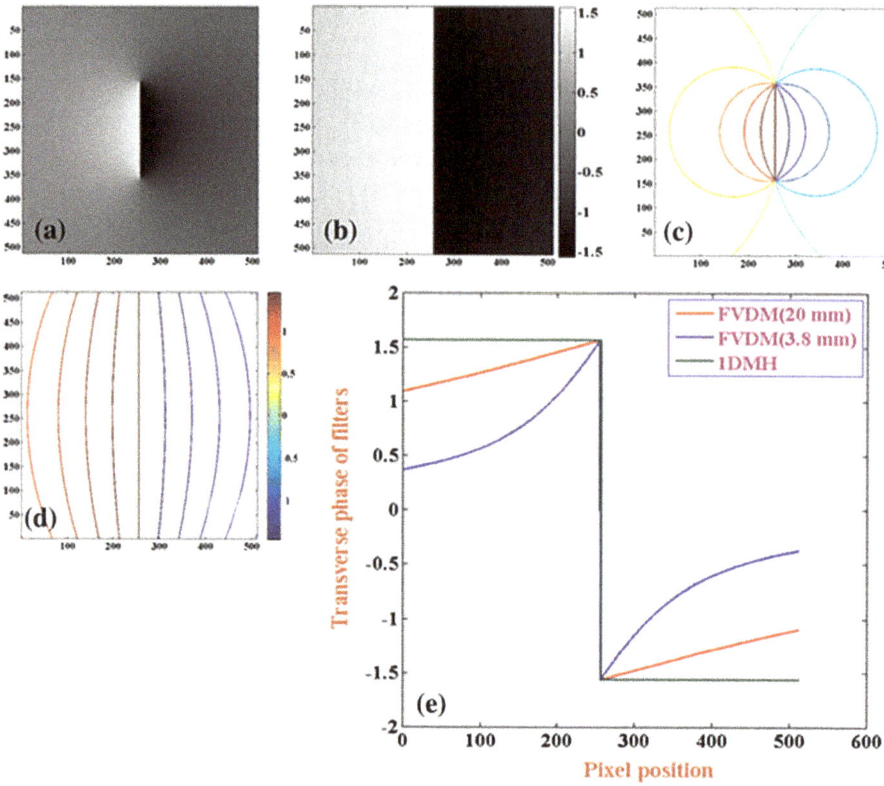


**Figure 8.7.** Selective edge enhancement using a fractional vortex dipole. (a) Phase distribution of a fractional vortex dipole. (b) The phase distribution of a typical 1D Hilbert transform mask. (c) Phase contours for a fractional vortex dipole with two smaller charge separation distance. (d) Phase contours when the dipole charges are separated by a large distance. (e) Horizontal scans across the phase filter masks for two fractional vortex dipole filters with two different charge separation distances are compared with that of a 1D Hilbert transform mask which has binary profile. Reprinted with permission from [37]. Copyright (2014) by Springer Nature.

variation is made visible. Instead of a phase shift to a zero frequency component of the object, a spiral phase plate introduced at the spatial frequency plane can result in isotropic edge enhancement of objects and for phase objects that provide very good phase contrast imaging [25, 38]. The spiral phase plate can be introduced in a normal bright-field microscope [39] and better images can be obtained for phase objects, with white light illumination as shown in figure 8.8.

Even when a spiral phase filter is used in the Fourier plane, having an absorptive or a transmissive centre makes the shadow effects look very different [40]. The one with a transmissive centre produces a realistic relief pattern during spatial filtering as shown in figure 8.9.

### Optical vortex coronograph

The optical vortex coronagraph helps direct imaging of extra solar planets in the neighborhood of their bright host star [41]. The telescope focuses the light of the star

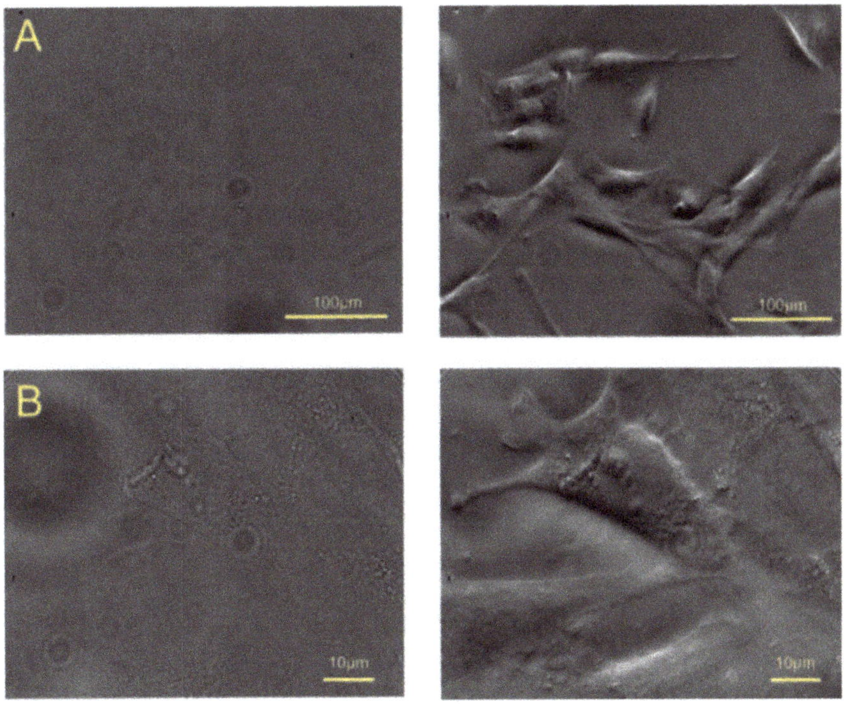


**Figure 8.8.** The images show fibroblast cells in row A as viewed with a 20 ×air objective, in row B recorded with a 100 ×oil immersions objective. The images at the left side show the bright-field images, whereas at the right are the corresponding spiral phase-filtered images. The centre of the spiral phase plate is placed 50 $\mu$m away from the optical axis. The recording time of the spiral phase-filtered images is only half as long as that of the bright-field images. Reproduced with permission from [39]. Copyright (2008) by John Wiley and Sons.

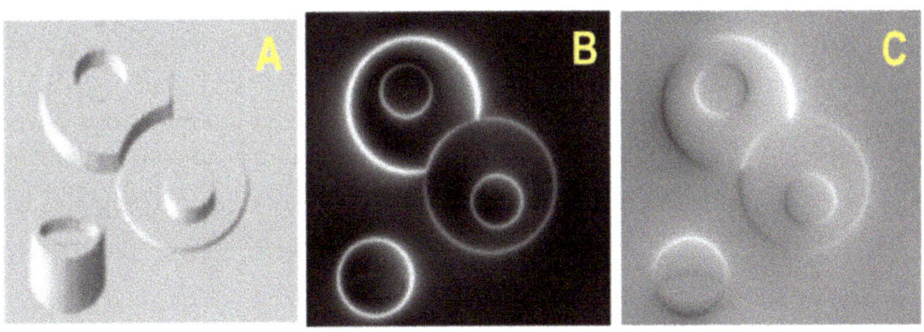


**Figure 8.9.** Simulation of a combination of amplitude and phase objects imaged by a spiral phase contrast filter. (A) shows the sample object to be imaged as a relief. (B) shows the numerically computed result for spatial filtering of the image wave with an ideal spiral phase element with a singular central point which is absorptive. (C) shows the result after filtering with a spiral phase element with a transmissive center, resulting in a relief-like view of the sample. Reprinted with permission from [40]. Copyright (2002) by the American Physical Society.

in the center of the vortex phase plate, transforming the stellar pattern into a ring embracing an optical vortex carrying a precise orbital angular momentum of light, which can be blocked by the donut shaped Lyot stop. Light from off-axis sources similarly will form asymmetric donut shaped ring patterns with their centres shifted. This means that the fainter nearby planet, being off-axis instead, does not acquire the same orbital angular momentum as the on-axis star, and its light produces an asymmetric pattern that passes through the center of the Lyot stop, revealing its presence to the observer. This approach promises superior performance compared to the standard Lyot coronagraph and other modern variants [42], both for space telescopes and ground-based telescopes with good seeing conditions in adaptive optical devices. Recent tests of the telescope, demonstrated the feasibility of this technique also with ground-based telescopes, by studying the effects of atmospheric turbulence in the generation of optical vortices from starlight beams [43]. Based on the asymmetric intensity distribution generated by an off-axis optical vortex, it is possible to resolve two sources at angular distances much below the Rayleigh criterion [44]. For the same reason, optical vortices can also be used to perform high precision astronomy and tip/tilt correction of the isoplanatic field [43].

**Observation of a weak star in the bright background**

The dark core of an optical vortex can be used as a window to examine a weak background signal hidden in the glare of a bright coherent source. Signal improvement of at least seven orders of magnitude is possible by the use of vortex masks [45].

To detect weak light from a source it is desirable to attenuate the disturbing intense glare of bright coherent beams from a nearly collinear source. Such a situation occurs in forward-scattered radiation in optical systems and sources adjacent to bright bodies. A vortex phase mask that can produce a dark focal spot can be used in a spatial filtering scheme.

Consider light from a distant single light source focused by a lens with a vortex phase mask as shown in figure 8.10. Due to the vortex phase mask a donut intensity profile with a dark focal spot results at the focal plane. The size of the darker region, called the core region, increases with increasing value of the magnitude of the charge $m$. The peak amplitude at the focal plane as a function of the radial coordinate for different charges of a vortex phase mask occurs at radial positions given by $R_{|m|=1} = 0.64R_{\text{diff}}$, $R_{|m|=2} = 1.03R_{\text{diff}}$, $R_{|m|=3} = 1.37R_{\text{diff}}$ and so on, where $R_{\text{diff}} = 1.22\frac{\lambda f}{D}$ is the radial position at which the first zero of the beam when the lens of focal length $f$ and diameter $D$ is used without mask for focusing.

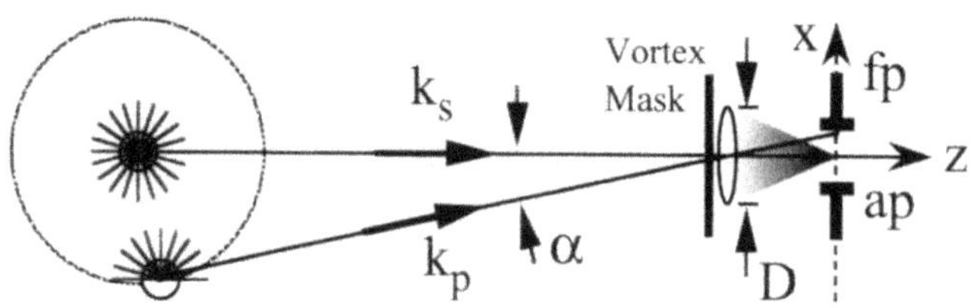


**Figure 8.10.** Vortex coronograph for observation of a faint object in the background of a bright nearby starlight. Angle $\alpha$ is between rays from the bright star and the reflected light from the planet at the objective. The focal length and diameter of the objective are $f$ and $D$ respectively. Reproduced with permission [45]. Copyright (2001) by OSA.

At the focal plane an on-axis bright source produces a dark core whose size is decided by $|m|$, $f$, $D$ and $\lambda$. The bright light from the source will come into focus not at the centre but away from it forming a donut structure and this can be filtered out by using an appropriate pin-hole at the focal plane. For any off-axis source emitting light, the lens with vortex mask will produce the donut focus with its center shifted. Because of this shift, a portion of the light from the second object can pass through the aperture and can be analyzed. Since the light captured is only from the off-axis light source and the light from the on-axis source is rejected, this method allows one to examine observation of weak sources in the vicinity of bright sources in an effective way.

The window (pin-hole/aperture) is placed to allow the dark core to pass through and reject the bright ring due to the primary strong source of light. The dark core region of light will remain dark if there are no other light sources in the vicinity and subtend at different angles at the lens. However, if there are other light sources, there will be light in the dark core and in this way the vortex opens a window, allowing us to see a background field without the glare of the original field.

## 8.5 Focal plane intensity manipulation

A plane wave focused by a lens with a circular pupil function creates a focal spot with the amplitude variation according to Jinc function (Airy function). But instead of a plane wave, if the same lens is illuminated by a beam with an optical phase singularity, its focal plane will have a donut shaped amplitude distribution. At the focal spot the intensity is zero, which is surrounded by a bright ring.

Under tight focusing the focal plane intensity distribution is susceptible to the polarization distribution of the incident beam. This can also be manipulated further by using azimuthal or radially polarized light. These beams are known to have optical phase singularities in them. The connection between these structured polarization distributions and phase singularities will become clear in the next chapter.

#### Polarization engineering

The helical phase structure and its role in the generation of inhomogeneous polarization distributions have drawn attention in recent years in the context of microscopy. Strong variation in the optical field around a singularity is highly sensitive to the changes found in its neighbourhood, and this property is very useful in the imaging of phase objects [46].

The radially polarized beam under tight focusing creates a focal spot, smaller than the diffraction limited focal spot [47–49]. The tight focusing of azimuthally polarized light produces a donut focus that can be used in stimulated emission depletion spectroscopy/microscopy.

## 8.6 STED microscopy

A singular beam with phase singularity is useful in the stimulated emission and depletion (STED) microscopy [50]. It is a fluorescence microscope in which the sample is initially excited by a beam (pump beam) with Gaussian profile but the depleting beams were not vortex beams in the initial designs [51]. The use of donut beams in STED was a later development. The excited sample can de-excite from higher states either by spontaneous (fluorescence) or stimulated emission. If a pump beam along with an erase beam of donut profile are focused onto the sample, the sample gets de-excited at regions where the donut intensity is as this induces stimulated emission. At locations where the erase beam does not have light, that is at the centre of the beam, there is no stimulated emission. Basically the stimulated emission depletes the excitation in the bright regions of the donut beam leaving the central spot in the excited state. In other words the fluorescent spot is depleted in the overlapping area of the beams by stimulated emission. This light is later detected from the shrunken fluorescent spot, the size of which is smaller than the diffraction limit, thereby detection at super-resolution level is possible. The nonlinear relationship between intensity of the erase beam and residual population of the fluorescent state plays a crucial role in breaking the diffraction barrier. This results in an effective point spread function (PSF) of a size smaller than the dimensions set by diffraction. STED-PSF is zero in the centre but strong at the periphery to ensure that the fluorescent spot is allowed only in the centre, and depleted at its periphery. The donut structure of the singular beam makes it suitable for inhibition of the fluorescent spot in the central region [52–54].

A schematic setup is shown in figure 8.11. A typical STED nanoscope is similar to a confocal microscope. It has a minimum of two co-aligned laser beams: one for excitation and a second called STED beam, for the depletion of fluorescence figure 8.11. The annular pattern of the STED beam is achieved using a vortex phase plate [55]. If resolution improvement is required along the optical axis, then the right choice is a bottle profile made by an axial phase plate [56].

In order to overcome the diffraction barrier, the effective fluorescence volume is squeezed by stimulated emission. The fluorophores at the periphery of the excited region is quenched by a second beam called the STED beam, which is a donut beam. The STED beam stimulates the emission of the fluorophores, instantaneously bringing them to the ground state. For unlimited resolution, the stimulated emission process should saturate and consequently broaden the effective donut area of depletion.

## 8.7 Optical trapping and tweezers

Ashkin *et al* have shown [57] that a single laser beam can be used to create a potential gradient which can be a 'force trap' also known as optical tweezer. The scattering force is proportional to the optical intensity and points in the direction of the incident light. The gradient force is proportional to the gradient of intensity and points in the direction of the intensity gradient. The single-beam gradient force trap is conceptually and practically one of the simplest radiation-pressure traps. In the earlier reported single-beam trap called an optical levitation trap [58], axial stability is achieved by balancing the scattering force and gravity [figure 8.12].

Gradient force optical traps are also known as optical tweezers. The limitations here are that particles that are trapped near the high-intensity focal region are

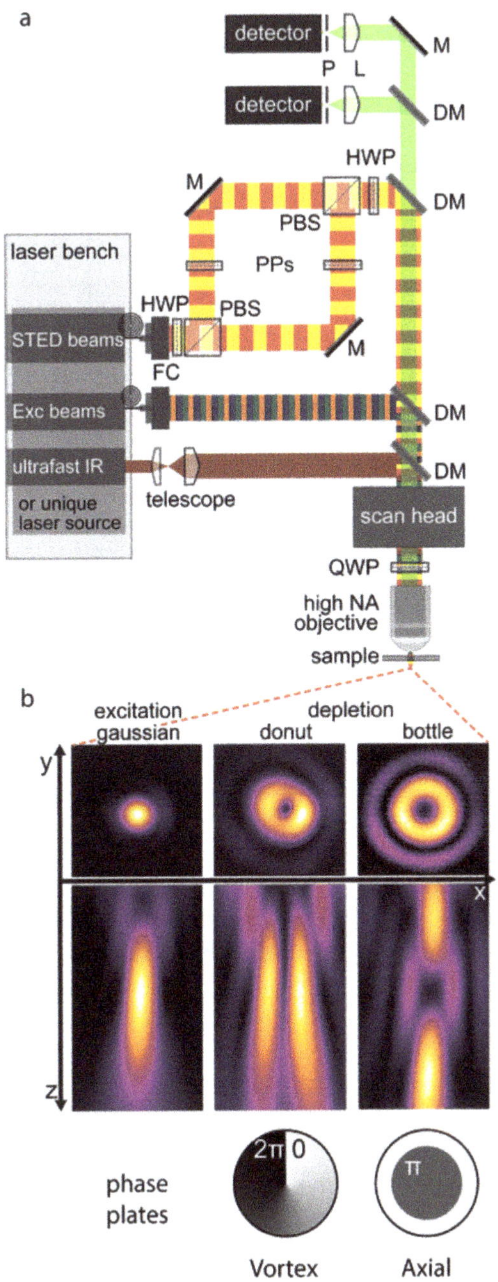


**Figure 8.11.** Schematic of the generalized stimulated emission depletion (STED) setup. (a) Several laser beams are used for multicolor STED. (b) Lateral and axial beam profiles of various beams as they appear at the focal plane where the sample is kept. Note the use of vortex plate for donut beam generation. HWP—half wave plate, QWP—quarter wave plate, PBS—polarizing beam splitter, PPs—phase plates, M—mirror, DM—dichroic mirror, L—lens, P—pin-hole. Reprinted by permission from [50]. Copyright (2015) by Springer Nature.

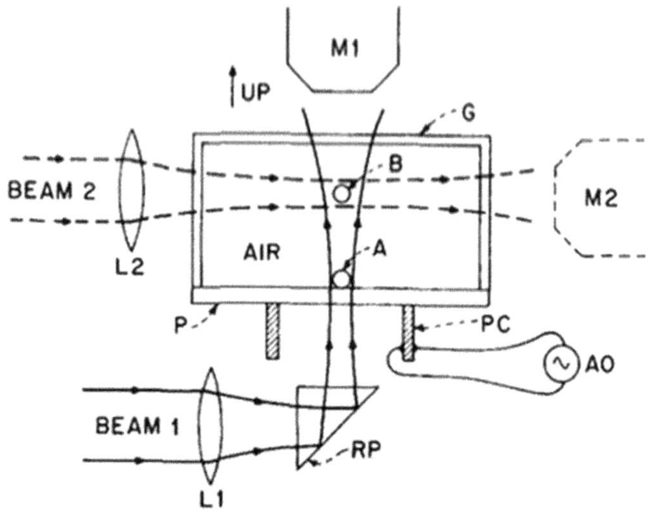


**Figure 8.12.** Levitation apparatus. In the axial direction gravity is balanced by the scattering force. A particle at A is shaken loose acoustically and lifted to B by $TEM_{00}$ mode beam 1. $TEM_{00}$ mode beam 2 is introduced later as a probe beam to study the strength of the trapping forces. Ll and L2 are lenses, P is a glass plate, G is a glass enclosure about 1.5 cm high, RP is a reflecting prism, PC is a piezoelectric ceramic cylinder driven by audio-oscillator AO, and Ml and M2 are microscopes. Reprinted from [58], with the permission of AIP Publishing.

susceptible to optical damage through absorptive heating. Isolating a single particle requires diluting the sample as multiple particles may be attracted into the same trap. What is more, the conventional stationary Gaussian-beam trap is not suitable for trapping spherical low-index particles such as bubbles and droplets. Hence, singular beams are useful for trapping low-index particles [59]. Optical rays with high convergence angles are more effective for trapping. A donut structure is more suitable for this purpose and an oil immersion objective can also be used to increase the angle of convergence. A dark core in the diffraction pattern of a singular beam is suitable for trapping and manipulation of living cells or biological objects [60–62]. A singular beam can trap both low-index, (in dark core) and high index particles (in bright regions) simultaneously [61].

Trapping by hollow beams has attracted attention due to the possibility of increased trapping efficiency. In conventional optical tweezers, radiation pressure force acts in a direction opposite to the axial gradient force, and thus lowers the trapping efficiency. Phase singular beams, having zero on-axis intensity lowers the radiation pressure and hence improves the trapping efficiency of optical tweezers.

Optical vortex traps are used to confine transparent particles as well as atoms, molecules, cells, colloidal suspensions, etc. Depending on the relative refractive index of a particle and the surrounding medium, the particles are trapped either in the beam's intensity minima or maxima. Trapping with the vortex beam is advantageous as the damage to the trapped particles is minimized due to the presence of a dark core at the center of the vortex beam. Typically, optical tweezers manipulate microscopic objects in planar geometry, and the particles are confined in a thin layer defined by the focal area of the beam.

It is well known that a periodic two-dimensional pattern can be created by the interference of laser beams. Recently these patterns were applied as optical traps for trapping dielectric particles such as polystyrene, $SrTiO_3$, and biologically interesting subjects. The advantage of using interference is that the period of the patterns may be as short as the optical wavelength. Periodic dielectric structures with lattice constants of micrometer or sub-micrometer dimensions play an essential role in photonic lattices. Multiple particle trapping has been demonstrated using three interfering beams which created a hexagonal pattern of high-intensity regions [63]. This simple configuration has been shown to trap hundreds of particles, typically in the high field regions. Grzegorczyk and Kong [64] have proposed a close-form analytical expression of the force on an infinite lossless dielectric cylinder due to multiple plane wave incidences. They have studied the curvature of the one-dimensional potential of an optical lattice created by the interference of three plane waves. It is shown that the points of zero curvature yield optical vortices which can be used to stably trap particles of particular size and index contrasts with the background. Under these circumstances, the trajectories of the particles can be accumulated to spirals whose centers correspond to the points of undetermined phase in the optical landscape.

## 8.8 Optically driven micro-motors

Optical vortices produce ring like optical traps. The orbital angular momentum carried by them can be transferred to the trapped particles and to the surrounding fluid [65, 66].

There is a need for new methods to pump and steer fluids in microfluidic channels. Current approaches based on hydraulic control and electroosmosis do not have real-time control and require external control apparatus. Optical methods using holographically generated vortices can offer dynamically reconfigurable micro-optomechanical pumps. Grier *et al* [67] have demonstrated that arrays of optical vortices created with the holographic optical tweezer technique can assemble colloidal spheres. The microoptomechanical pumps are due to optical vortices and they can be configured in real time using a spatial light modulator in which the phase patterns are generated and displayed. The light passing through it produces an array of micro-mechanical pumps and traps that use optical gradient forces for trapping and photon orbital angular momentum for rotating particles. This involves the force and torque exerted by optical vortices. Dielectric objects comparable in size to the wavelength of light are drawn by optical gradient forces toward an optical vortex's bright ring, and are driven around its circumference by the tangential component of the beam's momentum flux. Colloidal particles dispersed in a viscous fluid can be stably trapped near the focal plane.

In holographic optical tweezers, multiple optical vortices in arbitrary configurations can be realized [68]. Optical vortices can induce rotation in trapped particles and clusters of particles can be assembled into functional micro-mechanics. They can be used to pump fluids through micro-fluidics channels, control flows of fluids through these channels, to mix fluids within microfluidic channels, to transport and sort particles in small length scales.

## 8.9 Communications

The points in favor of using OAM for optical communication are: (i) it is possible to coaxially propagate OAM beams with different azimuthal OAM states that are mutually orthogonal, (ii) there is minimal inter-beam cross-talk and (iii) the beams can be efficiently multiplexed and demultiplexed. As a result, multiple OAM states could be used as different carriers for multiplexing and transmitting multiple data streams, thereby potentially increasing the system capacity [69, 70].

The viability of using the orbital angular momentum [70] of light to create orthogonal, spatially distinct data-transmitting channels that are multiplexed in a single fiber has been reported [70] recently. This gives the possibility of exploiting spatial modes of fibers to enhance the data capacity. This is important, especially when we have almost exhausted available degrees of freedom to orthogonally multiplex the data and the limits imposed by optical fiber nonlinear effects. As far as the angular momentum of light is concerned, the two degrees of freedom of using spin and infinitely many discrete orbital angular momenta make a singular beam an interesting candidate. The schematic of this concept is shown in figure 8.13. The different mode patterns of angular momentum states are depicted in the interferograms shown in figure 8.14.

Beams with OAM and the ability to multiplex such beams can potentially increase the system capacity. It has been shown in millimeter-wave wireless communication links with a single aperture pair by transmitting multiple coaxial data streams 32 Gb per second.

One property of electromagnetic waves that has been recently explored is the ability to multiplex multiple beams, such that each beam has a unique helical phase front. The amount of phase front twisting indicates the orbital angular momentum state number. Beams with different orbital angular momentum are orthogonal. Such

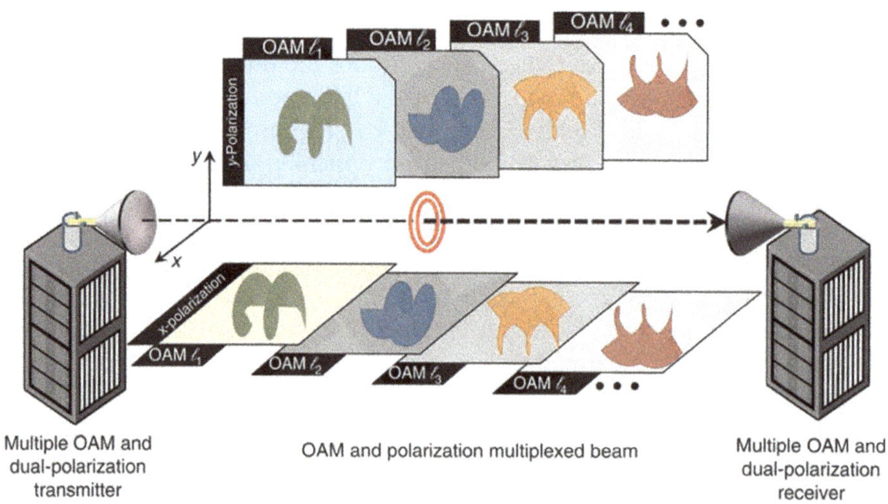


**Figure 8.13.** Concept of utilizing OAM and polarization multiplexing in a free-space mm wave communication link. This technique could have potential applications in places such as data centres, where large bandwidth links between computer clusters are required. Reprinted with permission from [71]. Copyright (2014) by Springer Nature.

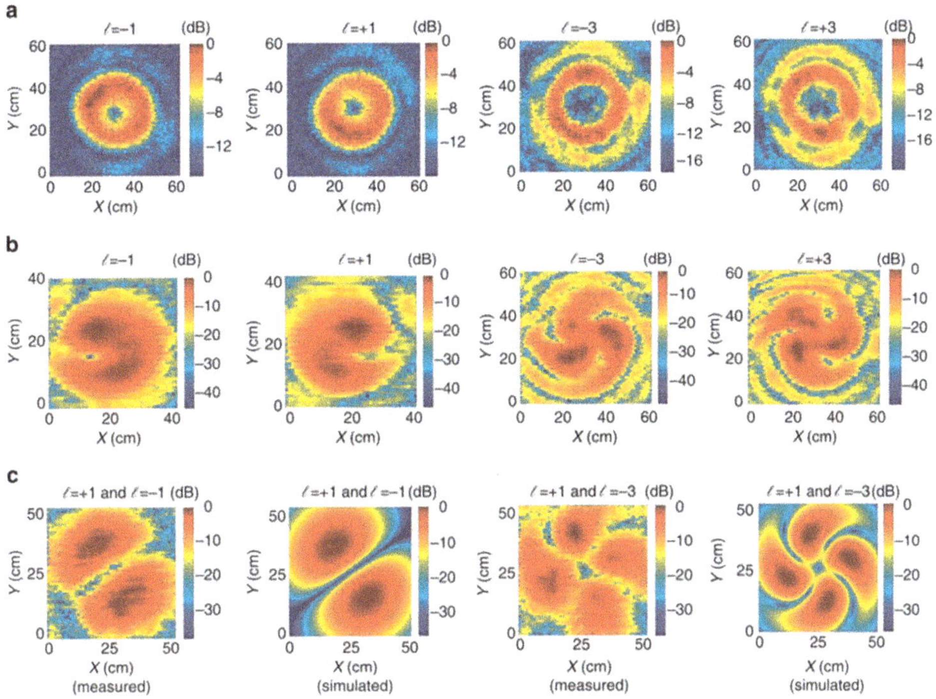


**Figure 8.14.** Normalized intensity and interferogram of mm wave OAM beams. (a) Normalized measured intensity of 4 mm wave OAM beams of charge $l = \pm 1$ and $l = \pm 3$. (b) Interferogram images of a Gaussian beam and OAM beams combined by a beam splitter. (c) Normalized measured and simulated intensity distribution of the multiplexed OAM beams after beam splitters. Reprinted with permission from [71]. Copyright (2014) by Springer Nature.

orbital angular momentum based multiplexing can potentially increase the system capacity and spectral efficiency of millimetre-wave wireless communication links with a single aperture pair by transmitting multiple coaxial data streams. A 32 Gbit per second millimetre-wave link over 2.5 m with a spectral efficiency of approximately 16 bits per second per Hertz, using four independent orbital angular momentum beams on each of two polarizations was demonstrated [71]. This group has also recovered all eight orbital angular momentum channels with bit-error rates below $3.8 \times 10^{-3}$. In addition, they have demonstrated orbital angular momentum mode demultiplexer to demultiplex four orbital angular momentum channels with cross-talk less than −12.5 dB and show an 8 Gbit per second link containing two orbital angular momentum beams on each of two polarizations.

In one demonstration [70], using over 1.1 km of a specially designed optical fiber that minimizes mode coupling, it has been shown 400 gigabits per second data transmission using four angular momentum modes at a single wavelength, and 1.6 terabits per second using two OAM modes over 10 wavelengths. These demonstrations suggest that OAM could provide an additional degree of freedom for data multiplexing in future fiber networks. A schematic of the OAM-MDM (MDM is mode division multiplexing) principle is shown in figure 8.15.

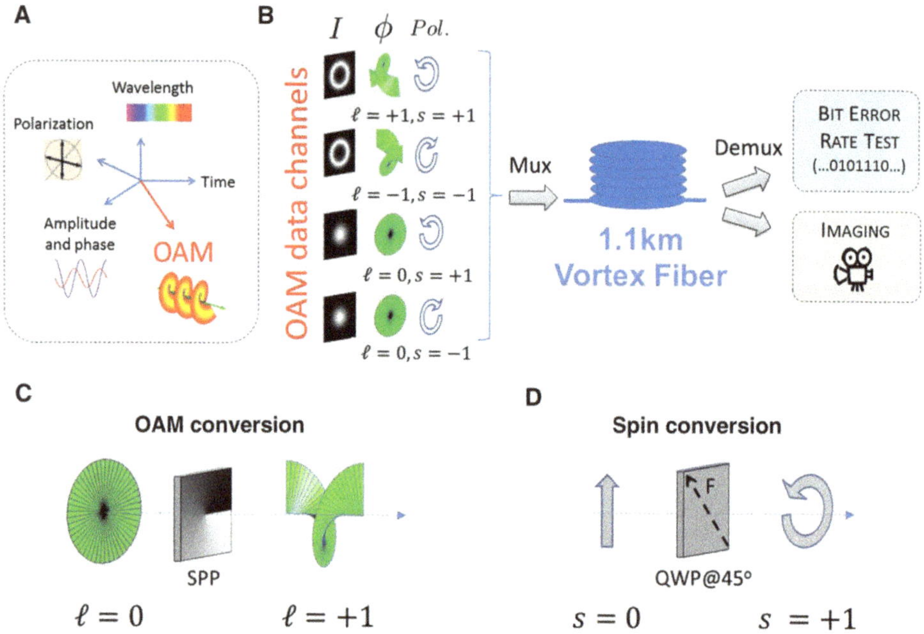


**Figure 8.15.** Schematic of the OAM-MDM principle [70]. (A) OAM may be considered as an orthogonal degree of freedom for data multiplexing. (B) Simplified OAM-MDM setup: four modes with distinct values of OAM (l) and spin (or circular polarization, s) are multiplexed into a specialty fiber, transmitted for 1.1 km, demultiplexed, and analyzed at the output by using big error rate (BER) testers and cameras. (C) Spiral phase plates (SPPs) can be used for OAM conversion from $l = 0$ to $l = 1$ state. (D) Likewise SAM states can be generated using a 45° oriented quarter wave plate from a linearly polarized light. From [70]. Reprinted with permission from AAAS.

In another demonstration [72], the transmitter and receiver units, based on spatial light modulators, prepare or measure a laser beam in one of eight pure OAM states. This is a free-space optical communication setup. They have shown that the information encoded in this way is resistant to eavesdropping in the sense that any attempt to sample the beam away from its axis will be subject to an angular restriction and a lateral offset, both of which result in inherent uncertainty in the measurement. This gives an experimental insight into the effects of aperturing and misalignment of the beam on the OAM measurement and demonstrates the uncertainty relationship for OAM.

## 8.10 Phase retrieval methods

The phase of a complex field can be extracted from the Fourier plane intensity measurements by iterative methods such as the Gerchberg–Saxton (GS) algorithm. The constraints in the complex field receiver (lens with aperture constraints) plane and the intensity measurement plane (Fourier plane) are used to recover the phase of the incident field. Instead of single intensity measurement at the Fourier plane, if multiple measurements, each of which contain different information which the earlier measurement may not contain, will result in efficient phase recovery. Therefore, what has been recently demonstrated involves two Fourier plane intensity distributions—one with an unknown complex field captured by the lens and the other with a vortex phase mask placed in front of the lens [73]. Using these two intensity measurements, it has been shown that better phase recovery is possible using iterative procedure, which is similar to the GS algorithm [74]. This subject has been explained in detail in chapter 6.

## References


[1] Masajada J 2004 Small angle rotations measurement with optical vortex interferometer *Opt. Commun.* **239** 373–81

[2] Masajada J, Popiolek A, Fraczek E and Fraczek W 2004 Vortex points localization problem in optical vortex interferometry *Opt. Commun.* **234** 23–8

[3] Masajada J and Dubik B 2001 Optical vortex generation by three plane wave interference *Opt. Commun.* **198** 21–7

[4] Vyas S and Senthilkumaran P 2007 Interferometric optical vortex array generator *Appl. Opt.* **46** 2893–8

[5] Vyas S and Senthilkumaran P 2007 Vortex array generation by interference of spherical waves *Appl. Opt.* **46** 7862–7

[6] Xavier J, Vyas S, Senthilkumaran P and Joseph J 2012 Tailored complex 3D vortex lattice structures by perturbed multiples of three-plane waves *Appl. Opt.* **51** 1872–8

[7] Popiolek-Masajada A, Kurzynowski P, Wozniak W A and Borwinska M 2007 Measurement of small wave tilt using optical vortex interferometer with the Wollaston compensator *Appl. Opt.* **46** 8039–44

[8] Borwinska M, Popiolek-Masajada A and Dubik B 2007 Reconstruction of a plane wave's tilt and orientation using an optical vortex interferometer *Opt. Eng.* **46** 073604

[9] Wozniak W A and Banach M 2009 Measurements of linearly birefringent media parameters using the optical vortex interferometer with the Wollaston compensator *J. Opt.* A **11** 094024

[10] Masajada J, Pipiolek-Masajada A and Wieliczka D M 2002 The interferometric system using optical vortices as phase markers *Opt. Commun.* **207** 85–93

[11] Fraczek W and Miroczka J 2008 Optical vortices as phase markers to wave-front deformation measurement *Metrol. Meas. Syst.* **15** 433–40

[12] Kurzynowski P, Borwinska M and Masajada J 2010 Optical vortex sign determination using self-interference methods *Opt. Appl.* **40** 165–75

[13] Mokhun I and Galushko Y 2008 Detection of vortex sign for scalar speckle fields *Ukr. J. Phys. Opt.* **9** 246–55

[14] Fraczek E, Fraczek W and Masajada J 2006 The new method of topological charge determination of optical vortices in the interference field of the optical vortex interferometer *Optik* **117** 423–5

[15] Fraczek E, Fraczek W and Mroczka J 2005 The experiemental method for topological charge determination of optical vortices in a regular net *Opt. Eng.* **44** 025601

[16] Malacara D 1978 *Optical Shop Testing* (New York: Wiley)

[17] Sirohi R S and Kothiyal M P 1987 Double wedge plate shearing interferometer for collimation test *Appl. Opt.* **19** 4054–6

[18] Kothiyal M P and Sirohi R S 1987 Improved collimation testing using Talbot interferometry *Appl. Opt.* **19** 4056–7

[19] Sriram K V, Kothiyal M P and Sirohi R S 1993 Self-referencing collimation testing techniques *Opt. Eng.* **32** 94–101

[20] Senthilkumaran P 2003 Optical phase singularities in detection of laser beam collimation *Appl. Opt.* **42** 6314–20

[21] Chang C W and Su D C 1991 Collimation method that uses spiral gratings and Talbot interferometry *Opt. Lett.* **16** 1783–4

[22] Sriram K V, Kothiyal M P and Sirohi R S 1994 Collimation testing with linear dual-field, spiral, and evolute gratings: A comparative study *Appl. Opt.* **33** 7258–60

[23] Furhapter S, Jesacher A, Bernet S and Ritsch-Marte M 2005 Spiral interferometry *Opt. Lett.* **30** 1953–5

[24] Jesacher A, Fürhapter S, Bernet S and Ritsch-Marte M 2006 Spiral interferogram analysis *J. Opt. Soc. Am.* A **23** 1400–8

[25] Furhapter S, Jesacher A, Bernet S and Ritsch-Marte M 2005 Spiral phase contrast imaging in microscopy *Opt. Express* **13** 689–94

[26] Goodman J W 2007 *Introduction to Fourier Optics* (Englewood, CO: Roberts)

[27] Khonina S N, Kotlyar V V, Shinkaryev M V, Soifer V S and Uspieniev G V 1992 The phase rotor filter *J. Mod. Opt.* **39** 1147–54

[28] Lohmann A W, Mendlovic D and Zaievsky Z 1996 Fractional Hilbert transform *Opt. Lett.* **21** 281–3

[29] Lohmann A W, Tepichin E and Ramirez J G 1997 Optical implementation of the fractional Hilbert transform for two-dimensional objects *Appl. Opt.* **36** 6620–6

[30] Davis J A, McNamara D E and Cottrell D M 1998 Analysis of the fractional Hilbert transform *Appl. Opt.* **37** 6911–3

[31] Davis J A, McNamara D E, Cottrell D M and Campos J 2000 Image processing with the radial Hilbert transform: theory and experiments *Opt. Lett.* **25** 99–101

[32] Crabtree K, Davis J A and Moreno I 2004 Optical processing with vortex producing lens *Appl. Opt.* **43** 1360–6

[33] Sharma M K, Joseph J and Senthilkumaran P 2011 Selective edge enhancement using anisotropic vortex filter *Appl. Opt.* **50** 5279–86

[34] Sharma M K, Joseph J and Senthilkumaran P 2013 Selective edge enhancement using shifted anisotropic vortex filter *J. Opt.* **42** 1–7

[35] Situ G, Pedrini G and Osten W 2009 Spiral phase filtering and orientation-selective edge detection/enhancement *J. Opt. Soc. Am.* A **26** 1788–97

[36] Kim G-H, Lee H J, Kim J-U and Suk H 2003 Propagation dynamics of optical vortices with anisotropic phase profiles *J. Opt. Soc. Am.* B **20** 351–9

[37] Sharma M K, Joseph J and Senthilkumaran P 2014 Fractional vortex dipole phase filter *Appl. Phys.* B **117** 325–32

[38] Situ G, Warber M, Pedrini G and Osten W 2010 Phase contrast enhancement in microscopy using spiral phase filtering *Opt. Commun.* **283** 1273–7

[39] Maurer C, Jesacher A, Furhapter S, Bernet S and Ritsch-Marte M 2008 Upgrading a microscope with a spiral phase plate *J. Miscrosc.* **230** 134–42

[40] Jesacher A, Furhapter S, Bernet S and Ritsch-Marte M 2005 Shadow effects in spiral phase contrast microscopy *Phys. Rev. Lett.* **94** 1–4

[41] Foo G D, Palacios D M and Swartzlander G A 2005 Optical vortex coronograph *Opt. Lett.* **30** 3308–10

[42] Palacios D M, Hunyadi S L and Saraha L 2006 Low-order aberration sensitivity of an optical vortex coronograph *Opt. Lett.* **31** 2981–3

[43] Anzoline G, Tamburini F, Bianchini A, Umbriaco G and Barbieri C 2008 Optical vortices with star light *Astron. Astrophys.* **488** 1159–65
[44] Tamburini F, Anzlin G, Umbrioco G, Bianchini A and Barbieri C 2006 Overcoming the rayleigh limit *Phys. Rev. Lett.* **97** 1–3
[45] Swartzlander G A Jr 2001 Peering into darkness with a vortex spatial filter *Opt. Lett.* **26** 497–9
[46] Spektor B, Normatov A and Shamir J 2008 Singular beam microscopy *Appl. Opt.* **47** A78–87
[47] Quabis S, Dorn R, Eberler M, Glocki O and Geuchs G 2000 Focusing light to a tighter spot *Opt. Commun.* **179** 1–7
[48] Dorn R, Quabis S and Leuchs G 2003 Sharper focus for a radially polarized light beam *Phys. Rev. Lett.* **91** 1–4
[49] Helseth L E 2006 Smallest focal hole *Opt. Commun.* **257** 1–8
[50] Bianchini P, Peres C, Oneto M, Galiani S, Vicidomini G and Diaspro A 2015 STED nanoscopy: a glimpse into the future *Cell Tissue Res.* **360** 143–50
[51] Well S W and Wichmann I 1994 Breaking the diffraction resolution limit by stimulated - emission - depletion fluorescence microscopy *Opt. Lett.* **19** 780–2
[52] Torok P and Munro P R T 2004 The use of Gauss-Laguerre vector beams in STED microscopy *Opt. Express* **12** 3605–17
[53] Willig K I, Keller J, Bossi M and Hell S W 2006 STED microscopy resolves nanoparticle assemblies *New J. Phys.* **8** 1–8
[54] Bokor N, Iketaki Y, Watanabe T, Daigoku K, Davidson N and Fujii M 2007 On polarization effects in fluorescence depletion microscopy *Opt. Commun.* **272** 263–8
[55] Schonle A, Keller J and Hell S W 2007 Efficient fluorescence inhibition patterns for resolft microscopy *Opt. Express* **15** 3361–71
[56] Klar T A, Jakobs S, Dyba M, Egner A and Hell S W 2000 Fluorescence microscopy with diffraction resolution barrier broken by stimulated emission *PNAS* **97** 8206–10
[57] Ashkin A, Dziedzic J M, Bjorkholm J E and Chu S 1986 Observation of a single-beam gradient force optical trap for dielectric particles *Opt. Lett.* **11** 288–90
[58] Ashkin A and Dziedzic J M 1971 Optical levitation by radiation pressure *Appl. Phys. Lett.* **19** 283–5
[59] Gahagan K T and Swartzlander G A Jr 1998 Trapping of low-index micro particles in an optical vortex *J. Opt. Soc. Am.* B **15** 524–34
[60] Gahagan K T and Swartzlander G A Jr 1996 Optical vortex trapping of particles *Opt. Lett.* **21** 827–9
[61] Gahagan K T and Swartzlander G A 1999 Simultaneous trapping of low-index and high-index microparticles observed with an optical vortex trap *J. Opt. Soc. Am.* B **16** 533–7
[62] Heckenberg N R, Nieminen T A, Friese M E J and Rubinsztein-Dunlop H 1998 Trapping microscopic particles with singular beams *Proc. SPIE* **3487** 46
[63] Burns M M, Fournier J M and Golovchenko J A 1990 Optical matter; crystallization and binding in intense optical fields *Science* **249** 749–54
[64] Grzegorczyk T M and Kong J A 2007 Analytical prediction of stable optical trapping in optical vortices created by three TE or TM plane waves *Opt. Express* **15** 8010–8
[65] Padgett M J and Allen L 1997 Optical tweezers and spanners *Phys. World* **10** 35–8
[66] Simpson N B, Allen L and Padgett M J 1996 Optical tweezers and optical spanners with Laguerre–Gaussian modes *J. Mod. Opt.* **43** 2485–92

[67] Ladavac K and Grier D G 2004 Microoptomechanical pumps assembled and driven by holographic optical vortex arrays *Opt. Express* **12** 1144–9

[68] Curtis J E, Koss B A and Grier D G 2004 Use of multiple optical vortices for pumping, mixing and sorting *US Patent* 6737634

[69] Willner A E *et al* 2015 Optical communications using orbital angular momentum beams *Adv. Opt. Photon.* **7** 66–106

[70] Bozinovic N, Yue Y, Ren Y, Tur M, Kristensen P, Huang H, Willner A E and Ramachandran S 2013 Terabit-scale orbital angular momenum mode division multiplexing in fibers *Science* **340** 1545–8

[71] Yan Y *et al* 2014 High-capacity millimeter-wave communications with orbital angular momentum multiplexing *Nat. Commun.* **5** 4876

[72] Gibson G, Courtial J, Padgett M J, Vasnetsov M, Pas'ko V, Barnett S M and Franke-Arnold S 2004 Free-space information transfer using light beams carrying orbital angular momentum *Opt. Express* **12** 5448–56

[73] Sharma M K, Gaur C, Senthilkumaran P and Khare K 2015 Phase imaging using spiral-phase diversity *Appl. Opt.* **54** 3979–85

[74] Khare K, Sharma M K and Senthilkumaran P 2018 Non-interferometric phase measurement *US Patent* 9947118

**IOP** Publishing

Singularities in Physics and Engineering
Properties, methods, and applications
**Paramasivam Senthilkumaran**

# Chapter 9

# Polarization singularities

## 9.1 Polarization of light

Polarization singularities in coherent optical fields are the points at which any of the parameters that define the state of polarization of light is indeterminate. In this chapter we deal with singularities that occur in completely polarized fields. We first give a brief introduction to polarization, representation of polarized light and go on to types of polarization distributions and singularities.

Polarization of light refers [1] to the variation in the magnitude and direction of electric field vector of an optical field in one cycle. Elliptically polarized light is the most general form of polarization—circular polarization and linear polarization are the two extremes. In linear polarization, the electric field oscillations are restricted to a plane, the magnitude of the electric field becomes zero twice in a cycle and the oscillation is simple harmonic for a coherent light. In elliptically polarized light the direction of the electric field vector changes by $2\pi$ in one cycle whereas the amplitude is always non-zero and oscillates between a maximum and a minimum. As the light propagates, the tip of the electric field draws an ellipse on a projected plane. In circularly polarized light the tip of the electric field changes by $2\pi$ in a cycle like elliptically polarized light and the magnitude of the electric field remains always constant.

The polarized light can be realized as a superposition of two orthogonal simple harmonic oscillations. Although there are an infinite number of ways to construct the two orthogonal states [2], the most commonly used orthogonal states are two linearly polarized states in which the oscillations for the two states occur in perpendicular planes. These two orthogonal states are chosen depending on the problem that is being dealt with. In reflection, one state is taken as parallel to the plane of incidence and the other perpendicular to it. In propagation through polarizing elements one state is taken parallel to the pass plane (for polarizers) or fast axis (for wave plates) and the other state is taken perpendicular to the first plane [3, 4]. This is how linear birefringence is dealt with. In problems involving circular birefringence, two circularly polarized fields, one right-handed and the other left-handed polarized light are considered as orthogonal states.

Any state of polarization (SOP) can be decomposed into its component states. Under linear decomposition, any change in the amplitude of one of the components results in a change in the azimuth (orientation angle of the major axis of the ellipse) and any phase difference between the orthogonal states leads to an ellipticity change. Under circular decomposition, in contrast, any change in the amplitude of one of the components results in a change in the ellipticity and any phase difference between the orthogonal states leads to an azimuth change.

In polarization optics for superposition of light beams, any change in the phase difference between the orthogonal linear states results in a SOP change whereas any change in the phase difference between the beams in the same SOP results in an intensity change. So, superpositions of beams in orthogonal states and in the same states are different. In fact there can be an infinite number of ways that orthogonal states can be constructed. But for the moment let us consider three sets of orthogonal states—two of them linear and one circular SOP.

## 9.2 Stokes parameters and Poincare sphere representation

Stokes parameters can be used to describe the state of polarization of a light [5, 6]. For a fully polarized beam they are defined as

$$
\begin{aligned}
S_0 &= I_x + I_y = |E_x|^2 + |E_y|^2 = I_0 \\
S_1 &= I_x - I_y = |E_x|^2 - |E_y|^2 = I_0\cos(2\chi)\cos(2\gamma) \\
S_2 &= I_{(+45^\circ)} - I_{(-45^\circ)} = 2\,\mathrm{Re}(E_x^* E_y) = I_0\cos(2\chi)\sin(2\gamma) \\
S_3 &= I_{(\mathrm{RCP})} - I_{(\mathrm{LCP})} = 2\,\mathrm{Im}(E_x^* E_y) = \sin(2\chi)
\end{aligned}
$$

where $I_x$, $I_y$, $I_{(+45^\circ)}$, $I_{(-45^\circ)}$, $I_{(\mathrm{LCP})}$ and $I_{(\mathrm{RCP})}$ are the component intensities when a given SOP is decomposed into linear states oriented along $x$, $y$, (+45°), (−45°) and circular states left circularly polarized (LCP) and right circularly polarized (RCP) respectively. $E_x$ and $E_y$ are the transverse electric field components. $\chi$ and $\gamma$ are the ellipticity and azimuth respectively. Using these Stokes parameters, a sphere of radius $S_0$ in which $S_1$, $S_2$, $S_3$ forming three orthogonal axis can be constructed. This sphere is the Poincare sphere and every point on the surface of the sphere represents a particular state of polarization (figure 9.1). Normalized Stokes parameters can be obtained by dividing each of the Stokes parameters by $S_0$. In this case the Poincare sphere has unit radius. The north and south poles of the Poincare sphere represent right and left circularly polarized light, equatorial points represent linearly polarized light with different azimuths and rest of the points represent elliptically polarized light. Points in the northern hemisphere are right-handed polarization states while points in the southern hemisphere are of left-handed.

All conceivable SOPs can be represented using the Poincare sphere in which each point on its surface represents a particular SOP with coordinates ($S_1$, $S_2$, $S_3$). Alternatively, any general polarization state can be described by a point $P(2\gamma, 2\chi)$ on the Poincare sphere, where $2\gamma$ and $2\chi$ are longitude and latitude, respectively. The

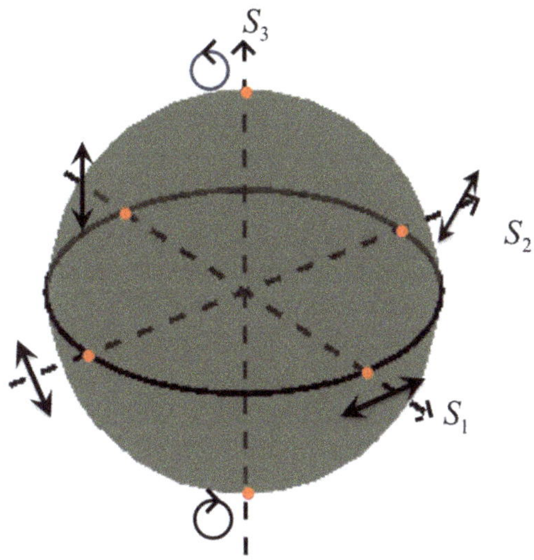


**Figure 9.1.** Poincare sphere. The north pole represents right circular polarization, the south pole represents left circular polarization, points on the equator represent linear states of polarization and the rest of the points represent elliptical polarization.

quantities $\gamma$ (azimuth) and $\chi$ (ellipticity) can be found from normalized Stokes parameters as $\gamma = \frac{1}{2}\tan^{-1}(\frac{S_2}{S_1})$ and $\chi = \frac{1}{2}\sin^{-1}(\frac{S_3}{S_0})$. Points along a given latitude represent SOPs of the same ellipticities but different azimuths, whereas points on a given longitude represent SOPs of the same azimuth but different ellipticities.

Consider a polarization state represented by a point A on the Poincare sphere. The intensity component of A in any other state (say B) can be found by [2]

$$I_{(A/B)} = I_A \cos^2\left(\frac{1}{2}c\right) \tag{9.1}$$

where $c$ is the length of the geodesic arc that connects states A and B. This is the angular separation of the states on the Poincare sphere. Using this definition the difference in the component intensities by decomposing a SOP into ($\pm1$, 0, 0) states gives $S_1$. Similarly, the component intensity difference between states (0, $\pm1$, 0) gives $S_2$ and between states (0, 0, $\pm1$) gives $S_3$.

### 9.2.1 Homogeneous polarization

Homogeneously polarized beams refer to beams having uniform polarization across their cross section. Circularly polarized beams and linearly polarized beams are examples of beams having homogeneous polarization. The SOP is the same at all points in the beam and hence the whole beam can be represented by a point on the Poincare sphere with coordinates ($S_1$, $S_2$, $S_3$). The interference of two beams A and B, in different states of polarization can lead to a resultant beam that can vary in the SOP as well as in intensity. By decomposing the state B into two orthogonal states—one being in state A and other orthogonal to it ($A'$), the interference of the A component of B and state A leads to intensity change. This intensity changed beam in state A and the beam in ($A'$ component of B) interfere to change the SOP of the final beam.

### 9.2.2 Inhomogeneous polarization

Inhomogeneous polarization refers to beams having spatially varying polarization distribution across the beam cross section. A radially polarized beam is an example of an inhomogeneously polarized beam. Since there is no single SOP associated with the beam, inhomogeneously polarized beams are represented by regions on the Poincare sphere. The Stokes parameters themselves are functions of position coordinates in the beam i.e. $S_0(x, y)$, $S_1(x, y)$, $S_2(x, y)$ and $S_3(x, y)$. Inhomogeneous polarization distributions may host polarization singularities. Polarization distributions that are inhomogeneous with polarization singularities are of interest and will be studied in great detail in this chapter.

There has been considerable interest in paraxial fields with slowly varying polarization distributions in recent years [7–11]. Ellipse fields have spatially varying elliptical SOPs. Likewise, vector fields have spatially varying linear SOPs.

**Ellipse fields**: Ellipse fields are inhomogeneously polarized fields in which the predominant SOPs are elliptical. Linear polarization states may occur at isolated points or on points in a line but not in a region. Two inhomogeneous polarization distributions in ellipse fields are shown in figure 9.2. In these figures the SOP distributions do not contain any polarization singularities. In such fields, at each of the position coordinate $S_3(x, y) \neq 0$. Ellipse fields can have right- or left-handed polarization states distributed on it and the linear polarization states appearing on a line called the L-line separate the two regions with opposite handedness.

**Vector fields**: Vector fields are inhomogeneously polarized fields in which the predominant SOPs are linear. In such fields $S_3 = 0$ at all points in the beam cross section. Two inhomogeneous polarization distributions in vector fields are shown in figure 9.3. These two example fields do not contain any polarization singularity.

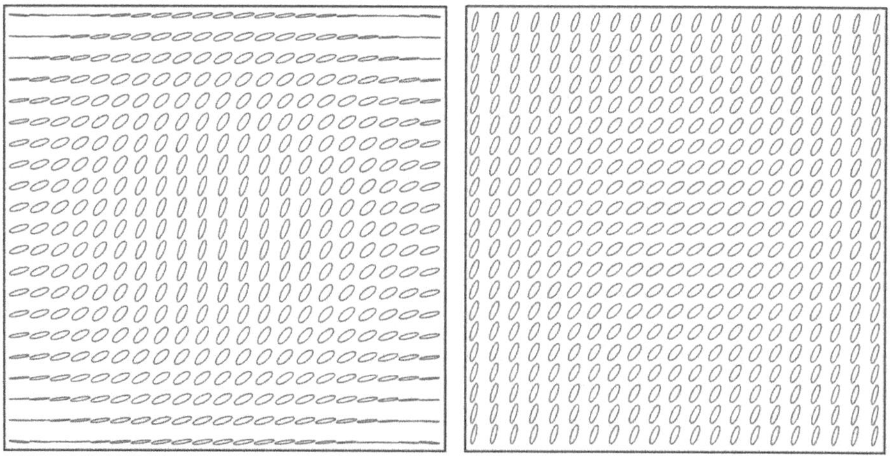

**Figure 9.2.** Inhomogeneous polarization distributions (left panel). Ellipse field distributions (right panel).

**Figure 9.3.** Inhomogeneous polarization distributions (left panel). Vector field distributions (right panel).

### 9.2.3 Encoding phase into polarization

The phase information of an optical wavefront can be represented in terms of polarization. We will give few examples here. Consider a wavefront with a dislocation line along $x = 0$ where the wave is polarized along $y$ direction.

$$\begin{aligned} U(x, y) &= \exp\{i\phi_0\} \quad x > 0 \\ &= \exp\{i\phi_0 + \pi\} \quad x < 0. \end{aligned} \tag{9.2}$$

Here $\phi_0$ is a constant phase. This wavefront with a line phase defect can be represented by a plane wave with SOPs marked as shown in figure 9.4(a). Note that the phase defect has disappeared by adopting this way of representing phase by polarization. As shown in the first example for linearly polarized light, it is easy to represent the phase difference of $\pi$ but it is difficult to encode other phase values. Using circularly polarized light, it is possible to encode any phase value into polarization. Consider again a similar example of a wavefront with a dislocation line along $x = 0$ where the phase difference between the two halves is not $\pi$

$$\begin{aligned} U(x, y) &= \exp\{i\phi_0\} \quad x > 0 \\ &= \exp\{i\phi_0 + \Delta\} \quad x < 0 \end{aligned} \tag{9.3}$$

Using circular polarization, different phase jumps, $\Delta = \frac{\pi}{4}, \frac{\pi}{2}$ and $\pi$ can be represented in a plane wave as depicted in figure 9.4(a), second, third and fourth rows. Note that the phase jump of $\pi$ is not represented by left- and right-handed circular polarization states. Likewise, rotation of the plane of polarization of plane polarized light can be construed as introducing phase difference between the two orthogonal circular polarization component states as depicted in figure 9.4(b).

The key to recording phase in terms of polarization is by using superposition of orthogonal states of polarization in which the phase that has to be recorded is given as the phase difference between orthogonal states of polarization. The superposition results in a spatially varying inhomogeneous polarization distribution. Later, one of the orthogonal states from the superposition can be retrieved which has the desired phase variation in the wavefront.

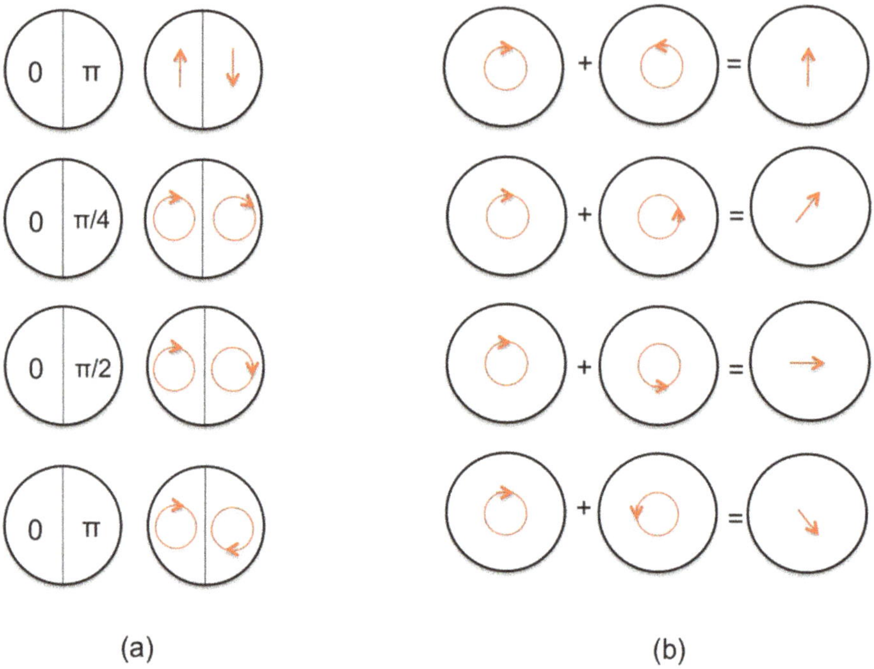


**Figure 9.4.** (a) Depiction of phase variation as polarization variation. (b) Superposition of equal amplitude left and right circular polarization state resulting in a linear polarization state.

Figure 9.5 shows an inhomogeneous polarization distribution which is decomposed into orthogonal linear states and figure 9.6 shows the same distribution is decomposed into orthogonal circular polarization states. It can be seen that depending on the type of decomposition the amplitude and phase distributions in different polarization basis states are different, but both the superpositions lead to the same resultant polarization distribution. It can be seen from these two examples that a plane wave or a helical wave can be extracted from the SOP distribution by the appropriate choice of polarized light.

## 9.3 Stokes fields

Using the generalized Stokes parameters $S_1$, $S_2$ and $S_3$, different Stokes fields can be generated. A Stokes field $S_{12}$ is a complex field and it can be formed by using $S_1$ and $S_2$ as $S_1 + iS_2 = A_{12}\exp\{i\phi_{12}\}$. Similarly one can construct Stokes fields $S_{23}$ using $S_2$ and $S_3$ or Stokes field $S_{31}$ using $S_3$ and $S_1$. Among these three Stokes fields, complex field $S_{12}$ constructed using otherwise real Stokes parameters $S_1$ and $S_2$, is of particular interest in the analysis of C-point and V-point polarization singularities. In the field $S_{12}$, the Stokes intensity is given by $A_{12}^2$ and Stokes phase is $\phi_{12} = \tan^{-1}[\frac{S_2}{S_1}]$. Since the information about $S_3$ is absent in $S_{12}$, the Stokes phase $\phi_{12}$ is the same for both left- and right-handed SOP distributions. Unlike phase singularities, ellipse field singularities can have handedness, which is not revealed by the Stokes field $S_{12}$. Similarly, distinction between C-points and V-points is also difficult using the Stokes field $S_{12}$ as they look same. This will be evident from many of the figures that follow.

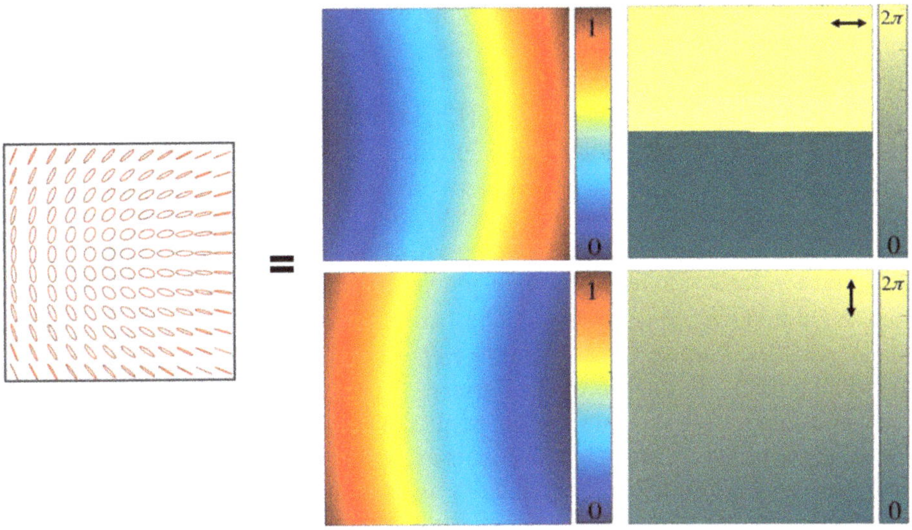


**Figure 9.5.** Decomposition of an inhomogeneous polarization distribution in the linear polarization basis. The amplitude and the phase variations in the component states are shown on the left and right respectively.

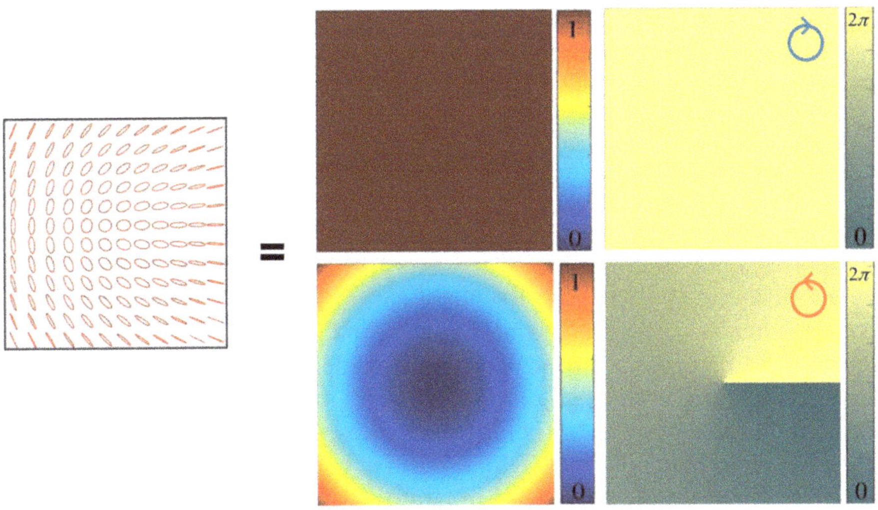


**Figure 9.6.** Decomposition of an inhomogeneous polarization distribution in the circular polarization basis. The amplitude and the phase variations in the component states are shown on the left and right respectively.

## 9.4 Ellipse field singularities

In polarization singularities the azimuth of the polarization state plays an important role [12]. In an inhomogeneous polarization distribution of ellipse fields, the point at which the azimuth is indeterminate is called a C-point singularity [13–15]. A C-point singularity is a point of circular polarization state. The neighborhood of this singular point has SOP distribution consisting of polarization ellipses where the orientation of these ellipses are arranged in an orderly fashion orienting in different directions.

This means, the major axes of the ellipses are arranged in an orderly fashion. These azimuths undergo rotation in a clockwise or anti-clockwise sense and in one complete closed path around the C-point the total rotation of azimuth can be $\pm\pi$ for the lowest order C-point singularity [13, 15, 16]. These C-point singularities are characterized by a C-point index given by

$$I_c = \frac{1}{2\pi} \oint \nabla\gamma \cdot dl \tag{9.4}$$

The C-point index $I_c$ can take both half-integer and integer values. The lowest order C-points have index values $I_c = \pm\frac{1}{2}$. They are named the lemon, monstar and star type singularities. Lemon and monstar have positive index whereas star has negative index. The polarization distributions corresponding to a lemon, monstar and a star are given in figure 9.7. At the singular point the field can have any intensity value. Hence it is possible to have bright C-points as well as dark C-points. Since at the C-point, the light is circularly polarized only the azimuth is undefined but the handedness is defined.

Let us consider the polarization distribution shown in figure 9.8(a) for a lemon. The circular polarization is at the center of the circle shown in the SOP distribution. The circular path shown is in the anti-clockwise sense and the SOPs along its path are elliptical. The major axes of all these ellipses at the numbered locations are extracted and are shown in figure 9.8(b). The major axis undergo a rotation of $\pi$ during the complete circular excursion around the singular point. Figure 9.8(c) shows rotation of the major axis by stretching the circular path into a straight line and in figure 9.8(d) these major axes are arranged with one end of each touching a common point to help visualize the sense of rotation of the major axis. In this figure we can see that the rotation is in the anti-clockwise sense and hence the index is taken to be positive and the amount of rotation is half of $2\pi$, hence the index for the C-point singularity here is $I_c = \frac{1}{2}$. In other words, rotation of the major axes of the surrounding ellipses and the sense of the circular path are the same. Similarly in figure 9.9, a negative index star is taken as an example and a similar depiction is made.

Since the magnitude and sign of the singularity indices are all based on tracking the rotation of the major axis of polarization ellipses or the rotation of the plane of polarization of a plane polarized light, a positive index lemon and a negative index star SOP distributions are considered in figures 9.10 and 9.11 respectively. In these

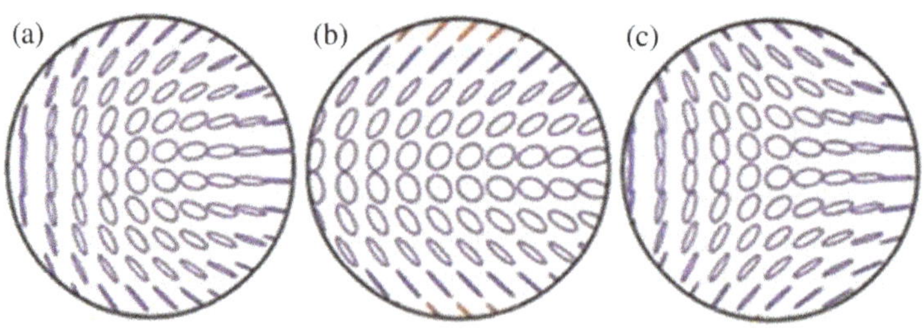


**Figure 9.7.** The SOP distributions for (a) lemon, (b) monstar and (c) star. Reproduced with permission from [26]. Copyright (2017) by OSA.

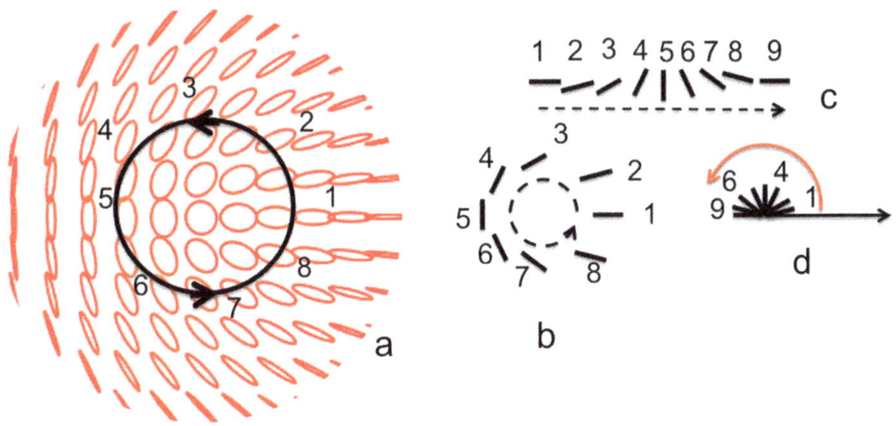


**Figure 9.8.** Lemon—positive half index ellipse field singularity. (a) SOP distribution. (b) The major axes of ellipses in the neighborhood of C-point plotted. (c) These major axes are arranged in a linear array and (d) arranged to indicate which way they rotate.

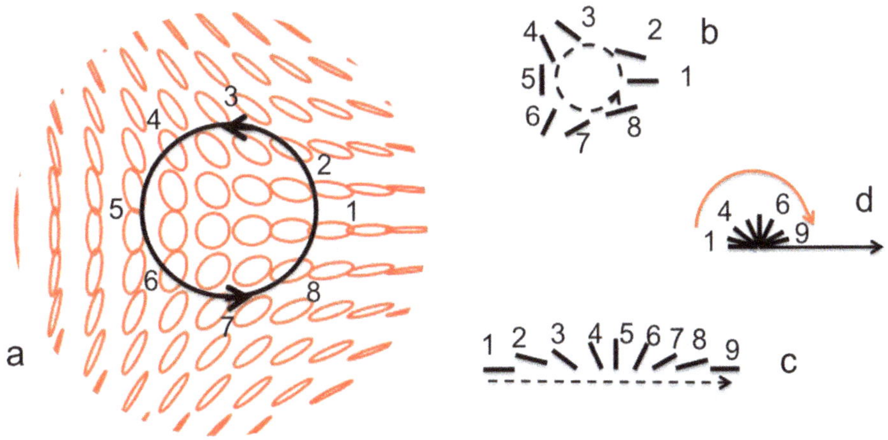


**Figure 9.9.** Star—negative half-index ellipse field singularity. (a) SOP distribution. (b) The major axes of ellipses in the neighborhood of C-point plotted. (c) These major axes are arranged in a linear array and (d) arranged to indicate which way they rotate.

figures, SOPs are drawn from the continuously varying polarization distribution by taking samples that are equally spaced both in the $x$ and $y$ directions. The sample is such that the figures represent the true spatial variations in the SOP. Samples at shorter intervals are needed if the variation is faster to capture the true picture. The concept of sampling theorem (signal processing) is applicable here in polarization sampling. In these two examples, around (neighborhood SOPs) a circular polarization point, the major axis of the ellipses on a positively oriented (anti-clockwise) closed curve, undergo rotation. If this rotation is anti-clockwise the index is positive and if it is clockwise, the index is negative.

The star, lemon or monstar, can all have right- or left-handed SOP distributions. Throughout this chapter the right- and left-handedness are distinguished by two

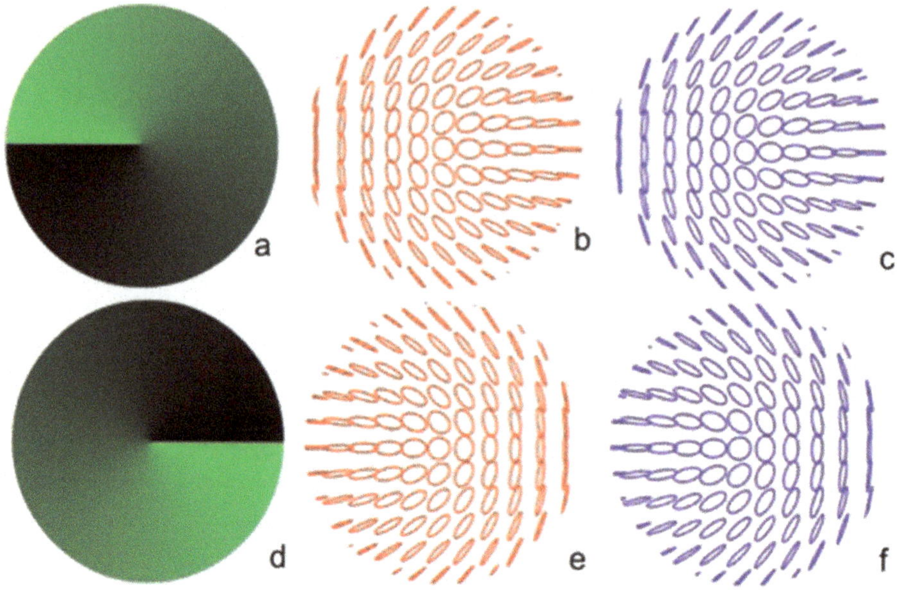


**Figure 9.10.** (a) and (d) Stokes phase distributions. (b) and (e) right-handed, (c) and (f) left-handed lemons. (b) is orthogonal to (f) and (c) is orthogonal to (e). Note that the Stokes phase is the same for both right- and left-handed singularity.

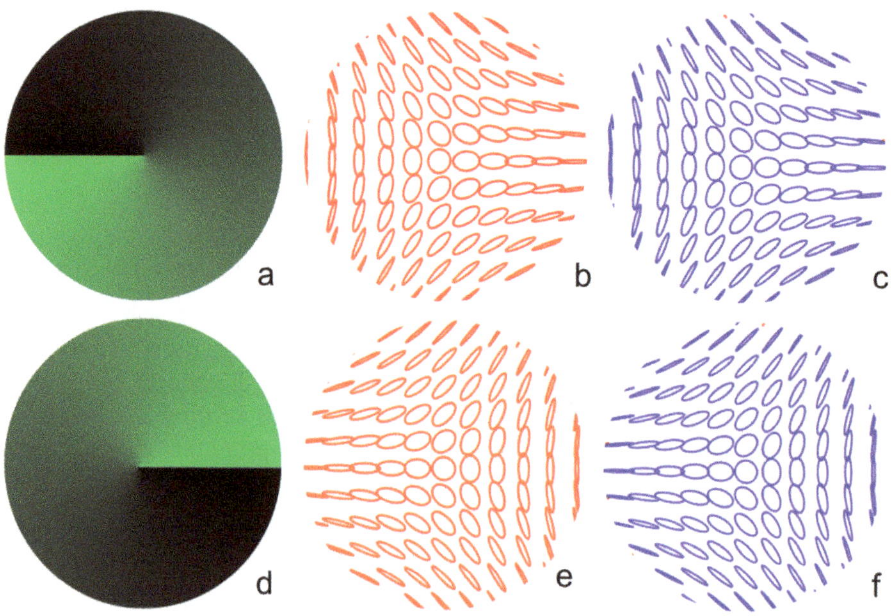


**Figure 9.11.** The same as in figure 9.10 but for a star.

different colors (red and blue respectively). For any given polarization distribution with Stokes parameter ($S_1$, $S_2$, $S_3$) as coordinates on the Poincare sphere, the state represented by a point on the Poincare sphere by ($-S_1$, $-S_2$, $-S_3$) is the orthogonal state. Similarly for polarization distributions also, we can define orthogonal polarization distributions [17, 18]. For the distribution shown in figure 9.10(b), the SOP distribution shown in figure 9.10(f) is orthogonal, for the one in figure 9.10(c) the orthogonal distribution is shown in figure 9.10(e), and so on. A similar pattern can be seen for star distributions shown in figure 9.11. For a (red) lemon at the north pole of the Poincare sphere the orthogonal state is a (blue) lemon (rotated) in the south pole. If you compare the two distributions, the SOP at every spatial point in one distribution is orthogonal in the other frame at the corresponding spatial point. Therefore, there is a point by point correspondence. For example a lemon with its separatrix rotated by different angles [12] occupies the same polar region, but the distribution within this region is different from one lemon to another. But each of them can be considered as a distinct polarization distribution. This is similar to the fact that two linearly polarized lights with their vibration planes oriented at 0° and 30° are considered as two distinct SOPs even though they both are linear. Orthogonal states in linear polarization are also linear and no-one brushs them aside as simple linear states. Similarly two lemons or stars with their separatrix rotated by a small angle are considered different distributions.

When the C-point index of the polarization singularity takes integer values, the major axis of each of the ellipse states is pointing radially or azimuthally as shown in figure 9.12 when the index $I_c = 1$. When the C-point index $I_c = -1$ the SOP distributions are as shown in figure 9.13. In figures 9.14, 9.15 and 9.16, C-point singularities with $|I_c| = \frac{3}{2}$, 2 and $\frac{5}{2}$ are shown. In each of these figures, the first rows represent positive index C-points and the second rows represent negative index C-points. The Stokes phase of a positive index and that of a negative index singularity is conjugate of each other. Similarly in figure 9.10 if the index is reversed one can get figure 9.11. Similarly by inverting the index, the distributions can be switched between figures 9.12 and 9.13.

## 9.5 Vector field singularities

In spatially varying linearly polarized fields, the point at which the azimuth is indeterminate and is surrounded by points having linear states of polarization with varying azimuth such that

$$\eta = \frac{1}{2\pi} \oint \nabla\gamma \cdot dl \tag{9.5}$$

is called a V-point singularity. The index $\eta$ which is called the V-point index or Poincare–Hopf index characterizes V-point singularities [15, 19]. Unlike ellipse field singularities, the vector field singularities can take only integer values. Since for a linearly polarized light the handedness is undefined, at the V-point singularity the azimuth as well as the handedness are undefined and hence the field itself is zero at the singular point. Hence a V-point singularity occurs at an intensity null point. The SOP distribution shown in figure 9.17(a) is called the radially polarized beam and in

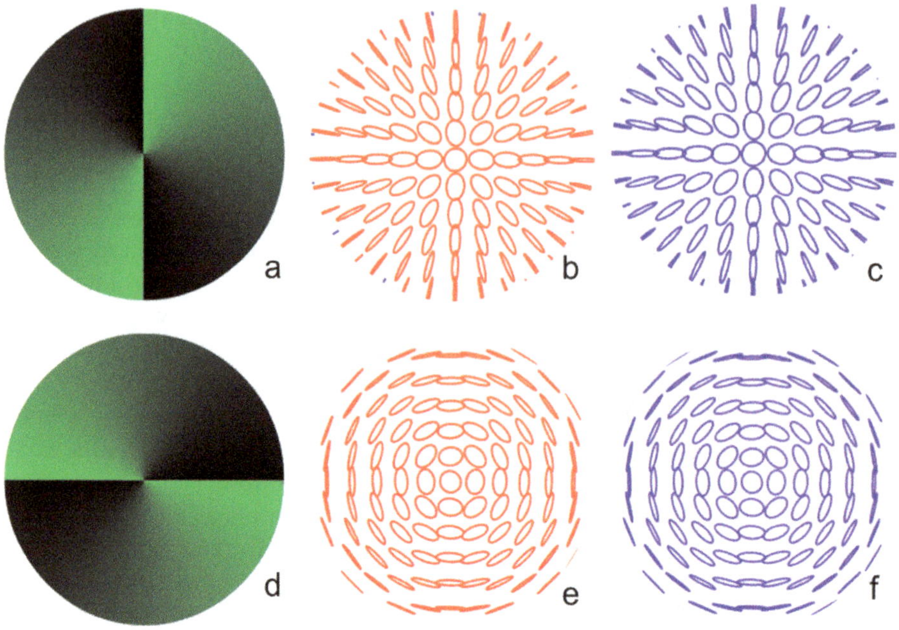


**Figure 9.12.** (a) and (d) Stokes phase distributions. (b) and (e) right-handed, (c) and (f) left-handed polarization singularity with $I_c = +1$. (b) and (f) are orthogonal to each other, (c) and (e) are orthogonal to each other.

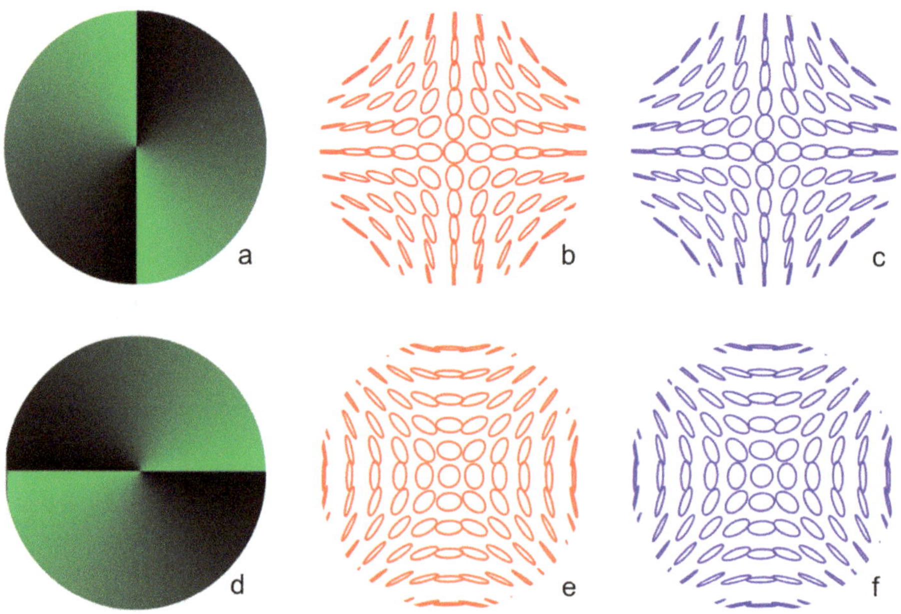


**Figure 9.13.** The same as in figure 9.12 but for a polarization singularity with $I_c = -1$.

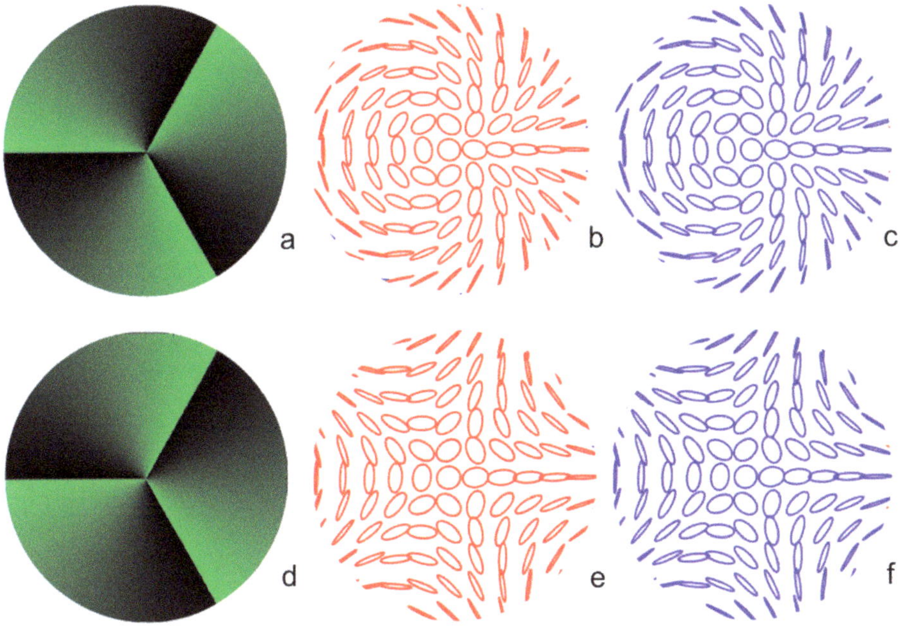


**Figure 9.14.** Stokes phase for right-handed and left-handed polarization singularity with $I_c = \frac{3}{2}$ in the first row (a)–(c) and $I_c = \frac{-3}{2}$ in the second row (d)–(f). The second row is the Stokes conjugate of the first row and vice versa.

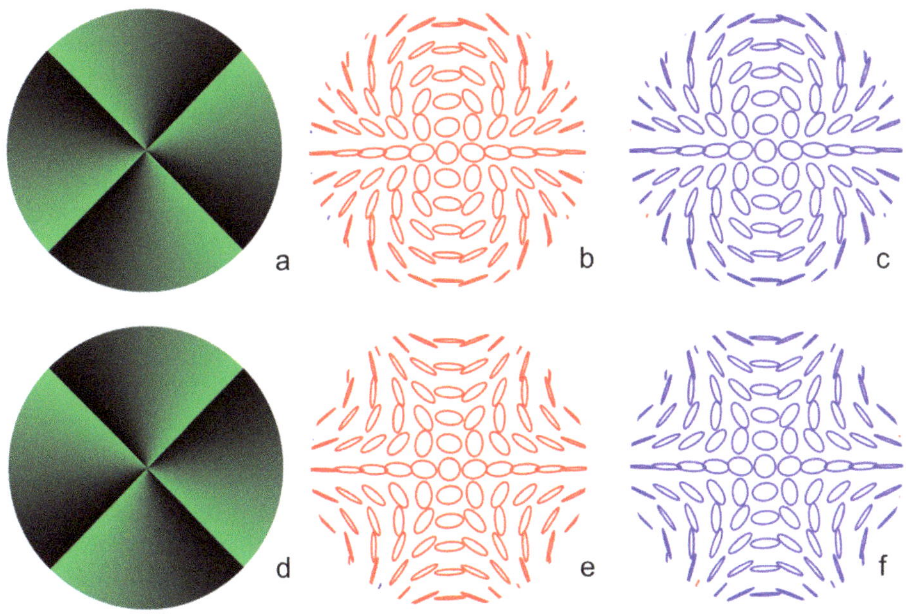


**Figure 9.15.** The same as in figure 9.14 but for a polarization singularity with $I_c = +2$ (a)–(c) and $I_c = -2$ (d)–(f).

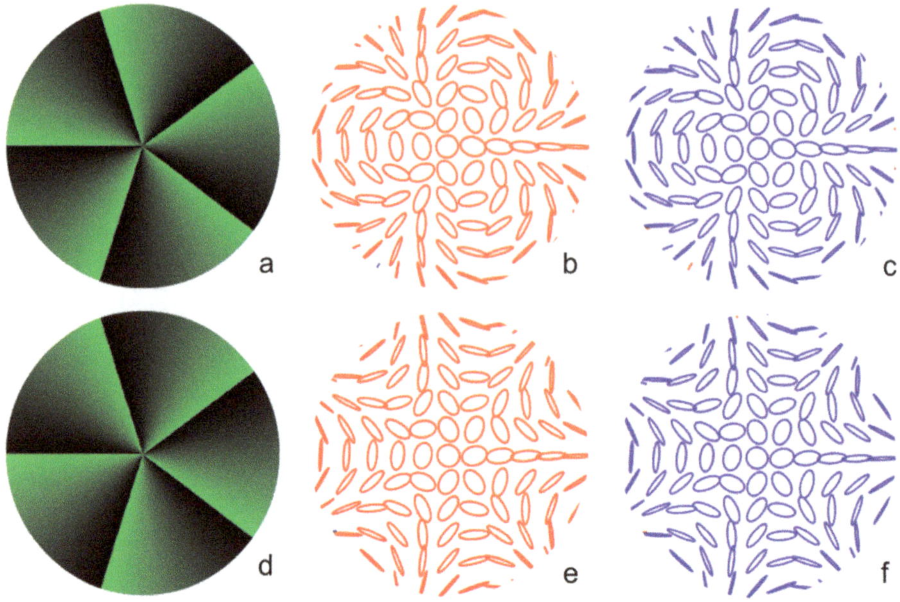


**Figure 9.16.** Stokes phase for right-handed, left-handed SOP distributions of a polarization singularity with $I_c = \frac{5}{2}$ are shown in first row (a)–(c) and for $I_c = \frac{-5}{2}$, are shown in second row (d)–(f).

figure 9.17(c) is called the azimuthally polarized beam. Both have the Poincare–Hopf index value of $\eta = 1$. The negative index $\eta = -1$ singularities are shown in figure 9.17(b) and (d) respectively.

V-point singularities with $\eta > 1$ have flower-like SOP patterns and $\eta < -1$ have spider web-like sectors since their polarization distributions have structures similar to flower petals and webs [19, 20]. Some of the positive and negative index V-points are shown in figures 9.18 and 9.19 and the flower and web structures can be seen clearly in them. The V-point singularity cannot be represented on a Poincare sphere like a C-point singularity. But the SOPs in the immediate neighborhood of the V-point singularity, can be mapped on to the Poincare sphere, where these states are given by equatorial points.

Having introduced singularities, a comparison of formulae would be appropriate at this juncture (table 9.1).

## 9.6 Stokes phase

Stokes phase $\phi_{12}$ is the argument of the Stokes field $S_1 + iS_2$ constructed from Stokes parameters. Among the three possible constructions of Stokes fields, the field $S_1 + iS_2$ is the most useful one. In this field both C-points and V-points appear as phase vortices. Stars, lemons and monstars appear as vortices of charge $\pm 1$ whereas lower index V-points appear as vortices of charge $\pm 2$. Also the C-point singularities appear at the intersections of the zero contours of $S_1$ and $S_2$. It can be seen that at C-points $S_3 = \pm 1$ and $S_1 = S_2 = 0$ and at V-points $S_1 = S_2 = S_3 = 0$. The Stokes phase distribution is related to azimuth distribution in the SOP distribution as

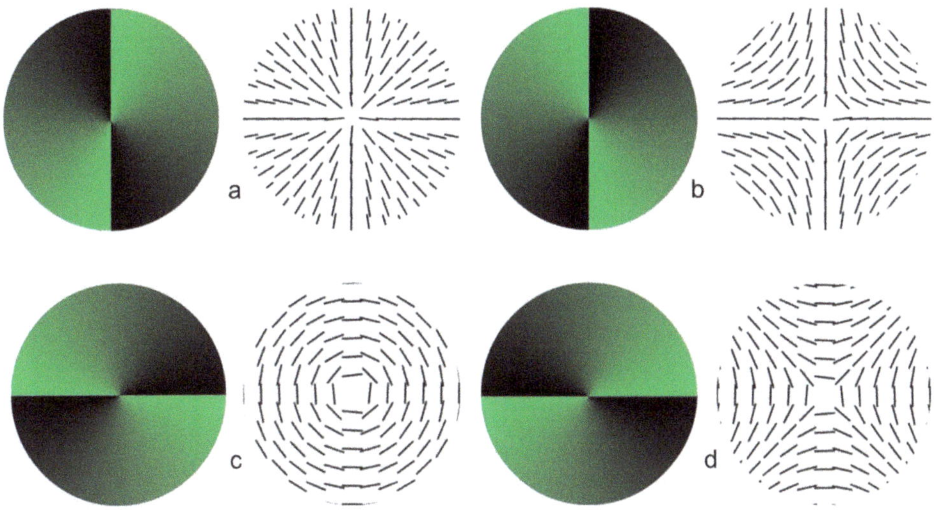


**Figure 9.17.** Stokes phase and polarization distributions for V-point singularities with $\eta = +1$ in (a) and (c) and with $\eta = -1$ in (b) and (d). The distributions shown in (a) and (c) are orthogonal to each other. Similarly the distributions shown in (b) and (d) are orthogonal to each other.

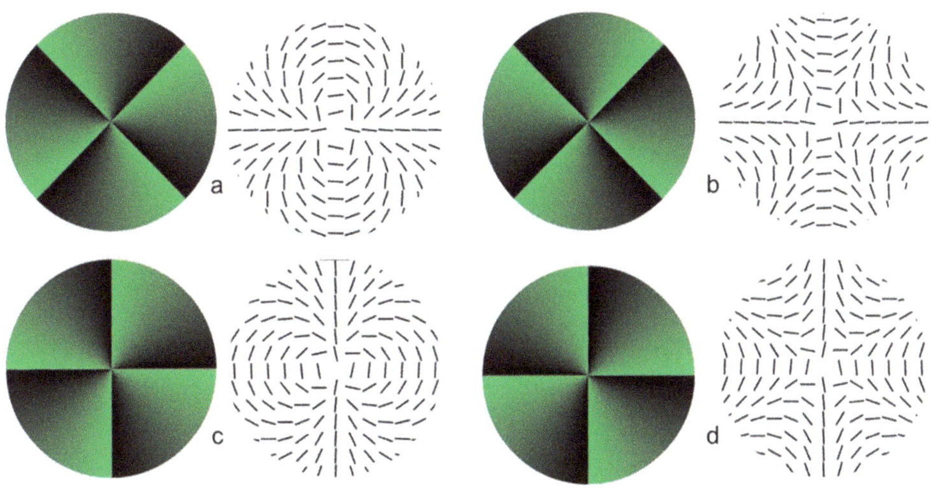


**Figure 9.18.** The same as in figure 9.17 but for a polarization singularity with $\eta = +2$ (a), (c) and $\eta = -2$ (b), (d).

$2\gamma(x, y) = \phi_{12}$. Figures 9.10–9.19 present $S_{12}$ Stokes phase distributions along with SOP distributions for various ellipse and vector field singularities.

Integer C-point index and V-point index have the same Stokes phase distribution. The phase distribution consists of a vortex of even integer charge in both cases. Hence the distinction between C-point and V-point has to be done by a limiting process based on the neighborhood SOPs. In the Stokes phase distribution the C-point index $I_c$ and the topological charge of the phase vortex $l$ are related by

**Figure 9.19.** The same as in figures 9.17 and 9.18, but for polarization singularity with $\eta = +3$ (a) and (c) and $\eta = -3$ (b) and (d).

**Table 9.1.** Comparison of various types of singularities. In the table $z$ is used to represent complex variable, $f(z)$ represents function of a complex variable, $\phi$ is the phase distribution and $\gamma$ is the azimuth distribution of the state of polarization.

| S.No | Type of singularity | Condition |
|---|---|---|
| 1 | Singularity in complex function | $\oint f(z)dz = 2\pi i$ |
| 2 | Phase singularity | $\oint \nabla\phi \cdot dl = 2\pi l$ |
| 3 | Polarization singularity (ellipse field) | $\oint \nabla\gamma \cdot dl = 2\pi I_c$ |
| 4 | Polarization singularity (vector field) | $\oint \nabla\gamma \cdot dl = 2\pi\eta$ |

$$2I_c = \frac{1}{2\pi}\oint \nabla\phi_{12} \cdot dl = \frac{1}{\pi}\oint \nabla\gamma \cdot dl = l \tag{9.6}$$

This topological charge of the phase vortex is also called the Stokes index $\sigma_{12}$. The Stokes index is related to both $I_c$ and $\eta$ by

$$\begin{aligned}\sigma_{12} &= 2I_c \\ \sigma_{12} &= 2\eta\end{aligned} \tag{9.7}$$

Since the phase distribution of the Stokes phase is that of a phase vortex, we can use the tools developed for phase vortex for polarization vortices. The phase contours in the Stokes phase are the contours of equal azimuth. At the vortex point, all the azimuth contours intersect. Also the sign rule of scalar field can be applied to vector field in terms of azimuth contours. The phase contours are related to the longitudes on the Poincare sphere.

Likewise in the Stokes fields $S_{31}$ or $S_{23}$, the $S_3$ contour lines are ellipticity contours. They are represented by latitudes on the surface of the Poincare sphere. Since on a sphere the longitudes are perpendicular to the latitudes, on the polarization distribution the azimuth contours and ellipticity contours cross each other perpendicularly. Hence the direction of the gradient of ellipticity is in the direction of azimuth contours and the direction of the gradient of azimuth is collinear to ellipticity contours. We also know that in the Stokes phase distribution, the phase contour and the circulating phase gradient corresponding to a vortex indicate that the polarization singularities are surrounded by closed ellipticity contours. But the converse is not true. This means that the closed ellipticity contour need not enclose a polarization singularity. But a polarization distribution in which the ellipticity contours are closed may induce a polarization singularity during propagation because of the presence of circulating azimuth gradients. Ellipticity contours can never cross each other while azimuth contours form star-like structures at the locations of polarization singularities. This is because, two different latitudes do not cross each other while longitudes merge at the poles. Hence, in an ellipticity distribution vortex-like structures cannot be found, but other critical points namely, extrema and saddles can be found. On the other hand, in azimuth distribution all three critical points—extrema, saddles and vortices can be found. Polarization vortices are azimuth vortices.

Among the three possible Stokes fields, the field $S_{12}$ is more important. A random polarization distribution is shown in figure 9.20 for example. The three Stokes fields corresponding to this polarization distribution are shown in figure 9.21. The

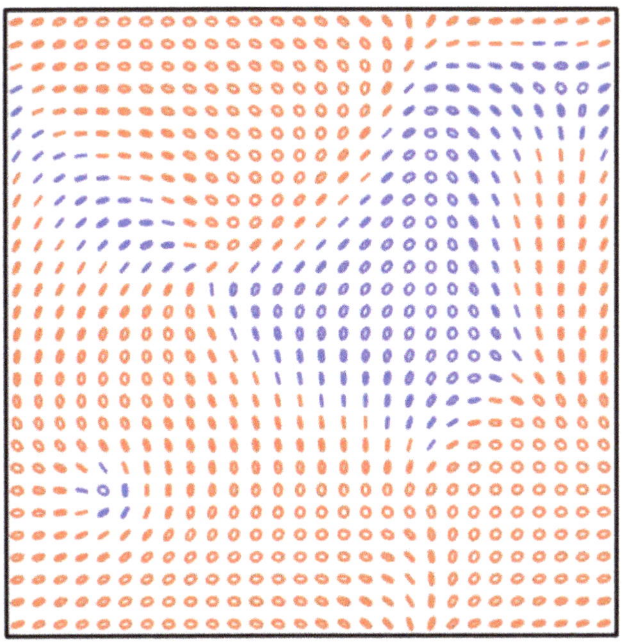

**Figure 9.20.** SOP distribution of a random field.

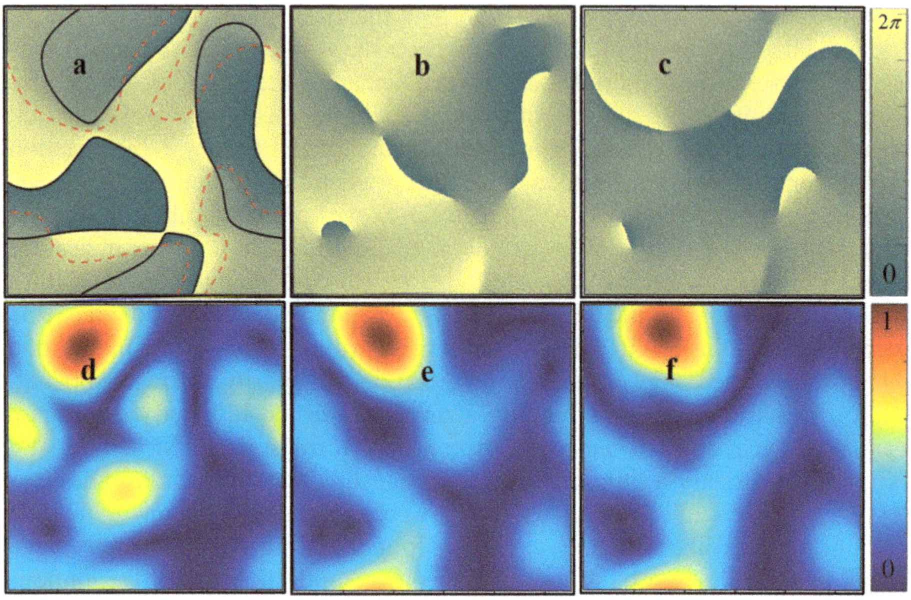


**Figure 9.21.** For the polarization field shown in figure 9.20, Stokes field phases $\phi_{12}$, $\phi_{23}$ and $\phi_{31}$ are shown in (a), (b) and (c) respectively. The corresponding Stokes amplitudes $A_{12}$, $A_{23}$ and $A_{31}$ are shown in (d), (e) and (f) respectively.

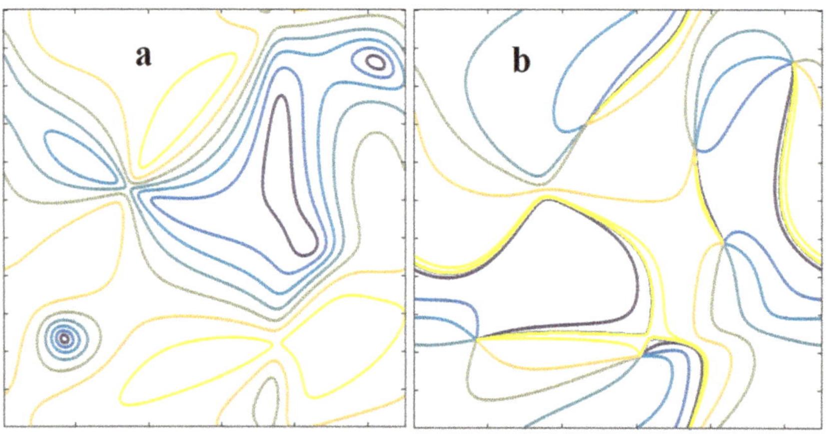


**Figure 9.22.** (a) Ellipticity contours and (b) azimuth contours for the polarization field shown in figure 9.20.

presence of vortices in the $\phi_{12}$ indicates the existence of polarization singularities in the distribution. Note from figure 9.22, at the vortex locations the azimuth contours merge (b) and ellipticity contours form closed loops (a).

## 9.7 Topological features of polarization singularities

Topological features such as extrema and saddles are there in real valued functions of two variables, such as intensity or amplitude distributions. In phase distributions, in addition to extrema and saddles, a new feature namely, a vortex is also present. But when it comes to polarization distributions, the number of topological features one can encounter can be many. Ellipticity, azimuth, handedness associated with polarized light leads to many different types of singularities namely, C-points, V-points, C-lines and L-lines. Further C-points have handedness associated with them. In addition we can have phase extrema and saddles, intensity/amplitude extrema and saddles in the three Stokes fields apart from intensity/amplitude maxima in the polarization distribution. The C-point singularities can occur at intensity extremum or at intermediate values of intensity, whereas the V-point singularities occur at intensity nulls. As a result, the occurrence of various combinations such as C-points with V-points or C-points with C-points or V-points with V-points are possible. Therefore, polarization singularity distributions have many interesting features.

**Sign rule**

Since the polarization singularities are phase vortices in Stokes phase distribution, there is also a sign rule in polarization singularity distributions. Adjacent singularities along a given Stokes phase contour line alternate the index sign [21]. This prohibits all singularities to have same index sign, i.e. a field cannot evolve with only lemons or only stars [22].

There should be an equal number of positive and negative index C-points in ellipse fields. In distributions there are also a mixture of C-points and V-points [12, 23]. Interference field distributions can be engineered to contain only V-points [24].

During diffraction it has been found that a V-point disintegrates into C-points [25] in such a way that there is index conservation. For example a V-point of Poincare–Hopf index $\eta = 1$ disintegrates into two C-points each of the same polarity ($I_c = \frac{1}{2}$) as that of the V-point. In many of the lattice fields produced by interference, both right- and left-handed singularities are seen in equal numbers. Such fields can have polarizations that can be mapped onto the Poincare sphere covering the entire area of the sphere. Such beams, that have diverse SOP distributions, are called Poincare beams. They are special since the SOP distributions of inhomogeneous fields often can be mapped onto one of the hemispheres itself and realizing field distributions in which both RCP and LCP (represented by both the poles of the Poincare sphere) are present is not easy. But it is not impossible to generate such fields.

C-point singularities have handedness $S_3 = \pm 1$. But V-points do not have any handedness. It has been observed that a disintegrating V-point produces C-points in which half of the C-points will have right-handedness and the other half of the C-points have left handedness [25].

## 9.8 Angular momentum in polarization singularities

C-point polarization singularity carries net orbital angular momentum whereas the net angular momentum carried by the V-point singularity is zero. This can be seen by noting that in the circular basis approach, the polarization singularities are expressed as the superposition of spin and orbital angular momentum states. Figures 9.23 and 9.24 sketch the basic mechanism of superposition of circular components in

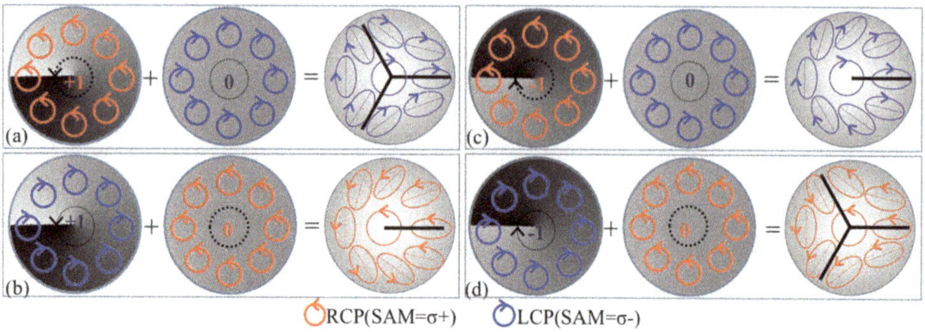


**Figure 9.23.** C-points: Decomposition in the circular polarization basis. Note that in the superpositions one beam has a vortex and the other does not have a vortex and both the beams have SOPs that are orthogonal to each other.

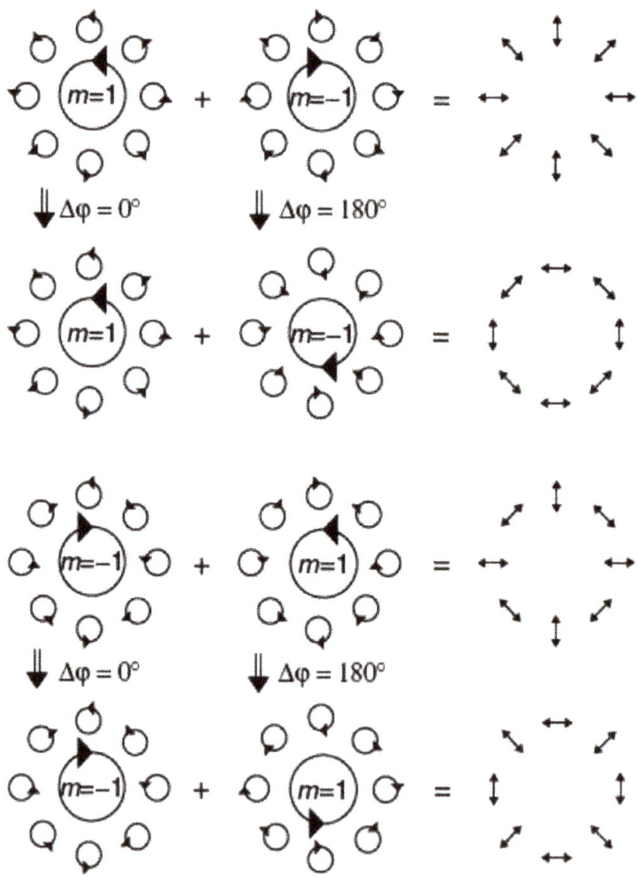


**Figure 9.24.** V-points: decomposition in the circular polarization basis. Reproduced from [19]. 

describing polarization singularities. This decomposition is useful in the study of their properties. For example, in figure 9.23(a) a star is shown as the superposition of a vortex beam ($m = 1$) in right circularly polarized (RCP) and a non-vortex beam ($m = 0$) in left circularly polarized (LCP), where $m$ is the topological charge of the beam. RCP and LCP states are shown in red and blue, respectively.

The polarization singularity as superposition can be expressed as

$$E(r, \theta) = r^m \exp(im\theta)\hat{e}_L + r^n \exp(i(n\theta + \theta_0))\hat{e}_R \tag{9.8}$$

In the above expression, $m \neq n$ and the amplitude variation in both the beams are different. Depending on $\theta_0$ different C-points with same index but different SOP distribution can be realized. When $m \neq n$ and any one of them is zero, the resulting C-point is a bright one, whereas when $m \neq n$ and both of them are not zero, the resulting C-point is a dark C-point. This is because, for $m, n \neq 0$ the phase vortices demand intensity nulls at the cores and as a result in the superposition a dark C-point occurs. Figure 9.24 shows circular decomposition of fields for V-point singularities. The general expression for the field consisting of a V-point in the circular decomposition [19] can be written as

$$E(r, \theta) = r^m \exp(im\theta)\hat{e}_L + r^m \exp(-im\theta)\hat{e}_R \tag{9.9}$$

The four types of vector beams are shown as superpositions of spin and orbital angular momentum states in figure 9.24.

**Spin–orbit beams:** Since polarization singularities are superpositions of spin and orbital angular momentum states, they are also called as spin–orbit beams.

## 9.9 Generation

### Interference methods

Polarization singularities can be generated interferometrically by overlapping two orthogonally polarized beams with phase singularities. The two beams should have appropriate amplitude and phase variations to realize polarization singularities. One possible interferometric setup is described in figure 9.25.

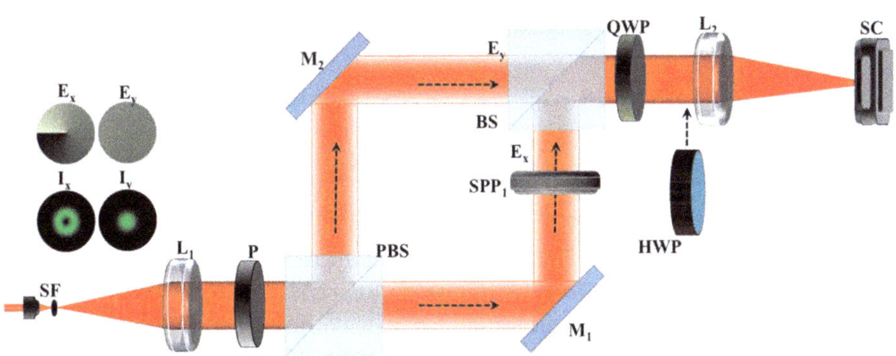


**Figure 9.25.** C-point generation by interference.

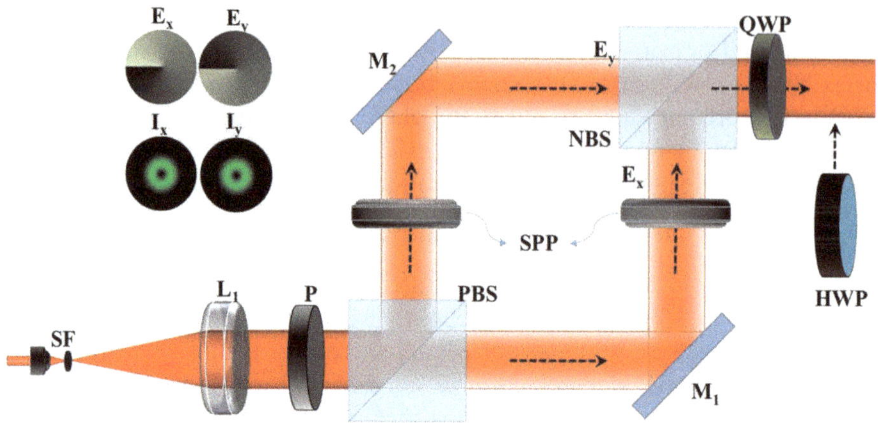


**Figure 9.26.** V-point generation by Mach–Zehnder interferometer.

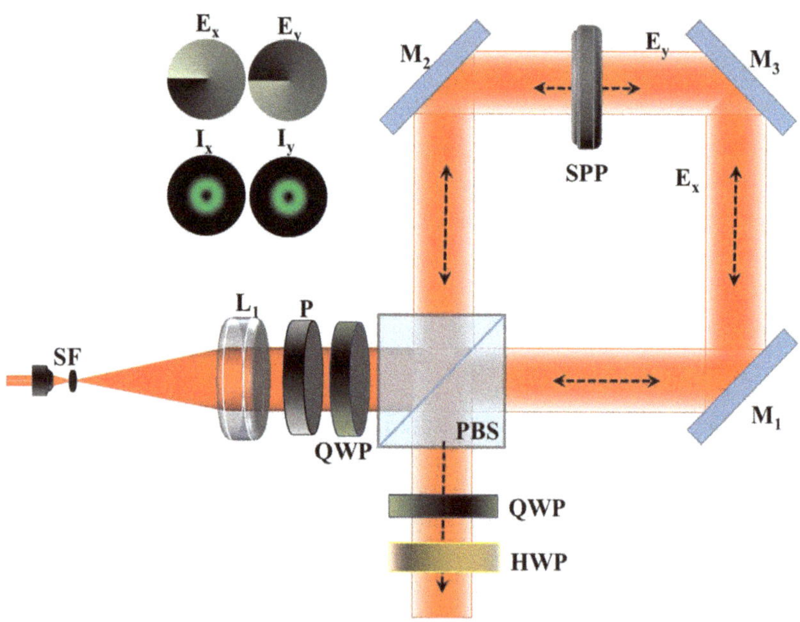


**Figure 9.27.** V-point generation by Sagnac interferometer.

In the C-point generator, a collimated beam launched into the Mach–Zehnder configuration is split into $\hat{x}$ and $\hat{y}$ components by a polarizing beam splitter (PBS). In one of the beams a vortex is planted by a spiral phase plate (SPP). These two beams are combined at the beam splitter (BS) and linear to circular SOP change by a quarter wave plate (QWP) makes the required superposition for the C-point. When C-point of opposite polarity is required, a half-wave plate (HWP) can be inserted into the beam. A HWP changes the sign of the index [26].

In figure 9.26, a Mach–Zehnder type interferometer configuration and figure 9.27, a Sagnac type interferometer configuration are shown for the generation of V-point singularity. The Mach–Zehnder configuration requires two SPPs, whereas in a Sagnac a single SPP will suffice. This is because the same SPP can produce positive as well as negative charge vortex beams depending on from which side light is passing through it. Two QWPs are used in the Sagnac configuration and their fast axes can be aligned appropriately to introduce a 0 or $\pi$ phase difference between the two circularly polarized beams at the output. A Sagnac type interferometer is much like a resonator configuration. There are many other ways the vortex phase can be invoked into the beam. Based on this fact, there are many resonator configurations possible as shown in figure 9.28

In all the interferometric methods [27, 28], the use of a spiral phase plate is inevitable and these methods however, require precise alignment of cores of the vortices and are very sensitive to vibrations. Other interference methods are by coherent superposition of orthogonal $TEM_{01}$ modes [29] and of many beams in image rotating resonator [30].

**Intra-cavity methods**

By introducing a polarization-selective mirror inside a laser resonator cavity, cylindrical vector beams can be achieved [31]. A binary dielectric diffraction grating etched in the backsurface of the mirror substrate makes the resonator polarization selective. Likewise, several intra-cavity methods (figure 9.28) which include the use of axicon [32], conical Brewster prism [33], windows [34], polarization selective grating mirror [35], calcite crystal [36], polarization based beam displacer [30], image rotating mirror arrangement [29], polarization selective GIRO (giant reflection to zero order) mirror [31] inside a resonator are there for the generation of cylindrical vector beams. Since in all the resonator configurations, the light enters through the inserted element in the cavity from both sides, only V-point generation is possible. Hence, generation of a C-point using resonator configuration is not there. All these methods discuss the generation of V-point singularities of positive Poincare–Hopf index beams. For negative index beams a HWP can be inserted [26] outside the cavity to change the polarity of the index.

Radial polarization can also be achieved by conical diffraction [37]. Spatially varying sub-wavelength grating structures can be designed and fabricated [38, 39] for the generation of radial polarization. Some of the structures of the grating are shown in figure 9.29. Diffractive optical elements [40], or an image rotator [29] or conical Brewster prism [33] or a GIRO mirror [31] can be used inside the resonator to achieve high power radial polarization. In another method spatially varying Brewster angles can be achieved by an appropriately designed element (figure 9.30) and this element can be inserted into a resonator to achieve radial/azimuthal polarization [34].

**Spatial light modulators**

Commercially available spatial light modulators (SLMs) respond only to one linear polarization state (say $\hat{y}$) while behaving like a simple reflecting mirror for the other state of polarization. This fact can be used to our advantage and the setup shown in figure 9.31 can be used. By displaying a vortex phase distribution on the reflective SLM, a plane wave with Gaussian profile falling on the SLM mirror assumes the charge of the

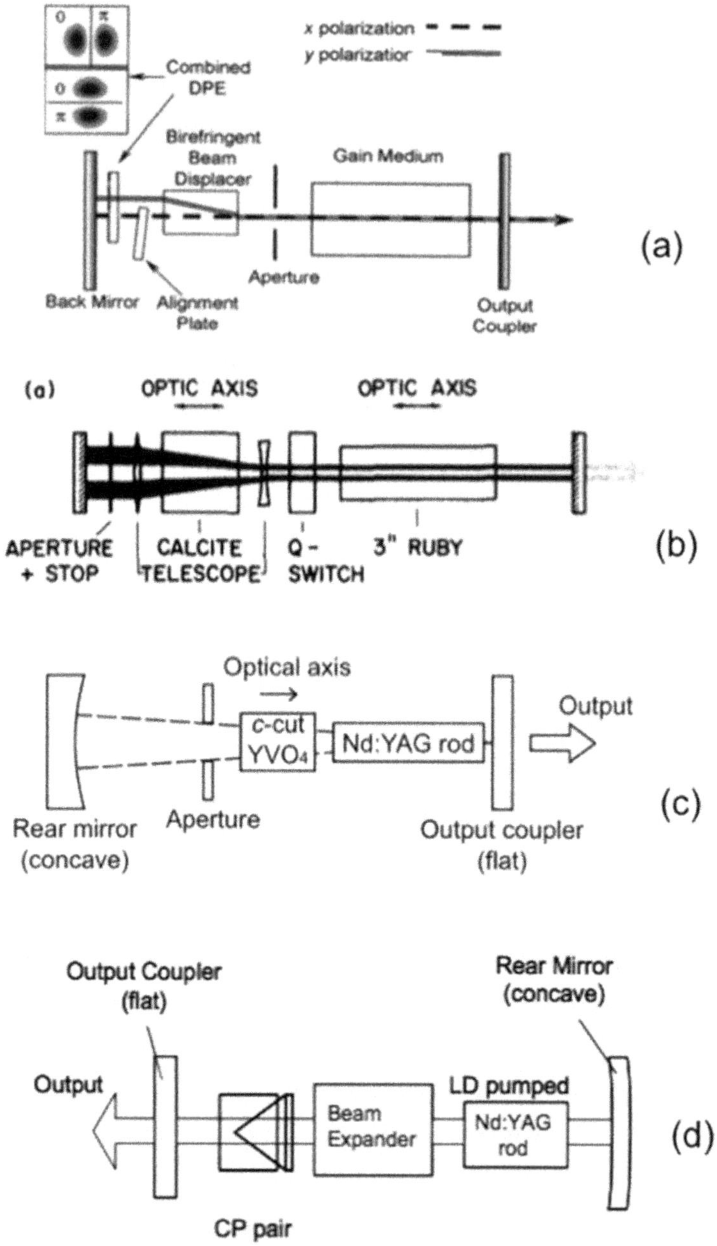


**Figure 9.28.** Various intra-cavity manipulations leading to vector vortex beam generation. In (a) a birefringent beam displacer [30], (b) a calcite crystal [36], (c) an undoped c-cut $YVO_4$ crystal [64]. (d) Conical prisms [65] are used inside the laser resonator cavity. Part (a) reprinted from [30], and part (b) reprinted from [36], with the permission of AIP Publishing. Part (c) reprinted with permission from [64], copyright (2007) by Springer Nature. Part (d) © 2005 IEEE, reprinted, with permission, from [65].

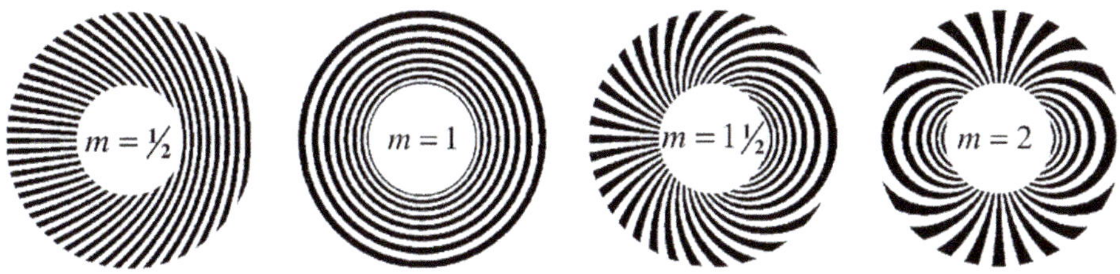


**Figure 9.29.** Magnified sub-wavelength grating structures for the generation of polarization singularity by diffraction. The charge of the phase vortex that the grating can produce is given by $m$. Reproduced with permission from [38]. Copyright (2003) by OSA.

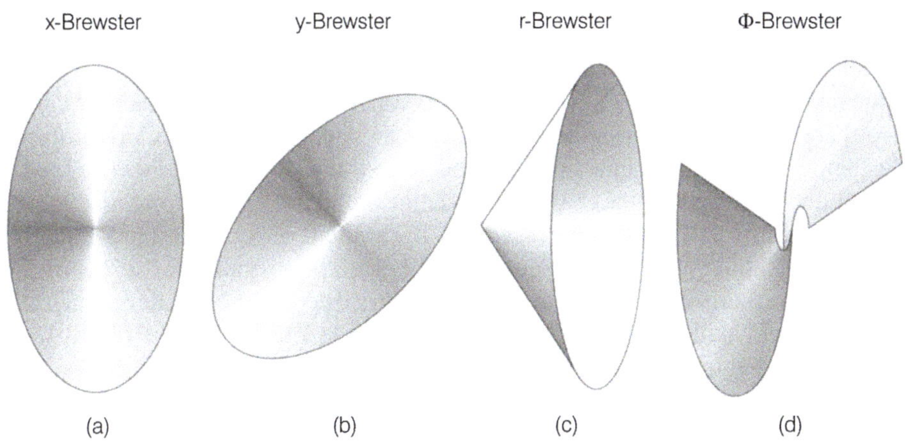


**Figure 9.30.** Different types of Brewster windows. The $r$ and $\phi$ type Brewster windows are of particular interest in radial or azimuthal polarization generation. Reproduced with permission from [34]. Copyright (1998) by OSA.

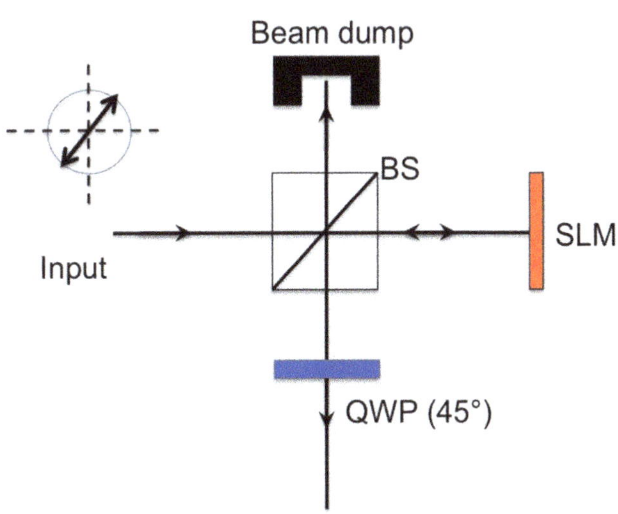


**Figure 9.31.** Schematic of the C-point generator. Input plane polarized light incident on the SLM, upon reflection possesses a vortex beam in $\hat{y}$ component and a plane wave in the $\hat{x}$ component. A C-point in the beam is generated after passing through the QWP.

vortex in $\hat{y}$ polarization state while preserving the plane wave nature in the $\hat{x}$ polarization state. These two beams are then combined to produce a superposition state and using a 45–135° oriented QWP, the linear states can be converted to circular polarization state. The resulting superposition is a C-point singularity.

## Spatially varying wave plates

Segmented spatially varying wave plates [41, 42] can be used for the generation of vector vortex beams. Another spatially varying structure that can be used is a sub-wavelength grating structure [38, 39], but this works on diffraction phenomena. The property of the half-wave plate is to rotate the azimuth of the polarization state. In the case of linear states the plane of polarization is rotated by twice the angle between the fast axis of the HWP and the plane of polarization. In the case of elliptically polarized light this rotation is also concomitant with handedness change. Since there is no proper orientation (of major axis) defined for circularly polarized light there is only a handedness change and since there is no handedness defined for linear states, there is only a change in orientation angle. Hence to realize radially polarized light from linearly polarized light, the amount of rotation that the plane of polarization should undergo depends on the spatial position in the beam. The orientation of the HWP is therefore spatially varied to get the desired polarization change. Such a spatially varying half-wave plate is called an S-wave plate. The S-wave plate presented here in figure 9.32 has sectors of half-wave plates, each one

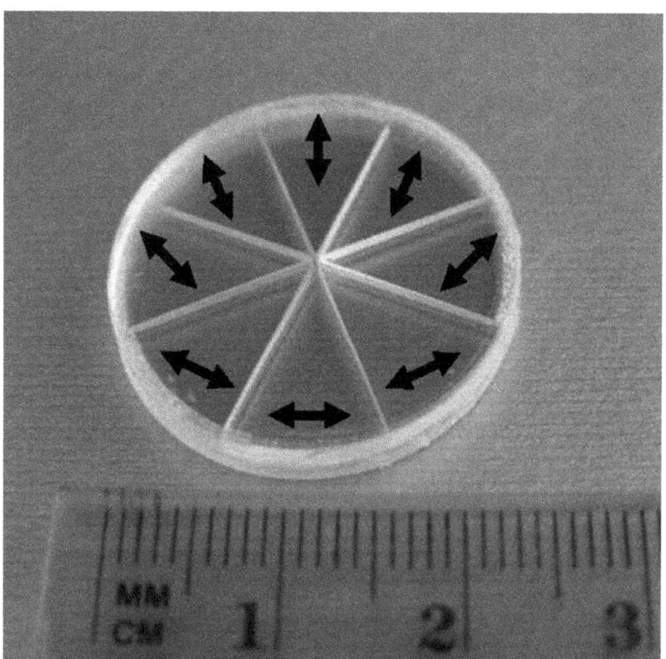


**Figure 9.32.** S-wave plate for V-point generation, made of segments of HWPs with their pass axes oriented as shown by arrows. Reprinted from [41]. Copyright (2008), with permission from Elsevier.

with different orientation to the crystal's optical axis. Depending on the plane of polarization of the incident plane polarized plane wave, radial or azimuthal SOP distribution can be realized using this plate. The superposition states of radial and azimuthal polarization states can also be realized.

In figure 9.32, an S-wave plate in which there are eight segmented half-wave plates whose fast axes oriented in different directions are shown. A plane polarized light which is vertically polarized after passing through the S-wave plate comes out as radially polarized, whereas a horizontally linearly polarized light after passing through it becomes azimuthally polarized. For a linearly polarized beam at any other angle, the output of the S-wave plate will be spirally polarized. This state is the superposition state of radial and azimuthal polarization states. The radial and azimuthal polarization distributions are orthogonal to each other.

### *q*-Plates

*q*-wave plates are introduced to generate and manipulate OAM states. A *q*-wave plate [43, 44] is essentially a birefringent wave plate with inhomogeneous patterned distribution of the local optical axis in the transverse plane. In these plates, $q$ is the topological charge of the plate which are multiples of half integers. A circularly polarized light passing through a *q*-wave plate of charge $q$ will produce a vortex beam of charge $2q$. The sign of the vortex beam is decided by the handedness of the incident circularly polarized light. Liquid crystal controlled *q*-plates [45] are also there. One such liquid crystal *q*-plate, that can be controlled electrically is shown in figure 9.33.

The spatial variation of the fast axis of the HWP in a *q*-wave plate is shown in figure 9.34. The number of times the fast axis undergoes a rotation of $\pi$ in one circular path around the center decides the charge $q$ of the plate. Accordingly figure 9.34(a) has charge $q = \frac{1}{2}$, figure 9.34(b) and (c) both have charge $q = 1$. A circularly polarized light passing through any of these plates will undergo a handedness change, which means that the change in SAM is $2\hbar$. The first plate produces a phase vortex of charge 1, whereas the other two will produce a phase vortex of charge 2. In general, a circularly polarized light passing through a *q*-plate of charge $q$ will become a beam with OAM equal to $2q\hbar$. For a *q*-plate of charge 1, the SAM change is $2\hbar$ and OAM change is also $2\hbar$. Therefore, the total variation of the angular momentum of light is 0, when $q = 1$, and there is no net transfer of angular momentum to the plate. Hence it acts as a coupler [44] of the two forms of optical angular momentum, allowing their conversion into each other.

This survey presented so far, shows that there are a large number of methods for the generation of V-points, while there are only a limited number of methods available for C-point generation.

## 9.10 Detection

One of the ways of detection of polarization singularities is to adopt methods that are used for the measurement of Stokes parameters of light. There are various schemes by which Stokes parameters can be detected. By using a polarizer

**Figure 9.33.** A $q$-plate device with topological charge $q = 3$ and electric tuning, as seen in a linear polariscope [43]. These plates are used for the generation of phase singular beams. Under circularly polarized light illumination they produce phase vortices. Under linear polarized light illumination they are capable of producing V-points. Reprinted from [43].

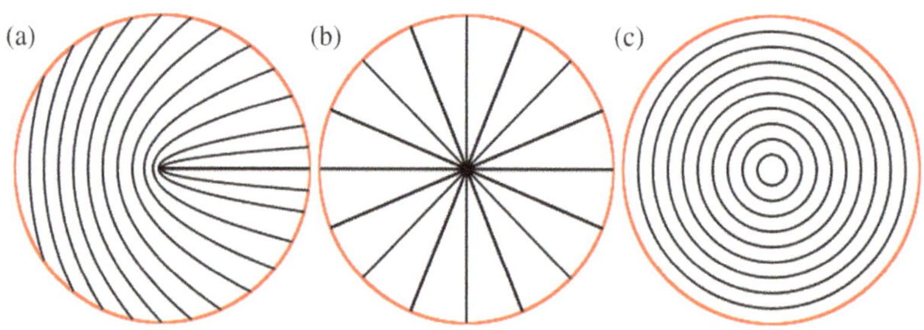


**Figure 9.34.** Spatially varying orientations of the fast axis (of the HWP) in the $q$-wave plate. $q$-wave plates of different charges ($+\frac{1}{2}$, $+1$, $+1$) are shown here. In reference [44] these plates are shown capable of producing phase vortices (OAM states). Reprinted with permission from [44] Copyright (2006) by the American Physical Society.

at 0°, 90°, 45° and 135° the intensity measurements $I_x$, $I_y$, $I_{45^\circ}$ and $I_{-45^\circ}$ are made. Then by using a quarter wave plate and a polarizer oriented at ±45° and 0° respectively intensity measurements $I_{\text{RCP}}$ and $I_{\text{LCP}}$ are made. After finding the Stokes parameters, each of which are functions of position coordinates for an inhomogeously polarized light, the polarization ellipse distributions can be plotted and the presence of C-points or V-points can be identified by looking at the SOP distribution.

### Stokes fields

The Stokes fields $S_{12}$ or $S_{13}$ can be constructed to identify the locations of polarization singularities. In the $S_{12}$ distribution, the polarization singularities appear as phase singularities and in $S_{13}$ distribution the polarization singularities occur at extremum or at saddle points in the phase distribution $\phi_{13}$. The phase distribution in the $S_{12}$ field is related to the azimuth distribution and the phase distribution in the $S_{13}$ is related to the ellipticity distribution. C-points occur at ellipticity extremum and V-points are known to occur at saddle points in the ellipticity distribution. Hence the phase contours in the $S_{12}$ field converge at/diverge from a polarization singularity whereas phase contours in the $S_{13}$ field either form closed loops or touch each other at the polarization singularity. In the $S_{12}$ field distribution, polarization singularities can be identified by looking for zero crossings of $S_1$ and $S_2$.

Interference methods for the detection of polarization singularities are not attractive. To obtain high interference fringe contrast, selection of a suitable SOP for the reference wave is difficult. For spatially varying SOP distribution, the interference patterns have spatially varying contrast as the contrast strongly depends on the SOP of the reference beam. Further, the fringe patterns themselves reveal different phase structures depending on the SOP of the reference beam in the interference [46]. The patterns therefore need careful examination.

### Interferometric method

The interferometric method of detection [47] of polarization singularity and s-contours (also referred to as L-lines) is based on capturing many interferograms and using the information about the presence of fork fringe structures in these interferograms, the location and path taken by s-contours (L-lines) are detected. Basically on an L-line, linear states of different azimuth are present and this s-contour separates right- and left-handed regions of polarization distribution. Hence, to detect linear states with different azimuths, the test beam which is inhomogeneously polarized is made to interfere with a circularly polarized beam. By using an analyzer, fork fringes in the interference pattern are obtained. The location where the fork is present, corresponds to a linear state with particular azimuth which is decided by the analyzer orientation. By rotating this analyzer, the location of the fork in the fringe pattern also changes as now a different azimuth is detected. The schematic of the interferometric setup used for drawing the L-lines is shown in figure 9.35. By subjecting the analyzer to continuous rotation, a continuous curve of s-contours on the polarization distribution can be drawn. This technique helps in detecting L-lines (s-contours) in the distribution.

To detect the C-points the same interferometer can be used. At C-points, the azimuth contours corresponding to different azimuth values converge. Hence now the aim is to draw azimuth contours on the polarization distribution. For this purpose, in the test beam, a QWP can be inserted. The action of the QWP is to bring all elliptical states with same azimuth to the linear state (figure 9.36). This can be understood by noting that a QWP brings all the points on the Poincare sphere along

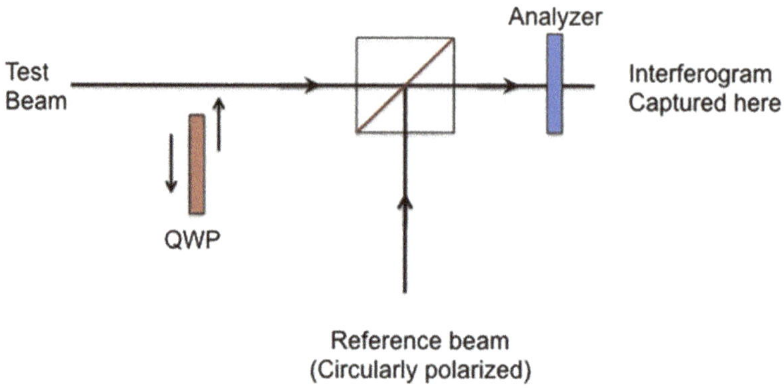


**Figure 9.35.** Circularly polarized reference beam and the beam with unknown SOP distribution produce an interference pattern after passing through the analyzer. The analyzer makes sure that dark point occurs, if the test beam at a particular location has a linear state which is orthogonal to the pass plane of the analyzer. That is identified by fork fringes. By rotating the analyzer pass plane, different linear SOPs can be found by tracking down the movement of fork fringes in the field of view. This way L-line detection by interference is possible. To detect the C-point, a QWP is introduced in the test beam and the exercise of drawing L-lines by rotating the analyzer is carried out. QWP is rotated and kept at different angles to draw different s-contours.

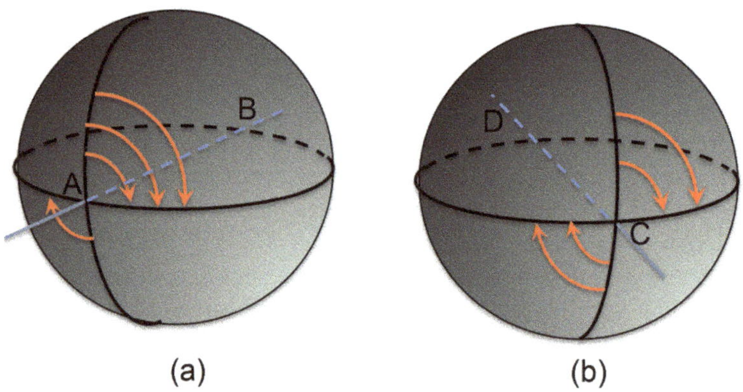


**Figure 9.36.** By using a QWP in the test beam, all the SOPs that have same particular value of azimuth are converted into linear states. Then using the L-line finding method described above, the new L-line (here we call it s-contour) is drawn. Actually this s-contour corresponds to azimuth contour in the polarization distribution of the test beam. To draw azimuth contour for another value of azimuth, the QWP is rotated so that another set of ellipses become linear states. The action of QWP on the azimuth contour (on the Poincare sphere it is the longitude) is explained in this figure. In (a) AB is the axis about which the $\frac{\pi}{2}$ rotation of points of Poincare sphere is carried out. In (b) the QWP axis is rotated so that CD represents new axis about which this rotation happens. This brings a different set of points on another longitude to the equator of the Poincare sphere.

a particular longitude to equatorial points. In this new SOP distribution obtained after using QWP, we know how to draw s-contours by rotating the analyzer (explained in the previous paragraph). But this time the s-contour corresponds to a particular azimuth in the SOP distribution of the test beam. Rotating the QWP by small angle will bring the points on another longitude of the Poincare sphere to the equatorial points. Now the job is clearly cut out. The s-contour in this new field will represent another azimuth contour in the test beam. By repeating this exercise, the location where many azimuth contours intersect is found to be the location of C-point.

### Polarizer

Many of the detection techniques in vector fields use a polarizer in the test beam and it is rotated to different angles to obtain different intensity patterns. Almost all singularities are being tested only in this way. For a V-point singularity a rotating polarizer yields a two lobe intensity pattern for the lowest order V-point. These patterns are depicted in figure 9.37. For different orientations of the polarizer pass

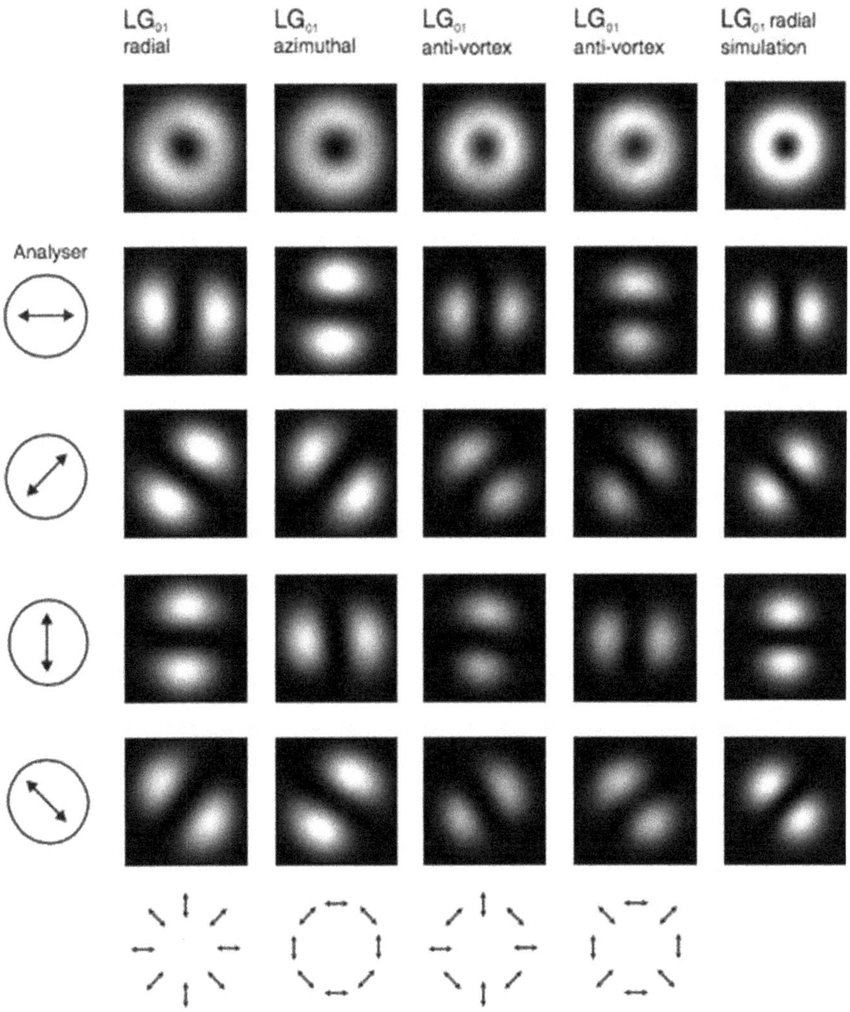


**Figure 9.37.** Intensity patterns for V-points ($|\eta| = 1$) passed through a polarizer, each time oriented in different directions. Reproduced from [19] © IOP Publishing Ltd and Deutsche Physikalische Gesellschaft.

plane, the two lobe intensity pattern gets aligned in different directions. But there is a problem in using this method. One of the patterns in the vector vortex always resembles one of the anti-vortex beams. Here the vortex and anti-vortex refers to positive and negative Poincare–Hopf index beams respectively. Hence this method poses a difficulty in distinguishing between positive and negative $\eta$ beams.

**Diffraction and polarization transformation—hybrid method for detection**

The first experiment on the diffraction of V-points through two types of triangular apertures was demonstrated recently [25]. Unlike using a polarizer, the diffraction pattern for all four types of V-point singularities are the same as shown in figure 9.38. This four level degeneracy in diffraction patterns and the doubly degenerate intensity patterns obtained using polarizer can be combined to evolve a method that can uniquely decide the type of the V-point singularity. The first experimental demonstration [48] used both diffraction and polarization transformation for uniquely distinguishing all the four states of a V-point singularity. It relied on the method that a positive and a negative phase vortex produce diffraction patterns that are mirror reflections about a plane decided by the diffracting aperture. Since V-points are superposition states of phase vortices in orthogonal circular polarization states, it is possible to make distinction between them in this experiment. The diffraction-polarization patterns for all four types of V-point are shown in figure 9.38. Since they all look same, to distinguish between the four different types of beams, the beam is sent through a polarizer as well as a diffraction setup. Before diffraction the left and

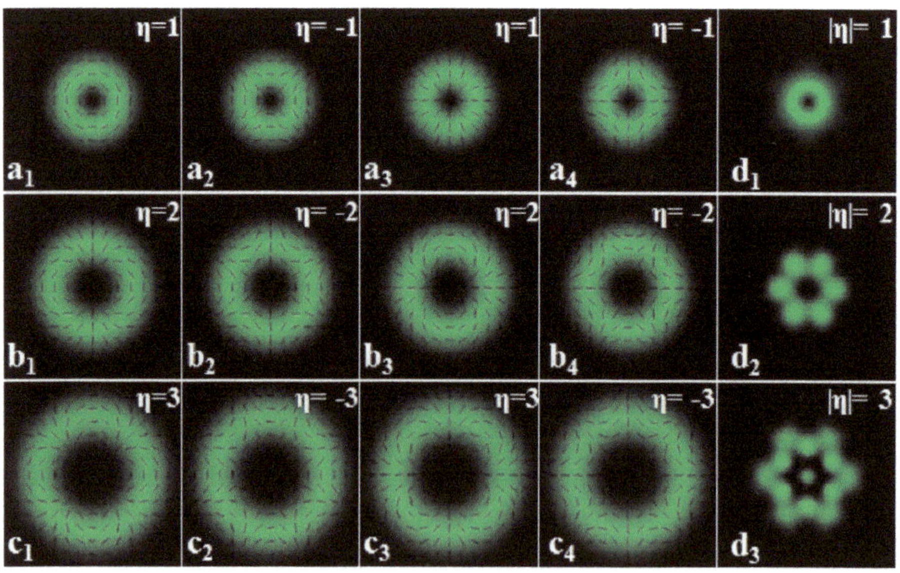


**Figure 9.38.** Four degenerates states of a V-point and the diffraction through a triangular aperture. The three rows show the patterns for three different $|\eta|$ values. Columns 1–4 show the free space propagation (or diffraction) of different types of vector vortex beams and the last column shows the diffraction pattern through the triangular aperture. Reproduced with permission from [48]. Copyright (2017) by OSA.

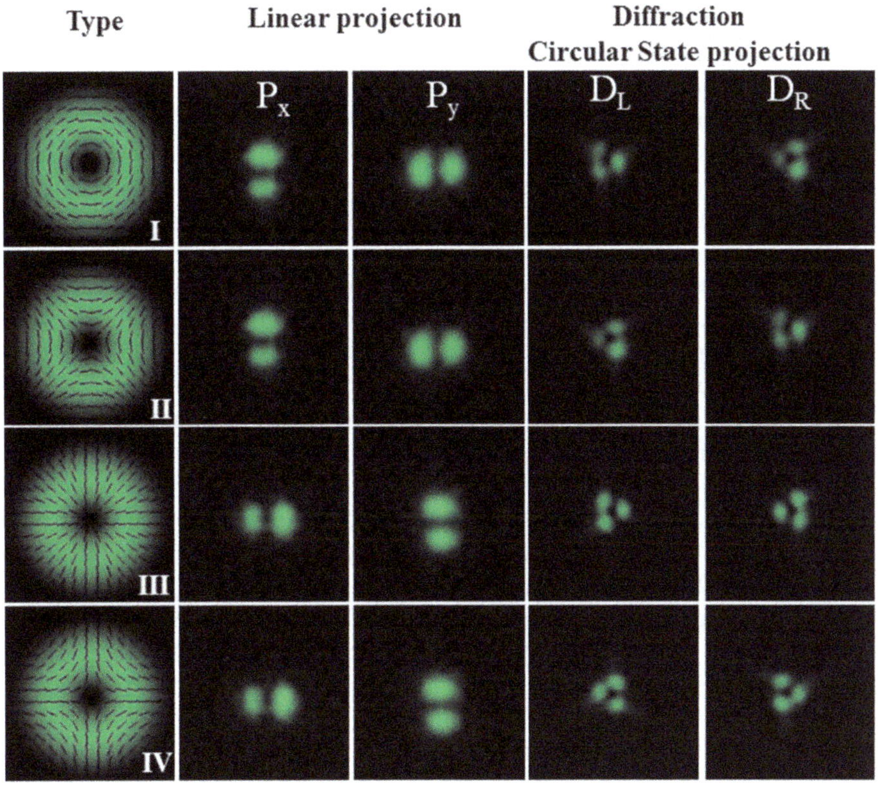


**Figure 9.39.** The first column shows the intensity and polarization distributions in the four types of V-point singularities. The second and third columns show the intensity distributions of the beam after passing through a polarizer with its pass plane along $x$ and $y$ respectively. Columns four and five show the diffraction pattern of the left and right circular polarization components of the V-point after passing through the triangular aperture. Reproduced with permission from [48]. Copyright (2017) by OSA.

right circular polarization component selection is made using a circular polarizer which is a polarizer-QWP combination. By analyzing the beam profile that passes through the polarizer and diffracting through the triangular aperture a distinction between the four types can be made as shown in figure 9.39.

## 9.11 Inversion and conversion methods

It is possible to change the sign of the polarization singularity or convert one type of singularity into another using some smart experiments. In the following section we discuss them.

### 9.11.1 Inversion methods

*Index sign inversion*

For the index sign inversion, a change in the SOP at a single point in the distribution will not suffice. A global change in the spatial distribution of SOPs is required. This simultaneous change in the distribution can be obtained by using a HWP [26]. The index sign inversion happens irrespective of the orientation of the fast axis of the HWP. This method is simple yet powerful, because the action of a HWP is to change not only the SOP of the incident light but also the topology of SOP distribution around the singular point. In this process, while a sign inverted C-point singularity does not change its position in the SOP distribution, it moves from one pole to the other on the Poincare sphere. For example, a star at the north pole after sign inversion appears as a lemon at the south pole of the Poincare sphere. The action of a half-wave plate has been explained elegantly using the Mueller matrix approach, orthogonal decomposition, Poincare sphere mapping and Stokes phase [26]. Different approaches bring new results and understanding which cannot be obtained otherwise. But here we present a simple and elegant approach to explain the mechanism of index sign inversion. In figures 9.23 and 9.24 we have seen that the C-point and V-point singularities can be expressed as superposition of OAM states in orthogonal circular polarization basis states. By passing a polarization singularity through a HWP, the sign of the handedness of the circularly polarization component states change. Since the components of the singularity have changed, the new superposition state can be seen as an index reversed polarization singularity. This is depicted in figure 9.40 for C-points and in figure 9.41 for V-points.

Polarization singularities are superposition of spin and orbital angular momentum states of light. Action of a HWP on a polarization singularity is to change the sign of the spin angular momentum components of the singularity and as a result the index sign inversion happens. This sign inversion by HWP comes with handedness inversion which cannot be avoided.

*Handedness inversion*

To change only the handedness of a C-point singularity it is now clear that a HWP cannot be used, as it also changes the sign of the C-point index along with the handedness. Interestingly a phase element and not a polarizing element can be used [17] to bring about this polarization parameter change. A spiral phase plate is a phase element that imparts phase delay in such a way that a plane wave passing through it becomes a helical wave. It's action has been described in chapter 2. These helical waves carry orbital angular momentum of light as explained in chapter 7. Depending on the amount of phase delay, the $m$th order SPP can impart a topological charge $m$ on the incident plane wavefront. Any incident phase singular beam with topological charge $m$ upon passing through a $n$th order SPP will produce a vortex of charge $(m + n)$. To switch the OAM states in the C-point singularity, the charge of the phase plate is chosen in such a way that it cancels the OAM in one SAM state while simultaneously planting oppositely charged OAM in the other SAM state. For example, consider a C-point that has +1 charge OAM in right circular polarization (RCP, SAM = $\sigma +$) and zero OAM in left circular polarization (LCP, SAM = $\sigma -$). The SPP is chosen to have −1 charge so that when the C-point passes through this SPP, it will acquire zero OAM in RCP, while getting −1 charge vortex in LCP. Such a combination of SAM and OAM states will show that the polarity of the C-point index is not changed, but the handedness of the SOPs in the distribution has inverted.

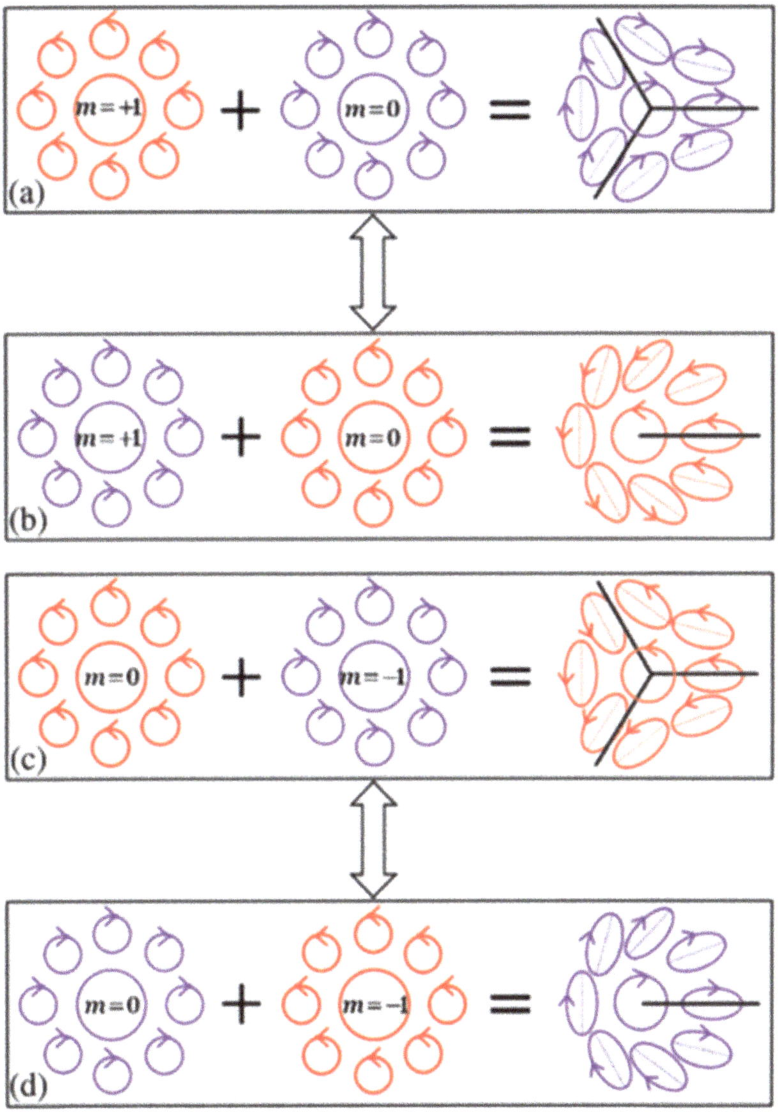


**Figure 9.40.** Circular basis representation of star (a) and (c) and lemon (b) and (d), respectively. By using the HWP, switching between (a) and (b) is possible. Similarly, switching between (c) and (d) is possible. Note that switching between (a) and (d) or between (b) and (c) is not possible using a HWP. Reproduced with permission from [26]. Copyright (2017) by OSA.

### 9.11.2 Conversion methods

The S-wave plate is a $q$-wave plate of charge $\frac{1}{2}$. A linearly polarized light incident on it becomes a radially polarized beam or azimuthally polarized beam depending on the orientation of the $q$-plate and the incident SOP of the linear state [49]. An S-wave plate that produces radially polarized light for the incident beam in vertically polarized state will produce an azimuthally polarized beam if the incident beam is

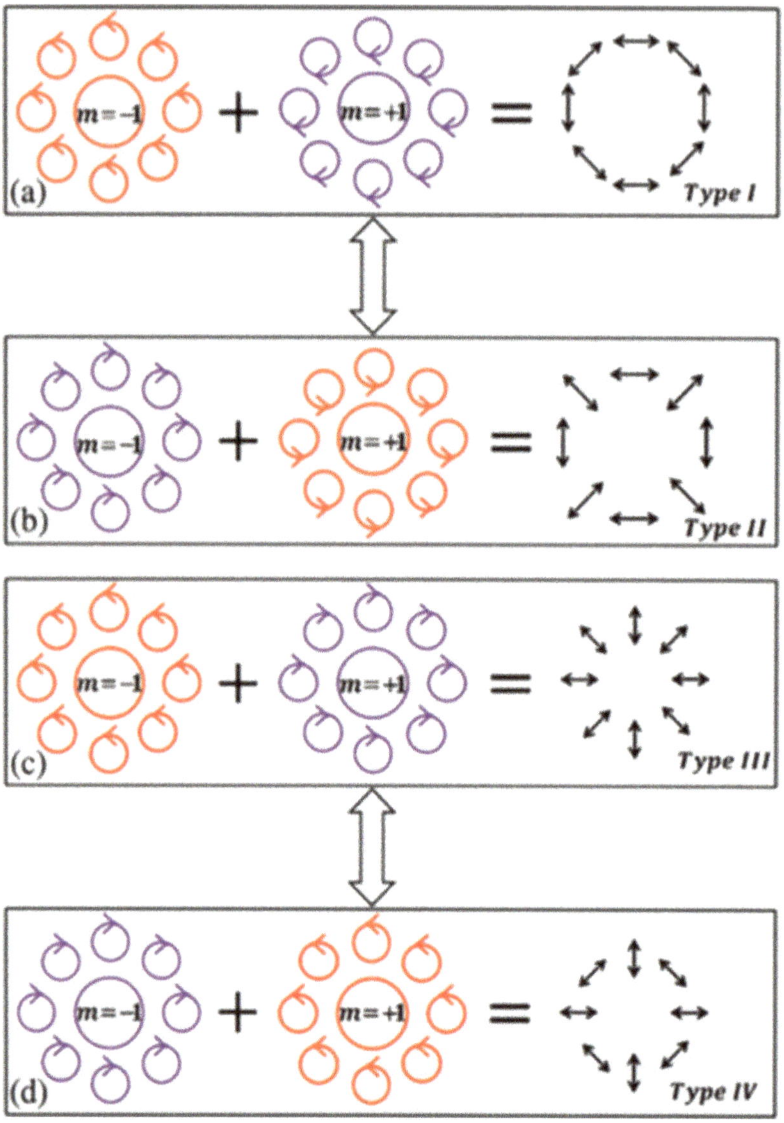


**Figure 9.41.** Circular basis representation of (a) Type I, (b) Type II, (c) Type III, and (d) Type IV vector beam, respectively. Switching by HWP between (a) and (b) or (c) and (d) is possible. Reproduced with permission from [26]. Copyright (2017) by OSA.

horizontally polarized. For other orientations of the linearly polarized beam, the output will be a superposition state of radial and azimuthal polarization. Similarly if a $q$-wave plate of charge 1 is used, the polarization distributions produced will be of higher Poincare–Hopf index beam $\eta = 2$. In fact all the polarization singularity generation methods, involve conversion of homogeneously polarized light with polarization singularity index zero to polarization singularities.

It is also possible to generate $q$-wave plates of different charges by using a sequence of optical elements that includes wave plates and $q$-wave plates. For example, it has been shown [50] that by using a $q$-wave plate, HWP and a $q$-wave plate the charge of the $q$-wave plate can be combined. Using this basic formula to produce a $q$-wave plate of charge 2 one can use one of the sequences: $q_{\frac{1}{2}}$ – HWP– $q_{\frac{1}{2}}$ – HWP – $q_{\frac{1}{2}}$ – HWP – $q_{\frac{1}{2}}$ or $q_1$ – HWP – $q_1$ This means that first the light enters through the first element $q$-wave plate and then passes through the second element HWP and so on. For negative charges the sequence starts and ends with HWP.

It is possible to convert one polarization singularity from one index to another by the transfer of angular momentum process [44]. This has been demonstrated in the case of OAM state generation by $q$-wave plates. In that process the incident light is linearly polarized with polarization singularity index 0 and after passing through the $q$-wave plate V-points of non-zero Poincare–Hopf index beams can be generated. For a circularly polarized light incident on the $q$-wave plate, the output is a vortex beam and in this process both input and output beams have zero polarization singularity index, but the phase singularity index has changed from zero to one. In this process spin angular momentum is converted into orbital angular momentum. Other methods of angular momentum transfers can be employed to convert one polarization singularity into another.

Recently it has been shown that by swapping the spin angular momentum states between the orbital angular momentum states of a polarization singularity the sign of the index can be inverted [26]. Likewise by switching OAM states between the angular momentum states of a polarization singularity, the handedness can be inverted [17].

## 9.12 Polarization singularity distributions

Apart from isolated singularities, there has been interest in polarization singularity distributions. In random fields, the distribution of polarization singularities follow the sign rule. Because of index conservation, the total index in the distribution is zero. For every positive index C-point there is a negative index C-point [51] or for every two positive index C-points, a V-point with negative index occurs in distributions [12, 22, 23]. In a generic field occurrence of C-points, or C-points with V-points are common. But by wave optical engineering, fields consisting of only V-points can be created [24]. These engineered fields can be created by multiple beam interference of polarized light.

Polarization structured light fields can be generated by plane wave interference methods. These generation methods and the fields realized have given much insight into the structures that exist in inhomogeneously polarized field distributions. Such lattice structures are envisaged to have applications in optical lattices, optical metrology, photonic crystals and can provide polarization structured illumination in microscopy.

## 9.13 Applications

### Edge enhancement

In spatial filtering experiments polarization can be used as an additional parameter apart from amplitude and phase. An S-wave plate that can convert a linearly polarized light into radially or azimuthally polarized light can be used as a spatial filter in a conventional Fourier 4*f* spatial filtering setup. At the output of the 4*f* processor a polarizer is used. The S-wave plate can be made to behave as a 1D Hilbert transform mask or radial Hilbert transform mask under linear and circular polarization respectively, of the illuminating beam. Linear SOP illumination results in anisotropic edge enhancement and circular SOP illumination results in isotropic edge enhancement. The key to understanding this is to realize the embedded phase structures in the inhomogeneous polarization and the subsequent projection by a polarizer at the output. An S-wave plate is a q-plate with charge $q = \frac{1}{2}$ and is made of HWP in which the fast axis of the HWP is spatially varying. In figure 9.42 it is depicted that illumination of the S-wave plate by light in two orthogonal linear states produce radial and azimuthal states respectively and for other orientations of the plane of polarization of the illuminating light, superposition states will be produced.

The spiral phase variation (point phase defect) and the binary signum phase variation (edge phase defect) are responsible for realization of radial and selective edge enhancement, respectively. The radial polarization hosts both point and edge phase dislocations in the beam [46]. When the radial polarization is viewed as a superposition state of two circularly polarized orthogonal states, the effect of the point phase defect comes out, and when the radial polarization is viewed as a superposition state of two linearly polarized orthogonal states, the effect of the line phase defect comes out. Hence, by analyzing the linear polarization component from radial polarization, it is possible to get selective edge enhancement, and by analyzing the circular polarization component, it is possible to get isotropic edge enhancement [52]. Similar arguments hold good for the azimuthal polarization variation. The radial and azimuthal polarization distributions are V-point singularities with positive Poincare–Hopf index beams. Beams with negative Poincare–Hopf index [53] also can demonstrate similar edge enhancement capabilities.

In the absence of a polarizer at the output, the 4*f* processor produces isotropic edge enhancement for all SOPs of the illuminating beam. The polarizer makes edge selection for display. With a fixed polarizer at the output, illumination with the radial or azimuthal SOP distribution gives selective edge enhancement—but the edges highlighted using radial SOP illumination are in the orthogonal direction to the edges that are highlighted by using azimuthal SOP illumination. This is because radial and azimuthal SOP distributions are orthogonal states. Hence their superposition states provides the flexibility to select the edges to be detected. Edge-enhanced images are shown in figure 9.43.

### C-points for optical activity measurement

When C-point polarization distribution is passed through a medium that is optically active, the polarization state at every point undergoes a rotation in its major axis and the resultant C-point distribution is a rotated one. For lowest order C-points with distinct separatrix this rotation is easily identifiable by looking at the SOP distributions. Rotated C-points are discussed in detail in singularity lattices [12].

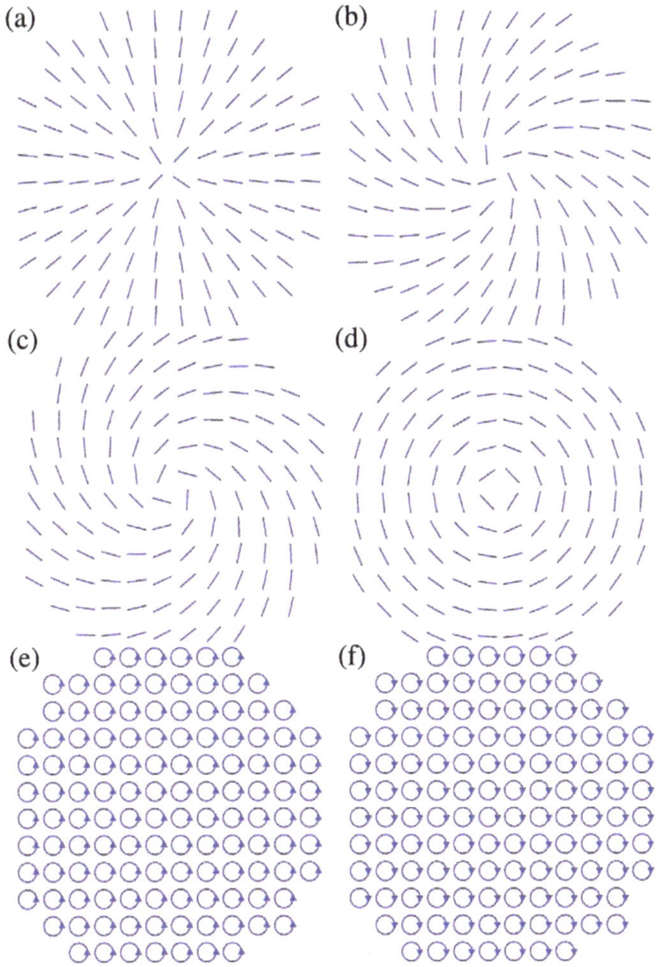


**Figure 9.42.** The polarization distributions generated by a S-wave plate when illuminated with (a)–(d) linearly polarized light and (e)–(f) circularly polarized light. SOP of the incident light on S-wave plate is (a) linear vertical, (b) linear at 30°, (c) linear at 60°, (d) linear horizontal, (e) LCP and (f) RCP. Since the S-wave plate is made of HWP, handedness change occurs while illuminating the plate with circular polarization. Among the generated states (a) and (d) are inhomogeneous and orthogonal to each other, whereas states (e) and (f) are homogeneous and orthogonal to each other. States (b) and (c) are superpositions states of (a) and (d). In (e) and (f) the phase information is not shown in the polarization distribution. Reproduced with permission from [52]. Copyright (2017) by OSA.

The optical activity exhibited by the stars and lemons in the double hopping process is explained and demonstrated [26, 54].

We have seen in edge enhancement, the use of a polarizer at the output of a V-point singularity provides angular selective edge enhancement [52] capabilities. In a similar way to when a polarizer is rotated in a C-point singularity the transmitted pattern has a single lobe of light whose orientation is decided by the angle between the direction of the separatrix and the pass plane of the polarizer. Note in a C-point

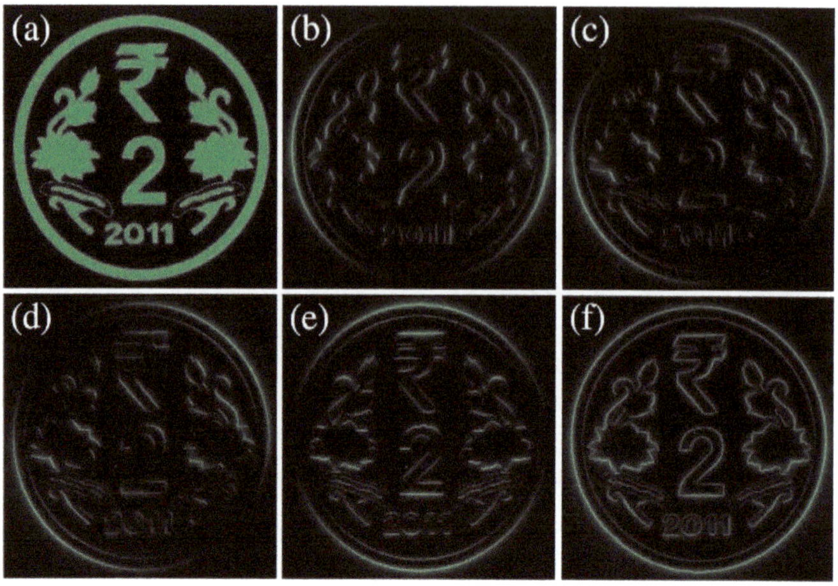


**Figure 9.43.** Simulation: the input amplitude object (a) directional edge-enhanced images when the filter (S-wave plate) is illuminated with plane polarized light at (b) 0°, (c) 30°, (d) 60° and (e) 90°. Isotropic edge enhancement (f) is possible when circularly polarized light or unpolarized light is used for illumination. Reproduced with permission from [52]. Copyright (2017) by OSA.

the azimuth undergoes $\pi$ rotation and in a V-point the azimuth undergoes $2\pi$ rotation about the singular point. Hence a polarizer in a V-point singularity gives rise to two lobes and a C-point gives rise to one lobe of transmitted intensity pattern. This concept is exploited in the determination of optical activity in chiral samples [55]. A lemon that is passed through an optically active medium undergoes rotation and the amount of rotation can be sensed by an analyzer.

**Robust beams**

Robust beam propagation through randomly fluctuating media is required for important applications such as free space communication, LIDAR systems, laser guided defense systems, imaging through biological tissues, etc.

Beams containing polarization singularities are robust in maintaining uniform intensity distribution while going through turbulent media. The problem of speckles that arises with coherent beams can thus be minimized by the use of polarization singularities. It has been shown that these singularities are superposition states of orbital angular momentum beams in orthogonal polarization states. Priyanka *et al* [56, 57] have shown that a laser beam engineered to carry $l = 0$ and $l = 1$ orbital angular momentum states in orthogonal polarization states can show robust behavior in maintaining a uniform intensity profile through turbulence. It is well known that a uniform amplitude plane wave and a vortex beam at the far field produce Airy and donut shape intensity profiles respectively. Comparison of the beam profiles in the two cases show that the intensity maximum in the Airy beam is

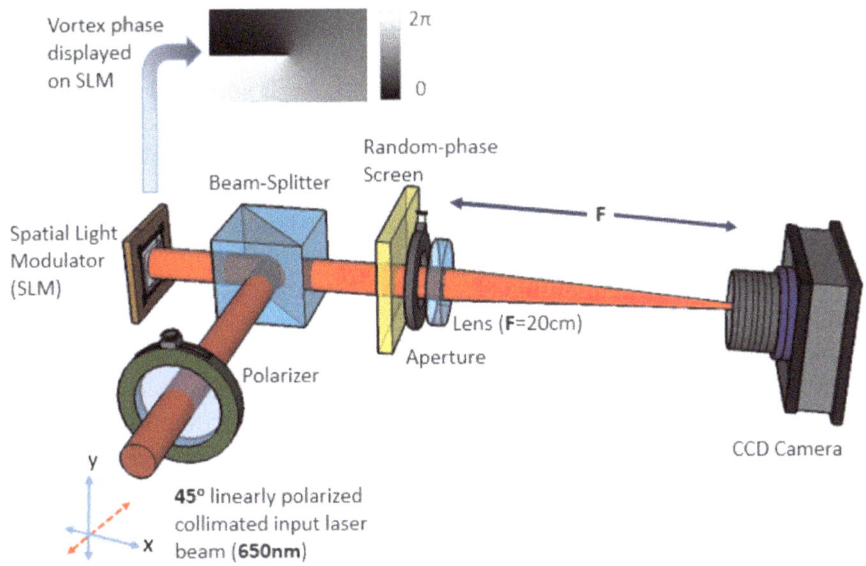


**Figure 9.44.** Experimental setup for generating the lemon–star dipole and observing effect of its propagation through a random phase screen. Reproduced with permission from [56]. Copyright (2017) by OSA.

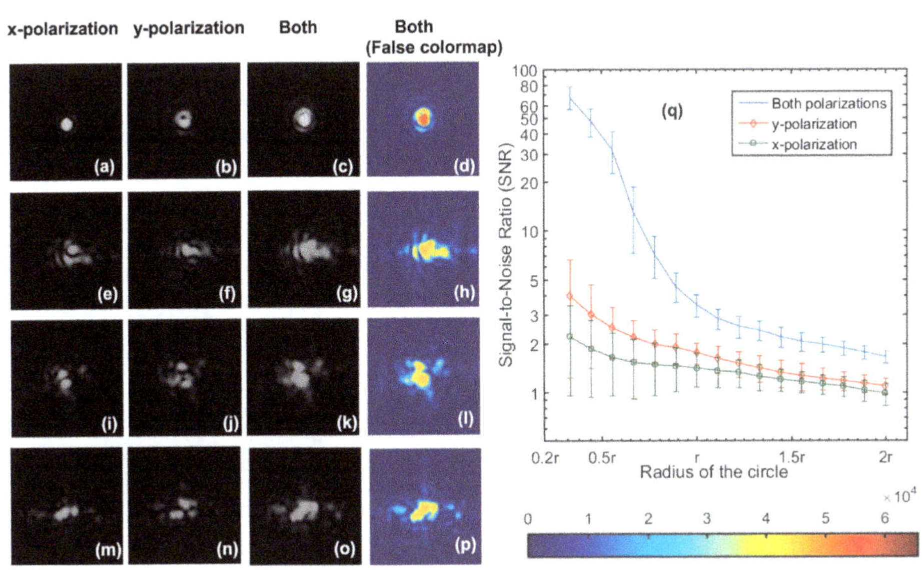


**Figure 9.45.** Experimentally recorded intensities (individual and combined polarizations) for the proposed beam at the far field diffraction pattern. Phase screens used are (a)–(d): open aperture, (e)–(p): three distinct areas of crumpled polythene sheet. (q) variation of signal-to-noise ratio (SNR) as a function of the radius of the circular area used for SNR calculation (error bars represent standard deviation for 50 realizations of the random phase screen), false colormap for displaying the recorded CCD images is also shown. The SNR of the beam is defined as mean intensity divided by standard deviation of pixel values in the circular region (detection area). Reproduced with permission from [56]. Copyright (2017) by OSA.

replaced by the intensity minimum in the donut beam. Added to this complementary nature, the fact that orthogonally polarized beams do not interfere suggest that beams made of orthogonally polarized beams with $l = 0$ and $l = 1$ orbital angular momentum states show significant negative correlation in diffraction. This concept of polarization and OAM diversity is considered valuable for robust laser beam engineering without the requirement for any active real-time correction methods.

Such superposition states result in C-point singularities. One C-point generator is shown in figure 9.44. When linear orthogonal states are taken for superposition the resulting beam is a star–lemon dipole [56] and when the circular orthogonal states are taken for superposition the resulting beam is a C-point singularity [57]. Instead of beams having $l = 0$ and $l = 1$ orbital angular momentum states, if the states $l = +1$ and $l = -1$ orbital angular momentum states are superposed the resulting singularities are V-point singularities. Speckles produced by orthogonal polarization basis states of a C-points are anti-correlated [57] whereas in the case of V-points they

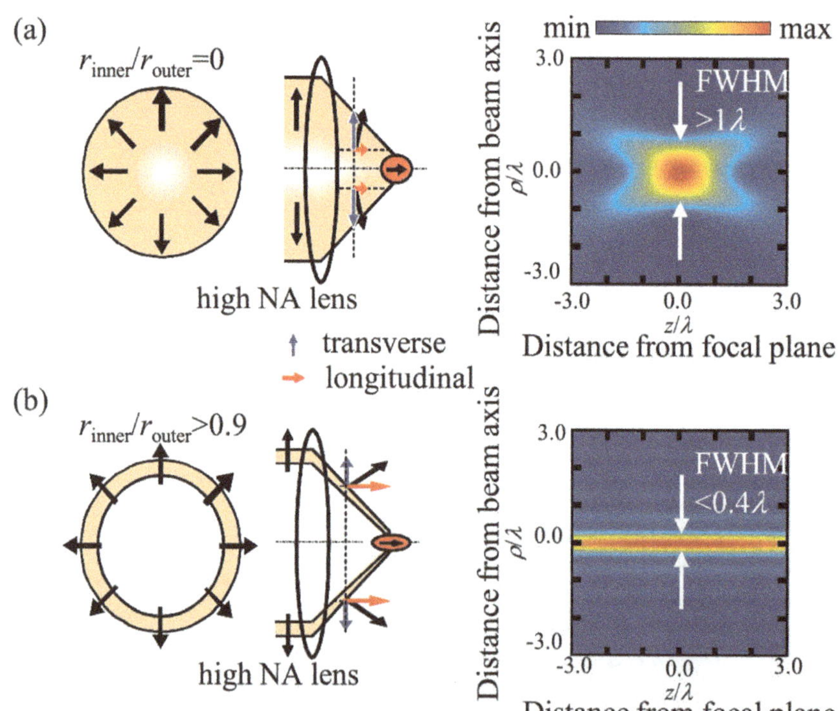


**Figure 9.46.** Schematic images of high numerical aperture (NA = 0.9) focused radially polarized beams and their calculated intensity profiles near the focal plane. Electric polarization is shown by black arrows. The electrical vectors passing through the lens can be resolved into transverse and longitudinal components. The longitudinal components constructively interfere at the beam axis near the focal plane. For a conventional donut-shaped beam (a), the inner part of the beam generates more intense transverse components than longitudinal components, whereas the outer part of the beam generates more intense longitudinal components than transverse components. For a halo-shaped beam (b) with a ratio of the inner to outer ring radius between 0.9 and 1, the longitudinal components dominate. Reprinted from [63], with the permission of AIP Publishing.

are not correlated [58]. Therefore C-points are more robust than the V-points. Robustness of the polarization structured beam for turbulence is demonstrated experimentally and the results are shown in figure 9.45.

### Smallest focal spot

Both the radially and azimuthally polarized light under loose focusing condition, produces a donut intensity profile. This is due to the presence of phase singularities in the V-point singularities. But when it comes to tight focusing, the radially polarized light produces a bright focal spot instead of a donut structure, which is smaller than the diffracted limited focal spot size [59–61]. A tightly focused azimuthally polarized light on the other hand produces a donut intensity structure that can be used in STED nanoscopy. Trapping metallic Rayleigh particles with radial polarization [62] has also been demonstrated. In figure 9.46 high numerical aperture focusing of radially polarized beam [63] is shown.

## References

[1] Born M and Wolf E 1980 *Principles of Optics* (New York: Pergamon)

[2] Pancharatnam S 1956 Generalized theory of interference and its applications *Proc. Indian Acad. Sci.* **XLIV** 247–62

[3] Hecht E and Zajac A 1974 *Optics* (Reading, MA: Addison-Wesley)

[4] Jenkins F A and White H G 1957 *Fundamentals of Optics* (New York: McGraw Hill)

[5] Goldstein D H 2011 *Polarized Light* (Boca Raton, FL: CRC Press)

[6] Collett E 1993 *Polarized Light* (New York: Marcel Dekker)

[7] Nye J F 1983 Lines of circular polarization in electromagnetic fields *Proc. R. Soc. Lond. Ser.* A **389** 279–90

[8] Hajnal J V 1987 Singularities in the transverse fields of electromagnetic waves. II theory *Proc. R. Soc. Lond. Soc.* A **414** 433–46

[9] Berry M V 2004 The electric and magnetic polarization singularities of paraxial waves *J. Opt. A: Pure Appl. Opt* **6** 475–81

[10] Berry M V and Dennis M R 2001 Polarization singularities in isotropic random vector waves *Proc. R. Soc. Lond. Ser.* A **457** 141–55

[11] Hajnal J V 1987 Singularities in the transverse fields of electromagnetic waves. II observations on the electric field *Proc. R. Soc. Lond. Soc.* A **414** 447–68

[12] Ruchi, Pal S K and Senthilkumaran P 2017 C-point and V-point singularity lattice formation and index sign conversion methods *Opt. Commun.* **393** 156–68

[13] Dennis M R 2002 Polarization singularities in paraxial vector fields: morphology and statistics *Opt. Commun.* **213** 201–21

[14] Freund I, Soskin M S and Mokhun A I 2002 Elliptic critical points in paraxial optical fields *Opt. Commun.* **208** 223–53

[15] Freund I 2002 Polarization singularity indices in Gaussian laser beams *Opt. Commun.* **201** 251–70

[16] Dennis M R, O'Holleran K and Padgett M J 2009 Singular optics:optical vortices and polarization singularities *Prog. Optics* **53**

[17] Bhargava Ram B S, Ruchi and Senthilkumaran P 2018 Angular momentum switching and orthogonal field construction of C-points *Opt. Lett.* **43** 2157–60

[18] Zhan Q and Leger J 2002 Focus shaping using cylindrical vector beams *Opt. Express* **10** 324–31
[19] Maurer C, Jesacher A, Furhapter S, Bernet S and Marte M R 2007 Tailoring of arbitrary optical vector beams *New J. Phys.* **9** 78
[20] Vyas S, Kozawa Y and Sato S 2013 Polarization singularities in superposition of vector beams *Opt. Express* **21** 8972–86
[21] Mokhun A I, Soskin M S and Freund I 2002 Elliptic critical points: C-points, $\alpha$-lines, and the sign rule *Opt. Lett.* **27** 995–7
[22] Pal S K and Senthilkumaran P 2016 Cultivation of lemon fields *Opt. Express* **24** 28008–13
[23] Pal S K and Senthilkumaran P 2018 Lattice of C-points at intensity nulls *Opt. Lett.* **43** 1259–62
[24] Ruchi, Pal S K and Senthilkumaran P 2017 Generation of V-point polarization singularity lattices *Opt. Express* **25** 19326–31
[25] Bhargava Ram B S, Sharma A and Senthilkumaran P 2017 Diffraction of V-point singularities through triangular apertures *Opt. Express* **25** 10270–5
[26] Pal S K, Ruchi and Senthilkumaran P 2017 Polarization singularity index sign inversion by half-wave plate *Appl. Opt.* **56** 6181–9
[27] Tidwell W C, Ford D H and Kimura W D 1990 Generating radially polarized beams interferometrically *Appl. Opt.* **29** 2234–9
[28] Tidwell S C, Kim G H and Kimura W D 1993 Efficient radially polarized laser beam generation with a double interferometer *Appl. Opt.* **32** 5222–9
[29] Armstrong D J, Phillips M C and Smith A V 2003 Generation of radially polarized beams with an image-rotating resonator *Appl. Opt.* **42** 3550–4
[30] Oron R, Blit S, Davidson N, Friesem A A, Bomzon Z and Hasman E 2000 The formation of laser beams with pure azimuthal or radial polarization *Appl. Phys. Lett.* **77** 3322–4
[31] Moser T, Balmer J, Delbeke D, Muys P, Verstuyft S and Baets R 2006 Intracavity generation of radially polarized $CO_2$ laser beams based on a simple binary dielectric diffraction grating *Appl. Opt.* **45** 8517–22
[32] Chang K C, Lin T and Wei M-D 2013 Generation of azimuthally and radially polarized off-axis beams with an intracavity large-apex-angle axicon *Opt. Express* **21** 16035–42
[33] Kozawa Y and Sato S 2005 Generation of a radially polarized laser beam by use of a conical brewster prism *Opt. Lett.* **30** 3063–5
[34] Tovar A A 1998 Production and propagation of cylindrically polarized Laguerre–Gaussian laser beamsaussian laser beams *J. Opt. Soc. Am.* A **15** 2705–11
[35] Moser T, Glur H, Romano V, Ahmed M A, Pigeon F and Parriaux O 2005 Polarization selective grating mirrors used in the generation of radial polarization *Appl. Phys.* B **80** 707–13
[36] Pohl D 1972 Operation of a ruby laser in the purely transverse electric mode $TE_{01}$ *Appl. Phys. Lett.* **20** 266–7
[37] Phelan C F, Donegan J F and Lunney J G 2011 Generation of a radially polarized light beam using internal conical diffraction *Opt. Express* **19** 21793–802
[38] Niv A, Biener G, Kleiner V and Hasman E 2003 Formation of linearly polarized light with axial symmetry by use of space-variant subwavelength gratings *Opt. Lett.* **28** 510–12
[39] Biener G, Niv A, Kleiner V and Hasman E 2002 Formation of helical beams by use of Pancharatnam-Berry phase optical elements *Opt. Lett.* **27** 1875–7

[40] Nesterov A V, Niziev V G and Yakunin V P 1999 Generation of high-power radially polarized beam *J. Phys. D: Appl. Phys.* **32** 2871–5
[41] Machavariani G, Lumer Y, Moshe I, Meir A and Jackel S 2008 Spatially variable retardation plate for efficent generation of radially and azimuthally polarized beams *Opt. Commun.* **281** 732–8
[42] Machavariani G, Lumer Y, Moshe I, Meir A and Jackel S 2007 Efficient extracavity generation of radially and azimuthally polarized beams *Opt. Lett.* **32** 1468–70
[43] Marrucci L 2013 The q-plate and its future *J. Nano Photon* **7** 1–4
[44] Marucci L, Manzo C and Paparo D 2006 Optical spin-to-orbital angular momentum conversion in inhomogeneous anisotropic media *Phys. Rev. Lett.* **96** 163905
[45] Slussarenko S, Murauski A, Du T, Chigrinov V, Marrucci L and Santamato E 2011 Tunable liquid crystal q-plates with arbirary topological charge *Opt. Express* **19** 4085–90
[46] Verma M, Pal S K, Joshi S, Senthilkumaran P, Joseph J and Kandpal H 2015 Singularities in cylindrical vector beams *J. Mod. Opt.* **62** 1068–75
[47] Angelsky O V, Mokhun I I, Mokhun A I and Soskin M S 2002 Interferometric methods in diagnostics of polarization singularities *Phys. Rev.* E **65** 1–5
[48] BhargavaRam B S, Sharma A and Senthilkumaran P 2017 Probing the degenerate states of V-point singularities *Opt. Lett.* **42** 3570–3
[49] Cardano F, Karimi E, Slussarenko S, Marucci L, de Lisio C and Santamato E 2012 Polarization pattern of vector vortex beams generated by q-plates with different topological charges *Appl. Opt.* **51** C1–6
[50] Delaney S, Sanchez-Lopez M M, Moreno I and Davis J A 2017 Arithmetic with q-plates *Appl. Opt.* **56** 596–600
[51] Kurzynowski P, Wozniak W A and Borwinska M 2010 Regular lattices of polarization singularities: their generation and properties *J. Opt.* **12** 035406
[52] Bhargava Ram B S, Senthilkumaran P and Sharma A 2017 Polarization-based spatial filtering for directional and nondirectional edge enhancement using an s-waveplate-waveplate *Appl. Opt.* **56** 3171–8
[53] BhargavaRam B S and Senthilkumaran P 2018 Edge enhancement by negative Poincaré-Hopf index filters *Opt. Lett.* **43** 1830–3
[54] Ruchi, Bhargava Ram B S and Senthilkumaran P 2017 Hopping induced inversions and Pancharatnam excursions of C-points *Opt. Lett.* **42** 4159–62
[55] Samlan C T, Suna R R, Naik D N and Viswanathan N K 2018 Spin-orbit beams for optical chirality measurement *Appl. Phys. Lett.* **112** 031101
[56] Lochab P, Senthilkumaran P and Khare K 2017 Robust laser beam engineering using polarization and angular momentum diversity *Opt. Express* **25** 17524–9
[57] Lochab P, Senthilkumaran P and Khare K 2018 Designer vector beams maintaining robust intensity profile on propagation through turbulence *Phys. Rev.* A **98** 023831
[58] Gateau J, Rigneault H and Guillon M 2017 Complementary speckle patterns: Deterministic interchange of intrinsic vortices and maxima through scattering media *Phys. Rev. Lett.* **118** 043903
[59] Quabis S, Dorn R, Eberler M, Glocki O and Geuchs G 2000 Focusing light to a tighter spot *Opt. Commun.* **179** 1–7
[60] Dorn R, Quabis S and Leuchs G 2003 Sharper focus for a radially polarized light beam *Phys. Rev. Lett.* **91** 1–4
[61] Helseth L E 2006 Smallest focal hole *Opt. Commun.* **257** 1–8

[62] Zhan Q 2004 Trapping of metallic Rayleigh particles with Radial polarization *Opt. Express* **12** 3377–82

[63] Kitamura K, Nishimoto M, Sakai K and Noda S 2012 Needle-like focus generation by radially polarized halo beams emitted by photonic-crystal ring-cavity laser *Appl. Phys. Lett.* **101** 221103

[64] Kozawa Y, Yonezawa K and Sato S 2007 Radially polarized laser beam from a Nd:YAG laser cavity with a c-cut YVO4 crystal *Appl. Phys.* B **88** 43–6

[65] Kozawa Y, Yoneyama T and Sato S 2005 Direct generation of cylindrical vector beam from nd:yag laser cavity *IEEE-Explore-CWAB3* P8

Lightning Source UK Ltd.
Milton Keynes UK
UKHW051337121221
395490UK00002B/94

9 780750 316965